METAMATERIALS - DEVICES AND APPLICATIONS

Edited by **Alejandro Lucas Borja**

Metamaterials - Devices and Applications
http://dx.doi.org/10.5772/62750
Edited by Alejandro Lucas Borja

Contributors

Yongjun Huang, Jian Wang, Jing Du, He-Xiu Xu, Yinchao Chen, Xin Zheng, Wensong Wang, Shuhui Yang, Daniel Shreiber, Mohammad Alibakhshikenari, Bal Singh Virdee, Mohammad Naser-Moghadasi, Ramazan Ali Sadeghzadeh, Ernesto Limiti, Andriy Serebryannikov, Abdelwaheb Ourir, Agnes Maurel, Carlos Javier Zapata-Rodríguez, Shuqi Chen, Zhancheng Li, Wenwei Liu, Hua Cheng, Jianguo Tian

Published by InTech
Janeza Trdine 9, 51000 Rijeka, Croatia

Publishing Process Manager Romina Skomersic
Technical Editor SPi Global
Cover InTech Design team

First published April, 2017
Printed in Croatia
Legal deposit, Croatia: National and University Library in Zagreb

Additional hard copies can be obtained from orders@intechopen.com

Metamaterials - Devices and Applications, Edited by Alejandro Lucas Borja
p. cm.
Print ISBN 978-953-51-3099-4
Online ISBN 978-953-51-3100-7

For EU product safety concerns:
IN TECH d.o.o., Prolaz Marije Krucifikse Kozulić 3, 51000 Rijeka, Croatia,
info@intechopen.com or visit our website at intechopen.com.

Meet the editor

Dr. Alejandro L. Borja received his MSc degree in telecommunication engineering and PhD degree from the Universidad Politécnica de Valencia, Valencia, Spain, in 2004 and 2009, respectively. From 2005 to 2006, he was with the Communication Group, University of Birmingham, where he was involved with the research and development of metamaterial-based antennas. He then joined, from 2007 to 2008, the Institut d'Électronique de Microélectronique et de Nanotechnologies (IEMN), Université des Sciences et Technologies de Lille 1; his research activity included the design of metamaterial-based structures with frequency-selective properties. Since 2009, he has been with the Departamento de Ingeniería Eléctrica, Electrónica, Automática y Comunicaciones, Universidad de Castilla-La Mancha, where he is an assistant lecturer. He has published more than 60 papers in peer-reviewed international journals and conference proceedings and frequently acts as a reviewer for several technical publications. In 2012, he served as a lead guest editor for a special issue of the *International Journal of Antennas and Propagation* related to metamaterials. His current research interests include EM metamaterials, substrate integrated waveguide devices, tunable structures, and their applications in microwave and millimeter-wave technologies.

Contents

Preface

Metamaterials have made a significant impact in all areas of electromagnetics. In the last decade, a wide range of theories, models, and novel devices have been published by researchers. It is likely that modern and future generations of communication systems may benefit from the main advantages of metamaterial technology and incorporate devices with improved properties. Metamaterials offer the possibility to engineer the fundamental characteristics of the propagating media made of subwavelength components. By these means, devices with enhanced responses in terms of impedance, phase constant, and dispersion control can be achieved. The use of metamaterials enables to develop new structures and applications beyond those realizable with natural materials.

This book aims to report on novel research and concepts established within the metamaterial field. It provides an opportunity for researchers to identify developments, compare performances with conventional devices, specify trends, and find potential applications where metamaterials could be employed. This book will be beneficial for the graduate students, teachers, researchers, engineers, and other professionals, who are interested to fortify and expand their knowledge related to metamaterial concepts and research.

The book is composed of 11 chapters from multiple contributors around the world, including China, Iran, the United Kingdom, Algeria, Spain, and the United States. It covers a wide variety of topics from different domains including polarization conversion, asymmetric transmission, transmission lines, filters, plasmonic lenses, tunable metamaterials, light manipulation, absorbers, and antennas, among others.

I would like to deeply thank all the authors for the submission of their encouraging chapters and their patience during the review processes, making thus possible the publication of this book in the present form. In addition, I take the opportunity to express my acknowledgment and gratitude to Dr. Ángel Belenguer, Prof. Joaquin Cascon, and Prof. Vicente Boria for their helpful discussions and comments related to the content of the chapters. Finally, I offer my special thanks to InTech Open Access Publisher for the opportunity to edit this book and Ms. Romina Skomersic, Publishing Process Manager, for her assistance and help during the whole publication process.

Dr. Alejandro Lucas Borja
Departamento de Ingeniería Eléctrica, Electrónica, Automática y Comunicaciones
Escuela Politécnica de Albacete
Universidad de Castilla-La Mancha
Spain

Chapter 1

Effects of Dielectric Substrate on Polarization Conversion Using Coupled Metasurfaces With and Without Tunneling

Andriy E. Serebryannikov, Mehmet Mutlu and Ekmel Ozbay

Additional information is available at the end of the chapter

http://dx.doi.org/10.5772/65950

Abstract

Dielectric substrates are technologically necessary components of various microwave and optical structures and devices, and may strongly affect their performance. For metasurfaces composed of subwavelength resonators, placing dielectric components in the proximity of resonators can lead to strong modification of subwavelength resonances and related transmission regimes. We focus on the effects exerted by material and geometrical parameters of such a dielectric substrate on linear-to-linear polarization conversion that appears in quasiplanar structures containing two coupled metasurfaces and enabling chirality. It is shown that spectral locations of the polarization conversion resonances and transmission efficiency at these resonances are strongly sensitive to the substrate parameters, whereas the ability of polarization conversion and related asymmetry of transmission can be preserved in wide ranges of parameter variation. The effects of a substrate are considered in detail for the mechanisms with and without tunneling, indicating a route to compact designs of quasiplanar structures for single- and multiband polarization conversion.

Keywords: metasurface, polarization conversion, subwavelength resonator, permittivity

1. Introduction

It is well known that material parameters of the components of an entire structure can strongly affect dispersion, transmission, and scattering characteristics. For some classes of the structures, such as the idealized lossless cavities and waveguides, effects of variation of the material parameters can be easily quantified. In particular, the rule of $\varepsilon^{-1/2}$, being applicable to eigenfrequencies of the closed cavities filled with a linear isotropic dielectric, is commonly known. The quantifying of resonances becomes much more complicated in the case of open resonance structures, where the resonance fields can be strong beyond the resonators [1]. In this case, an a priori estimate of the strength of the effect of dielectric is very complicated and, moreover, one cannot predict the principal possibility of obtaining of a desired transmission regime. It can be even more difficult to preserve the same strength of some effects, e.g., efficiency of conversion of the incident wave energy to a certain diffraction order or polarization state, simultaneously with the shift of resonance frequencies, while only varying characteristics of the dielectric layer(s) adjacent to the resonators. Nevertheless, stacking resonant or nonresonant arrays with dielectric layers and placing dielectric components inside individual resonators are known as effective tools to control resonance frequencies and transmission and scattering characteristics [2–5].

Theory and technology related to subwavelength resonators have extensively been developed since the early 2000s. Initially, the interest to them has been stimulated by the possibility of obtaining artificial magnetism and negative refraction [6]. Later, a chiral way to negative refraction has been proposed, which is realized with the aid of planar metamaterials [7]. Artificial bianisotropic [8] and, in particular, electrically thin chiral structures based on metasurfaces composed of subwavelength resonators [9] have extensively been studied. The latter suggest efficient solutions for polarization conversion problem. Circular dichroism and polarization rotation belong to the most distinguished properties of the artificial chiral materials. One of the basic features is that electric and magnetic dipoles are strongly coupled and excited simultaneously, so that chirality originates from the collinear excitation of the effective electric and magnetic responses. Moreover, these responses can be switched at a fixed frequency by changing polarization of the incident electromagnetic wave. It is noteworthy that manipulation by polarization states can be achieved, in addition to the chiral structures, which are based on two coupled metasurfaces, with the aid of various other schemes and classes of the structures. In particular, high-contrast gratings, advanced quarter-wave plates, and structures with a single or multiple (anisotropic) metasurfaces can be mentioned [10–15].

In the coupled arrays of subwavelength resonators, conversion of the incident linear polarization to the circular polarization [16–21] and to the orthogonal linear polarization [22, 23] has been demonstrated. These conversions are directly related to *asymmetric transmission*, a Lorentz reciprocal phenomenon that enables strong forward-to-backward transmission contrast between two opposite incidence directions [22, 24–28]. Since reciprocity forbids one-way transmission in two-port reciprocal systems, asymmetric transmission requires breaking of spatial inversion symmetry and, hence, new transmission and reflection channels. For the twisted metasurfaces, they can be obtained by involving polarization states different from the incident one [16–18, 22, 29]. Diodelike asymmetric transmission with reflections vanishing for one of two opposite incidence directions has been theoretically predicted in the zero-loss approximation for both diffraction [30] and polarization conversion [1, 23] inspired mechanisms. The principal possibility of the perfect polarization conversion in a generalized double-layer structure has been demonstrated [31].

In one of the recent studies, linear-to-linear polarization conversion and diodelike asymmetric transmission in a thin chiral structure with 4-U unit cells have been obtained [23]. It was shown that the problem of their *perfectness* is an eigenstate, phase propagation, and impedance matching. It can be reached due to the coexistence of tunneling, optimization of the axial ratio of the eigenwaves, and optimization of transmission phases of the eigenwaves that destructively interfere in one direction and constructively interfere in the opposite one. Recently, similar mechanisms have been realized in the structures with U-type unit cells, with and without an evanescent-wave metallic mesh and, thus, with and without tunneling [1]. Moreover, perfect multiband linear-to-linear polarization conversion and asymmetric transmission can be obtained with the aid of the coupled arrays of U-shaped apertures [32]. Besides, many recent studies should be mentioned which are dedicated to dual-band and broadband conversions of polarization that involve linear and circular states [1, 20, 21, 33–36].

In this chapter, the emphasis is put on the effects exerted by variations of the dielectric substrate parameters on subwavelength resonances and related regimes of linear-to-linear polarization conversion and asymmetric transmission. Consideration is restricted to quasiplanar structures containing two coupled metasurfaces that represent arrays of subwavelength split-ring resonators (SRRs) enabling simple U-type unit cells with resonant behavior at microwave frequencies, and to the case of normal incidence. Zero-loss approximation is utilized to clarify the main components of the underlying physical mechanisms. Two mechanisms will be discussed, which are realized in the structures with and without a metallic mesh (small-hole array) placed between the SRR arrays, i.e., tunneling either appears and contributes or does not. The focus will be on sensitivity of the subwavelength resonances in the studied structures with chirality to the variations in permittivity and thickness of the dielectric substrate layers located between the SRR arrays and the metallic mesh, or simply between two SRR arrays. Indeed, in the both cases, the resonance characteristics of individual resonators and those of the coupled metasurfaces can be strongly affected, because the resonators are placed directly on the dielectric layer(s). Perfect transmission that originates from the perfect matching of real impedances can be obtained in the mesh-free structures in some ranges of parameter variation. These ranges either coincide or do not coincide with the ranges, in which matching can appear for the structures with a mesh as a result of fulfillment of certain phase conditions. The role of the choice of permittivity and thickness of the dielectric layers can be very important for preserving the features related to the matching and perfect polarization conversion. While variations of substrate characteristics give big freedom in design without changes in SRR arrays, it can be further extended by a proper selection of an array period. Simulation results are obtained by using CST Microwave Studio, a full-wave commercial solver based on the finite integration method (see www.cst.com for software details). The same methodology can be applied to study the effects of a substrate for linear-to-circular and circular-to-circular polarization conversion in various structures based on metasurfaces. This chapter presents a review of the basic effects achievable by variations of substrate parameters in the structures with U-type assembly, while a comparative study of the existing and suggested performances for multi-/broadband polarization conversion is beyond its scope.

2. Polarization conversion, asymmetry in transmission and tunneling

First, let us briefly describe the structures to be studied that may enable polarization conversion and related asymmetry in transmission. Schematic of a unit cell and perspective view of the studied periodic structure are presented in **Figure 1**. Each of two SRR arrays has period of a

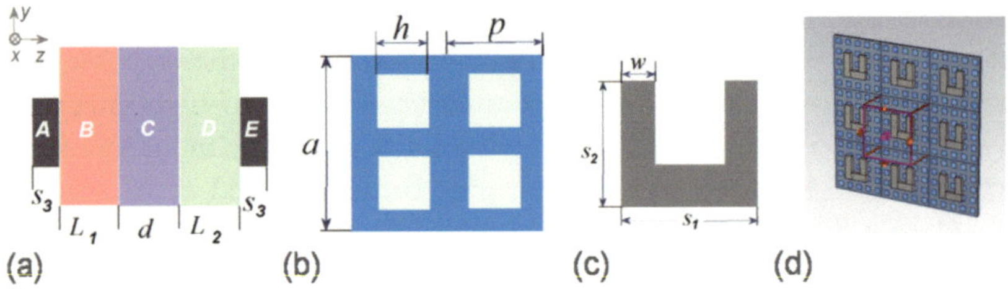

Figure 1. (a) Schematic of a complex unit cell (side view): A—SRR, B and D—dielectric layers with permittivity ε_1 and ε_2, respectively, C—metallic mesh, E—rotated SRR. (b) Metallic mesh with square holes seen from the frontside and backside at $a = 2p$. (c) SRR seen from the backside (denoted by E in the side view). (d) $(3a) \times (3a)$ fragment of the basic configuration at $a = 5p$ (perspective view).

in the x and y directions. The metallic mesh with thickness d and period p is obtained by periodic arrangement of the square holes, i.e., $a > p$; p is assumed to be the same over x and y. For the *basic configuration* with the metallic mesh [1], we take a = 22 mm, p = 4.4 mm, h = 2.2 mm, $L_1 = L_2 = 1.25$ mm, d = 0.5 mm, $s_1 = s_2 = 10$ mm, $s_3 = 1.5$ mm, and w = 3 mm. Hence, the total thickness of the structure is $S = d + 2s_3 + L_1 + L_2$. The back-side SRR array represents the front-side SRR array rotated by 90° in the clockwise direction. Such a location of the SRR arrays can create chirality and, thus, the ability of polarization conversion. The stronger the conversion, the stronger asymmetry in the transmission. In fact, the ability of asymmetric transmission originates from the fact that conversion of the incident wave is distinguished at front-side illumination and at back-side illumination, so that the different channels (e.g., that associated with the orthogonal polarization in transmission mode for the former, and that associated with the same polarization in reflection mode for the latter) are the main acceptors of the incident wave energy.

The use of the layers with $\varepsilon < 0$ (metallic mesh) and $\varepsilon > 0$ (dielectric substrate) in the same structure should allow one obtaining destructive interferences of the waves reflected at the interfaces between the layers from A to E and those between air and the layers A and E. This can result in zero reflection and, accordingly, in perfect tunneling. Generally, the operation is based on wave interference, and, thus, the phase is a critical parameter. However, a negative-ε layer is not necessary for obtaining of zero reflections, if input impedances at the interfaces of the individual layers are properly adjusted. This is a reason why similar regimes of polarization conversion can be obtained in the structures with and without a metallic mesh and, thus, the mechanisms with and without tunneling are worth comparing [1].

In the general case, a linearly polarized incident wave changes its polarization state when passing through such a coupled system. The complex amplitudes of the incident ($E_{xi}^{f,b}$ and $E_{yi}^{f,b}$) and transmitted ($E_x^{f,b}$ and $E_y^{f,b}$) waves are expressed through each other with the aid of the T-matrix as follows [22, 23]:

$$\begin{pmatrix} E_x^{f,b} \\ E_y^{f,b} \end{pmatrix} = \begin{pmatrix} T_{xx}^{f,b} & T_{xy}^{f,b} \\ T_{yx}^{f,b} & T_{yy}^{f,b} \end{pmatrix} \begin{pmatrix} E_{xi}^{f,b} \\ E_{yi}^{f,b} \end{pmatrix}, \tag{1}$$

where $E_{xi}^{f,b}$ and $E_{yi}^{f,b}$ are the x and y components of the incident wave; $T_{xx}^{f,b}$ and $T_{yy}^{f,b}$ are the *copolarized* transmission coefficients, and $T_{xy}^{f,b}$ and $T_{yx}^{f,b}$ are the *cross-polarized* transmission

http://dx.doi.org/10.5772/65950

coefficients, f and b stand for the forward and the backward transmission cases, which correspond to front-side and back-side illumination, respectively.

The amplitudes in these two cases are related to each other, depending on the structural symmetries. For the structures showing the same symmetries as the studied one, $T^f_{xy} = -T^b_{yx}$, $T^f_{yx} = -T^b_{xy}$ and $T^f_{xx} = T^b_{xx} = T^f_{yy} = T^b_{yy}$, provided that $\varepsilon = \varepsilon_1 = \varepsilon_2$. According to references [23, 29], if the structure parameters can be adjusted so that $T^{f,b}_{xx} = 0$, $T^{f,b}_{yy} = 0$, $T^f_{yx} = T^b_{xy} = 0$ and $|T^f_{xy}| = |T^b_{yx}| = 1$ at some resonances, then perfect linear-to-linear polarization conversion is achieved. This regime is schematically illustrated in **Figure 2(a)**, where transmission, reflection, and polarization conversion are explained by using a four-port system. Perfect polarization conversion and perfect asymmetry in transmission may occur when either y-polarized plane wave is incident from the frontside, or x-polarized plane wave is incident from the backside. In turn, if either y-polarized wave is incident from the backside, or x-polarized wave is incident from the frontside, perfect reflection takes place. As a result, transmission may be strong at either front-side or back-side illumination, depending on which of two incident linear polarizations is utilized. At the same time, it vanishes at the opposite-side illumination when the same polarization state of the incident wave is used. Clearly, this reversibility of the direction of diodelike transmission occurs when the incident polarization is changed (for the both incidence directions) to the orthogonal one, because Lorentz reciprocity forces the scattering matrix to be symmetric.

It may occur that at the other polarization conversion resonances the above-described operation regime appears when $T^{f,b}_{xx} = 0$, $T^{f,b}_{yy} = 0$, $T^b_{yx} = T^f_{xy} = 0$ and $|T^b_{xy}| = |T^f_{yx}| = 1$. In this case, the schematic in **Figure 2(a)** should be modified, i.e., the ports with perfect reflection and those with perfect transmission replace each other.

For the circularly polarized (CP) waves, we have, instead of Eq. (1),

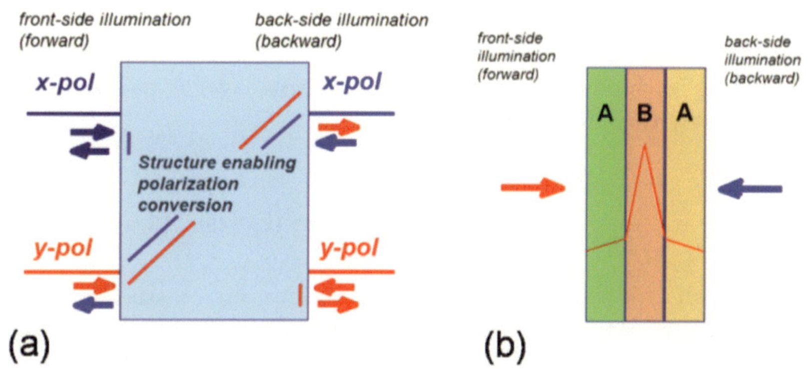

Figure 2. (a) Schematic of perfect polarization conversion and asymmetric transmission regime. (b) Schematic of tunneling in ABA-type stack with propagating-wave layers A and evanescent-wave layer B; red line shows possible (simplified) variation of the field magnitude in the propagation (normal) direction.

$$\begin{pmatrix} E_{+}^{f,b} \\ E_{-}^{f,b} \end{pmatrix} = \begin{pmatrix} T_{++}^{f,b} & T_{+-}^{f,b} \\ T_{-+}^{f,b} & T_{--}^{f,b} \end{pmatrix} \begin{pmatrix} E_{+i}^{f,b} \\ E_{-i}^{f,b} \end{pmatrix}, \tag{2}$$

where $E_{+i}^{f,b}$ and $E_{-i}^{f,b}$ are the components of the incident wave that correspond to right- and left-handed CP waves; $T_{++}^{f,b}$ and $T_{--}^{f,b}$ are the transmission coefficients of right- and left-handed CP waves, and $T_{+-}^{f,b}$ and $T_{-+}^{f,b}$ are the cross-coupling transmission coefficients, superscripts f and b indicate the forward and backward transmission cases; $T_{++}^{f,b} = T_{xx}^{f,b} + T_{yy}^{f,b} + i(T_{xy}^{f,b} - T_{yx}^{f,b})$, $T_{--}^{f,b} = T_{xx}^{f,b} + T_{yy}^{f,b} - i(T_{xy}^{f,b} - T_{yx}^{f,b})$, $T_{+-}^{f,b} = T_{xx}^{f,b} - T_{yy}^{f,b} - i(T_{xy}^{f,b} + T_{yx}^{f,b})$, $T_{-+}^{f,b} = T_{xx}^{f,b} - T_{yy}^{f,b} + i(T_{xy}^{f,b} + T_{yx}^{f,b})$, $T_{-+}^{f} = T_{+-}^{b}$ and $T_{+-}^{f} = T_{-+}^{b}$. Efficient polarization conversion and asymmetry in transmission can also be obtained for the CP waves [18, 19, 21, 36]. However, the focus in this chapter is related to the effects arising for the linearly polarized waves.

For a part of the structures considered here, tunneling is an important component of the resulting polarization conversion mechanism that has initially been suggested in reference [23]. Previously, a similar tunneling-based mechanism has been utilized for obtaining perfect transmission through the sandwiched structures comprising evanescent-wave layers, which do not enable polarization conversion [37–39]. In particular, it was shown that arrays of subwavelength resonators may play two roles: either work as positive-ε layers in the tunneling mechanism or contribute to the effective negative-index behavior.

The general idea of matching in case of the presence of evanescent-wave components can be explained in terms of transmission through isotropic AB type (bilayer) and ABA type (trilayer) stacks, in which A and B stand for propagating-wave and evanescent-wave layers, respectively, see **Figure 2(b)**. According to reference [37], the criterion of the perfect transmission for AB stack can be written as follows:

$$(k_1/k_0 - k_0/k_1)tan(k_1 d_1) - (\alpha_2/k_0 + k_0/\alpha_2)tanh(\alpha_2 d_2) + i(k_1/\alpha_2 + \alpha_2/k_1)tan(k_1 d_1)tanh(\alpha_2 d_2) = 0, \tag{3}$$

where k_1 and d_1 are wave number and thickness of the layer A, $k_2 = i\alpha_2$ and d_2 are wave number and thickness of the layer B. This criterion cannot be satisfied, since the second term is imaginary, i.e., the waves scattered by the layers A and B, being in different phase planes, cannot entirely cancel each other. Then, if we add one more layer A, as shown in **Figure 2(b)**, the criterion can be rewritten as follows:

$$\begin{aligned}(k_1/k_0 - k_0/k_1)2tan(k_1 d_1) - (\alpha_2/k_0 + k_0/\alpha_2)tanh(\alpha_2 d_2) - [k_1^2/(\alpha_2 k_0) \\ + (\alpha_2 k_0)/k_1^2]tan^2(k_1 d_1)tanh(\alpha_2 d_2) = 0.\end{aligned} \tag{4}$$

Now, both the first and the second term are real and, thus, Eq. (4) has a solution that corresponds to the perfect transmission. This model enables a qualitative but physically correct prediction of the perfect transmission also for complex anisotropic structures with the polarization conversion ability, including the case of coupled metasurfaces with a metallic mesh [23]. For the structures without a mesh and, hence, without tunneling, conditions of the perfect transmission can easily be obtained from matching of impedances. For an ABA-type stack, this yields

$$tan(k_1d_1) = 0 \text{ and } tan(k_2d_2) = 0, \tag{5}$$

where k_2 is real wave number of the layer B. In the next sections, the existence, efficiency, and generality of polarization conversion achievable in ABA-type stacks at variations of substrate parameters will be discussed.

3. Varying characteristics of dielectric substrate

3.1. Basic effects of substrate permittivity

Here, we demonstrate how the choice of a dielectric substrate material can affect the ability and manifestations of polarization conversion in the structures based on coupled metasurfaces. Let us consider the case of an intermediate distance between metasurfaces and vary ε from 1 (substrate-free case) to 40 (some types of ceramics, e.g., see reference [40]). In **Figure 3**, results are presented for the structures with and without a mesh. In the latter case, it is assumed that the mesh is simply removed from the mesh-containing structures, and a homogeneous dielectric layer fully occupies the region between two metasurfaces. Then, its thickness is $L = d + L_1 + L_2$. To better illustrate the basic features, we first consider the case of $\varepsilon = 1$, see **Figure 3(b, c)**. For the mesh-free structure in **Figure 3(b)**, polarization conversion is observed between 6 and 8 GHz. However, all of the diagonal and nondiagonal components of the *T*-matrix are of the same order, so that a nearly perfect one-way linear-to-linear polarization conversion and relevant diodelike transmission cannot be obtained. At the same time, conversion into polarization states being different from a linear one can be quite strong. The twin maxima like those shown in **Figure 3(b)** are typical for the coupled SRR arrays. They may appear regardless of whether polarization conversion is possible or not, and can be explained by using Lagrange (hybridization) formalism [9, 41–43]. The strength of coupling can be utilized to control optical activity and polarization conversion. In particular, either two narrow bands of polarization conversion, which are well separated from each other, or a wide band with two weak maxima might be obtained.

Each SRR can approximately be presented as an LC circuit, with inductance L and capacitance C, and resonance angular frequency $\omega_0 = (LC)^{-1/2}$. Accordingly, the Lagrangian of a single SRR can be written as $\Gamma = (L/2)(\dot{q}^2 - \omega_0^2 q^2)$, where q stands for a charge that is considered as a generalized coordinate. In the case of two coupled SRRs that belong to different metasurfaces, the Lagrangian represents the sum of the Lagrangians of two individual SRRs plus the coupling term, i.e.

$$\Gamma = (L/2)(\dot{q}_1^2 - \omega_0^2 q_1^2) + (L/2)(\dot{q}_2^2 - \omega_0^2 q_2^2) + M\dot{q}_1\dot{q}_2, \tag{6}$$

where q_1 and q_2 correspond to the first and the second SRR, respectively, and M is mutual inductance arising due to the magnetic coupling of two SRRs. Magnetic coupling is considered

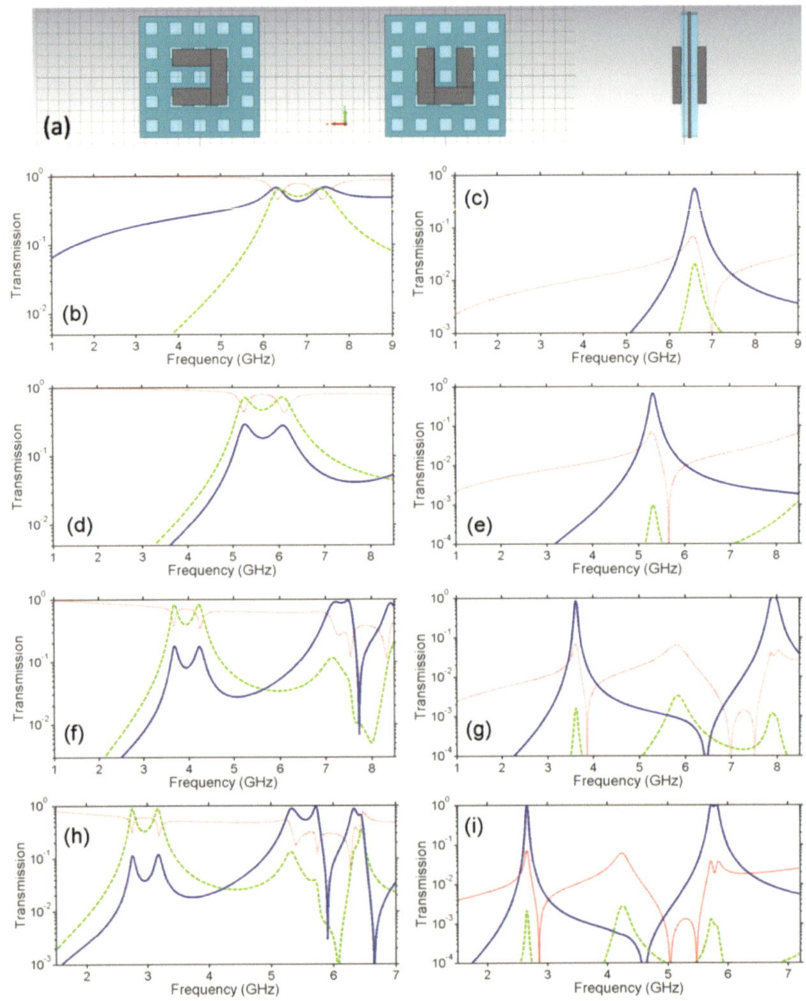

Figure 3. (a) Front, back, and side view of a unit cell, and transmission for the basic configuration (b, d, f, h) without and (c, e, g, i) with metallic mesh; (b, c) $\varepsilon = 1$, (d, e) $\varepsilon = 2.1$, (f, g) $\varepsilon = 5.8$, (h, i) $\varepsilon = 11.4$; solid blue line – $|T^f_{xy}| = |T^b_{yx}|$, dashed green line – $|T^f_{yx}| = |T^b_{xy}|$, and dotted red line – $|T^f_{xx}| = |T^b_{xx}| = |T^f_{yy}| = |T^b_{yy}|$.

to be independent of the angle of rotation of the second SRR [43]. The eigenfrequencies are given by $\omega_{\pm} = \omega_0[1 \mp (M/L)]^{-1/2}$, where ω_+ and ω_- correspond to an antisymmetric and a symmetric charge distribution, respectively [41]. Hence, a spectral distance between the eigenfrequencies is

$$\Delta\omega = \omega_+ - \omega_- \approx (M/L)\omega_0. \tag{7}$$

Mutual inductance quickly decreases while the distance between SRRs is increased and, thus, the spectral separation becomes smaller. Electric coupling of two SRRs, which depends on the rotation angle, can be easily involved into this model [9, 43].

For the structure with the metallic mesh in **Figure 3(c)**, the diagonal components and one of the nondiagonal components of the T-matrix are suppressed, as desired. As a result, one-way polarization conversion can be obtained. However, a nearly perfect transmission is again not reached: $max|T^f_{xy}| = 0.54$ and $max|T^b_{yy}| = 0.067$ at $f = 6.59$ GHz. On the contrary to the mesh-containing structure with 4U-type unit cells [23] and the mesh-free structure in **Figure 3(b)**, the unwanted components of the T-matrix are well suppressed, enabling strong directional selectivity in a wide frequency band. Note that the mesh alone only allows one obtaining a very weak transmission in this band, i.e., evanescent-wave regime is evident. In this case, Lagrange formalism in the above-presented form is not applicable, since additional modifications are required to properly take into account the effects of the mesh.

Placing a low-ε layer between two metasurfaces in the mesh-free configurations and between each metasurface and the mesh in the mesh-containing configurations results in that all of the basic transmission and polarization conversion features observed in **Figure 3(b, c)** are preserved, while the resonance frequencies are redshifted. This is quite expectable, although the obtained shift is weaker than it would be when the rule of $\varepsilon^{-1/2}$ is valid. In **Figure 3(d, e)**, we use a material with $\varepsilon = 2.1$ that belongs to the ε-range, to which many materials that are widely used at microwave and optical frequencies (e.g., polytetrafluoroethylene/Teflon, SiO_2) do belong. In particular, the first maximum of $|T^f_{xy}|$ in the case of the mesh-containing structure is redshifted from $f = 6.59$ to $f = 5.29$ GHz, compare **Figure 3(c, e)**. In turn, two first maxima of $|T^f_{yx}|$ in the case of the mesh-free structures are redshifted from $f = 6.35$ and $f = 7.3$ GHz to $f =$ 5.27 and $f = 6.09$ GHz, respectively, see **Figure 3(b, d)** for comparison. Moreover, the maximal values of $|T^f_{yx}|$ shown in **Figure 3(d)** are several times larger than the values of $|T^f_{xy}|$ at the same frequencies. The maximal values of $|T^f_{xy}|$ in **Figure 3(d, e)** become closer to unity, so the nearly perfect conversion regime is approached.

The above-discussed features can be enhanced at a further increase of ε. In **Figure 3(f, g)**, the results are presented for $\varepsilon = 5.8$ (e.g., some types of glass and diamond). Now, we obtain $max|T^f_{yx}| > 0.8$ at $f = 3.67$ and $f = 4.22$ GHz, and $max|T^f_{xy}| > 0.89$ at $f = 7.22$ and $f = 7.49$ GHz for the mesh-free structure in **Figure 3(f)**, and $max|T^f_{xy}| > 0.8$ at $f = 3.6$ GHz, and $max|T^f_{xy}| > 0.995$ near $f = 7.9$ GHz for the mesh-containing structure in **Figure 3(g)**, while the second cross-polarized component is well suppressed. Higher resonances are redshifted so that dual-band operation with a large distance between the bands is possible at $f < 8.5$ GHz. Note that $S/\lambda < 0.17$ and $a/\lambda < 0.62$ at $f < 8.5$ GHz, where λ is the free-space wavelength.

What can be even more important for practical applications is that the high-frequency polarization conversion band at $f > 7$ GHz is wider than those connected with the maxima near $f = 4$ GHz, in **Figure 3(f, g)**. If single-band operation is sufficient, design can be optimized based on the trade-off between the bandwidth and electrical size. It is worth noting that a part of the near-unity transmission scenarios is realized by using the y-polarized incident wave, while the other part does with the aid of the x-polarized incident wave. In particular, for the mesh-free structure in **Figure 3(f)** we obtain different directions of near-unity transmission for the first two maxima and for the high-frequency band, while this direction is the same for the both bands in the mesh-containing structure in **Figure 3(g)**.

While the maxima twinning is a general feature shown by the mesh-free structures, there is no twinning for the mesh-containing structures. The absence of twinning may originate from the merging of the maxima, which correspond to different but very weakly separated resonances. An additional study based on Lagrange formalism or other models of resonance coupling is required to clarify this difference. Due to the Lorentz reciprocity, a nearly perfect polarization conversion is a one-way effect. Indeed, if the same polarization state is used at both frontside and backside illumination, we obtain the perfect reflection for illumination from the side being opposite to that one, at which a nearly perfect polarization conversion is obtained. A key feature is that the copolarized components are well suppressed in a wide frequency range, including the polarization conversion bands. Up to now, the mesh free and the mesh-containing performances that contain the dielectric layers with ε = 5.8 have been the best among the discussed ones in terms of conversion efficiency, suppression of the unwanted components, and number of the achievable conversion bands.

Next, we investigate whether we have more flexibility in choice of ε, i.e., whether the basic features observed in **Figure 3(f, g)** can be kept when a higher-ε material (i.e., with $\varepsilon > 5.8$) is used for the dielectric layers. Indeed, since the resonance frequencies are expected to be further redshifted, the desired values of impedances, phases, and efficiency of one-way polarization conversion are not guaranteed. **Figure 3(h, i)** present the transmission results for the two structures at $\varepsilon = 11.4$. The chosen value of ε corresponds to the range, to which many materials such as graphite, Si, and GaAs are belonging. By comparing to **Figure 3(f, g)**, the spectra shown in **Figure 3(h, i)** are very similar but are much denser. For instance, the dependences in **Figure 3 (f)** at $f < 8.5$ GHz are similar to those in **Figure 3(h)** at $f < 6.3$ GHz, so that changing a scale at the abscissa axis would make these dependences almost identical. The same remains true for the dependences in **Figure 3(g)** at $f < 8.5$ GHz and those in **Figure 3(i)** at $f < 6.3$ GHz. In **Figure 3(h)**, we obtain $max|T^f_{yx}| = 0.885$ and $max|T^f_{yx}| = 0.88$ at f = 2.76 and f = 3.14 GHz, respectively, whereas $|T^f_{xy}| = 0.87$ and $|T^f_{xy}| = 0.96$ for two maxima in the vicinity of f = 5.5 GHz. Thus, the cross-polarized components with magnitudes higher than 0.9 can be obtained even without a metallic mesh and tunneling, i.e., only due to matching of the real input impedances. In spite of that the copolarized components are still significant, high-efficiency transmission and polarization conversion are possible for this structure at two different incident polarizations.

From the *broadband* operation perspective, bringing the neighboring resonances together might be useful, as shown in **Figure 3(i)** around 5.76 GHz for the structure with the mesh. Indeed, such a coupling of two resonances can be obtained that the dip between the neighboring maxima is very weak and two neighboring bands are merged. In this regime, $|T^f_{xy}| > 0.99$ at the maxima, whereas $min|T^f_{xy}| = 0.88$ between them. Thus, the vicinity of f = 5.76 GHz can be suggested for broadband operation. One more high-efficiency polarization conversion band occurs at f = 2.66 GHz, where $max|T^f_{xy}| \approx 0.94$. Note that $a/\lambda \approx 0.2$ and $S/\lambda \approx 0.05$, and $a/\lambda \approx 0.42$ and $S/\lambda \approx 0.12$ at f = 2.66 GHz and f = 5.76 GHz, respectively.

Based on the obtained results, one can conclude that the main functions of the high-ε layers include redshift of the resonances, at which polarization conversion may occur, and improvement of the phase and/or real impedance matching, which results in that the regime of perfect one-way polarization conversion is approached. The functions of the mesh include suppression of the unwanted components of the *T*-matrix and the collaborative effect with the high-ε layers and SRR arrays in the tunneling mechanism. In turn, the coupled SRR arrays enable polarization conversion, on the one hand, and contribute to the resulting transmission mechanism due to either tunneling or real impedance matching, on the other hand. The role of the SRR arrays here is similar, in principle, to that in the structures with 4U-type unit cells [23]. However, wideband suppression of the unwanted components has not been reached therein, in contrast with the results shown in **Figure 3**. To compare, transmission for the structures obtained from those in **Figure 3(c, e, g, i)** by removing the SRR arrays remains low, i.e., $|T^f_{xx}| = |T^f_{yy}| < 0.1$ and $|T^f_{xy}| = |T^b_{yx}| = 0$. Hence, the roles of the different structural components in obtaining of the nearly perfect matching and tunneling are clear.

The above-discussed features are kept even in wider ranges of variation of the problem parameters, including very high values of ε. For example, for the mesh-free structure similar to those in **Figure 3** but with ε = 35.4, we obtain the maxima of $|T^f_{yx}| = 0.96$ and $|T^f_{yx}| = 0.94$ at f = 1.63 and f = 1.87 GHz, respectively. The values of $|T^f_{xy}|$ that are equal to 0.86, 0.97, 0.88, 0.86, 0.92, and 0.98 correspond to the well-separated peaks located at 3.11, 3.4, 3.72, 4.32, 4.49, and 4.69 GHz, respectively. Among them, there are four narrow bands, in which either $|T^f_{xy}| > 0.99$ ($|T^b_{xy}|{\approx}0$) or $|T^f_{yx}| > 0.99$ ($|T^b_{yx}|{\approx}0$). More than five bands of high-efficiency, one-way polarization conversion can be easily obtained in one structure at subwavelength scale, i.e., at $a/\lambda < 0.5$ and $S/\lambda < 0.5$. This structure is really very thin as compared to λ, e.g., $S/\lambda{\approx}0.03$ at f = 1.63 GHz and $S/\lambda{\approx}0.09$ at f = 4.69 GHz. Note that $\varepsilon^{1/2}L/\lambda{\approx}0.095$ and $\varepsilon^{1/2}L/\lambda{\approx}0.28$, respectively. Thus, classical Fabry-Perot resonances in the dielectric layer are not expected to contribute to the resulting transmission mechanism that distinguishes our structures from those studied by Markovich et al. [31]. For the mesh-containing structure with ε = 35.4, we obtain four maxima, at which either $|T^f_{xy}| > 0.99$ ($T^b_{xy}{\approx}0$), or $|T^f_{yx}| > 0.99$ ($T^b_{yx}{\approx}0$), whereas $S/\lambda < 0.11$ and $a/\lambda < 0.4$. Although the total number of the polarization conversion bands (within the same frequency range) is usually larger for the mesh-free structures, those with the mesh often allow a better approaching to the case of perfect polarization conversion.

The resonance frequencies corresponding to the first maxima of $|T^f_{xy}|$ in **Figure 3(c, e, g, i)** are approximated by

$$f = f_{(0)}\varepsilon^{-0.36}, \tag{8}$$

where $f_{(0)}$ stands for the resonance frequency at $\varepsilon = 1$. If a wider range of ε variation is considered (i.e., $\varepsilon_{\max} > 11.4$), 0.36 in Eq. (8) must be replaced with a larger value, e.g., with 0.39 [1]. For a more accurate approximation, subranges of larger ε and smaller ε should be considered separately. In turn, for the both maxima of $|T^f_{yx}|$ from the first pair in **Figure 3(b, d, f, h)**, the following approximation can be used:

$$f = f_{(0)} \varepsilon^{-0.33}. \tag{9}$$

Thus, comparing with the classical scaling rule of $\varepsilon^{-1/2}$, one obtains similar rules for the basic configuration, so that the effect of variations in ε can be quantified. Surprisingly, the appropriate exponent values differ from the classical rule not strongly, although a large part of the resonance field energy may correspond to the exterior of subwavelength resonators. Moreover, not only locations of the maxima but also separation between them, Δf, can be predicted for different values of ε. In terms of the Lagrange model, a decrease in Δf at larger ε indicates a weaker coupling that might be connected with a larger electrical thickness of the dielectric substrate and a larger phase difference at the opposite sides of a dielectric layer.

A simultaneous widening of two polarization conversion bands can be achieved in the basic configuration just by modification of the metallic mesh parameters. **Figure 4** presents an example, in which the mesh has a larger period than that shown in **Figure 3**. In fact, this structure differs from that in **Figure 3(i)** only in the values of p and h. One can see that the second band of $|T^f_{xy}| = |T^b_{yx}| \approx 1$ in the vicinity of f = 6.4 GHz is similar to such bands shown in **Figure 3(g, i)**. However, the first band (at f = 2.8 GHz) has now two neighboring maxima with a very weak dip between them, so that our guess regarding a possible nature of the single maxima observed in the structures with the mesh in **Figure 3** may be correct. Note that the values of $|T^f_{yx}| = |T^b_{xy}|$ can also be quite large within the bands of $|T^f_{xy}| = |T^b_{yx}| \approx 1$, as shown in **Figure 4(b)** at f = 6.5 GHz. In this case, nearly perfect two-way polarization conversion is obtained, i.e., it occurs for both incident linear polarizations. Thus, both one-way and two-way polarization conversion bands can appear in one structure. Besides, it is interesting to compare locations of the maxima in **Figures 3(i)** and **4(b)**, while p is varied but the total volume occupied by a metal is fixed. The first band is shifted from f = 2.66 GHz to the vicinity of f = 2.8 GHz (two close maxima at 2.78 and 2.84 GHz, with a weak dip between them). The second band is shifted from the vicinity of f = 5.75 GHz to the vicinity f = 6.4 GHz. Hence, the introduced modification of the mesh geometry leads to the enhancement of the capacitive effect for the both broadbands.

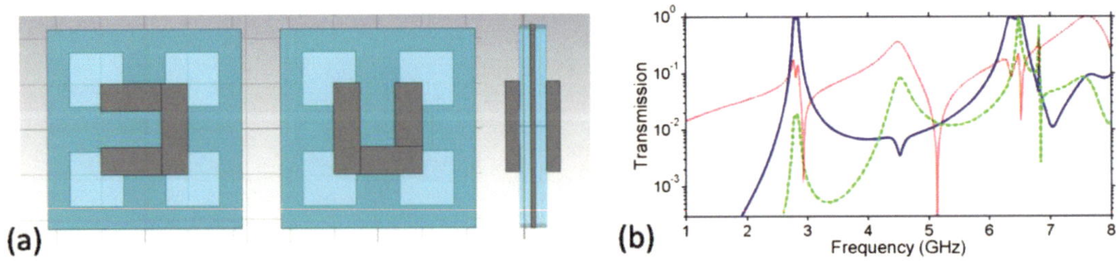

Figure 4. (a) Front, back, and side view of a unit cell, and (b) transmission for the basic configuration with a modified metallic mesh, p = 11 mm and h = 6 mm: solid blue line – $|T^f_{xy}| = |T^b_{yx}|$, dashed green line – $|T^f_{yx}| = |T^b_{xy}|$, and dotted red line – $|T^f_{xx}| = |T^b_{xx}| = |T^f_{yy}| = |T^b_{yy}|$; ε = 11.4.

http://dx.doi.org/10.5772/65950

3.2. Thin and thick configurations

Next, we consider the structures with a smaller and a larger distance between the coupled metasurfaces. Results for the structures with $L_1 = L_2 = 0.25$mm, $s_3 = 0.5$mm, and the same remaining parameters as for the basic configuration are presented in **Figure 5**. Now, we have $S = 2$mm, so that we refer to it as *thin configuration*. First, one should notice that separation of the resonances is here stronger pronounced than for the basic configuration. Generally, it may occur because each metasurface strongly affects the resonance field of the other in the thin configuration, leading to a stronger coupling. For example, the difference in location of the maxima observed in **Figure 5(b)** for the mesh-free structure with $\varepsilon = 2.1$ is larger than 2 GHz. In this case, the maximal values for two cross-polarized components are close to each other, while the copolarized components are not suppressed at the resonances.

Introducing a metallic mesh between the metasurfaces results in that the distance between the maxima dramatically decreases. Indeed, the center frequency $f_0 = (f_u + f_l)/2$, where f_u and f_l are the upper and the lower frequencies in a pair of the coupled (neighboring) resonances, is

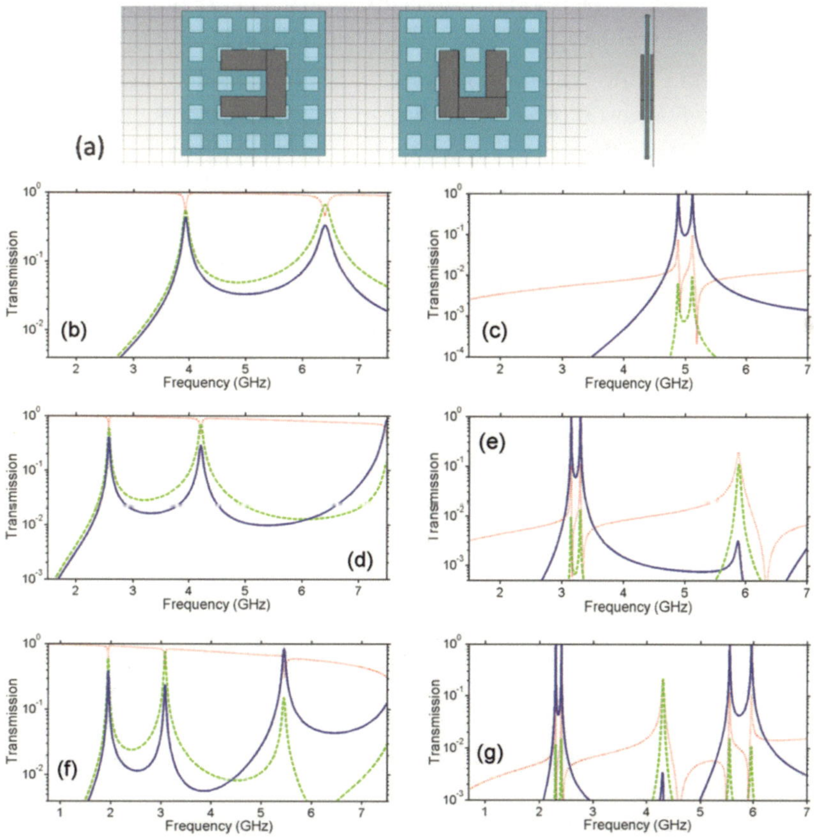

Figure 5. (a) Front, back, and side view of a unit cell, and transmission for the thin configuration (b, d, f) without and (c, e, g) with metallic mesh; (b, c) $\varepsilon = 2.1$, (d, e) $\varepsilon = 5.8$, (f, g) $\varepsilon = 11.4$; solid blue line – $|T^f_{xy}| = |T^b_{yx}|$, dashed green line – $|T^f_{yx}| = |T^b_{xy}|$, and dotted red line – $|T^f_{xx}| = |T^b_{xx}| = |T^f_{yy}| = |T^b_{yy}|$.

equal to 5.16 and 4.99 GHz in **Figure 5(b)** and **Figure 5(c)**, respectively, i.e., the shift is not very strong. At the same time, $\Delta f = f_u - f_l$ is 2.45 and 0.23 GHz, i.e., the difference between two structures in terms of resonance separation is very strong. For the mesh-free structures in **Figure 5(b, d, f)**, these features are in agreement with the predictions based on the Lagrange model (f_u and f_l correspond to ω_+ and ω_-): stronger coupling appears at smaller distances between the SRRs [41, 42]. On the other hand, assuming that the spectral separation of the neighboring maxima is connected with the strength of resonance coupling for a wider class of the structures, one may expect that in the mesh-containing structures in **Figure 5(c, e, g)**, the coupling is substantially weaker because of the effect exerted by the mesh. This feature probably originates from the fact that the mesh prevents direct coupling of nonevanescent waves at two sides of the entire structure. Similarly to **Figure 3**, nearly perfect polarization conversion and related diodelike transmission are obtained for the mesh-containing structures in **Figure 5**. Moreover, on the contrary to **Figure 3**, $|T^f_{xy}| > 0.99$ is achieved in a very wide range of variation of ε, while other components of the *T*-matrix are weak.

An increase of ε only leads to the resonance redshift, so that a larger number of narrow bands of polarization conversion might appear at the subwavelength range. Four nearly perfect maxima of $|T^f_{xy}|$ are observed in **Figure 5(g)** at ε = 11.4. Accordingly, Δf is decreased for each pair of the coupled resonances. For instance, for the first pair of resonances, we obtain $\Delta f = f_u - f_l$ = 0.15GHz and 1.64GHz in the structures with and without mesh at ε = 5.8, see **Figure 5(d, e)** and Δf = 0.11GHz and 1.14GHz for similar structures at ε = 11.4, see **Figure 5(f, g)**. Hence, a strong effect of the metallic mesh on coupling and location of high-*Q* resonances is typical at small distances between metasurfaces. Whereas obtaining nearly perfect polarization conversion does not need a special choice of substrate materials in the mesh-containing structures with small distance between metasurfaces, obtaining a near-unity cross-polarized component in similar mesh-free structures is a more challenging task.

The resonance frequencies corresponding to the first maxima of $|T^f_{yx}|$ in **Figure 5(b, d, f)** can be approximated by Eq. (9) but with the exponent of -0.375 instead of -0.33. For the first and second maxima of $|T^f_{xy}|$ (in the first pair) in **Figure 5(c, e, g)**, it is recommended to use Eq. (8) with -0.42 instead of -0.36. Note that the difference in the exponent required in the case of mesh free and mesh-containing structures is increased as compared to the basic configuration. Qualitatively, a better approaching to the value of -1/2 that is observed for the thin structures with the mesh means that the resonance fields are strongly localized and, thus, they may correspond to larger values of *Q*. Indeed, one can see that the resonances are sharper in **Figure 5** than in **Figure 3**. One should keep in mind that high-*Q* resonances can be more sensitive to the Ohmic losses, leading to undesired absorption enhancement and suppression of the transmission maxima [16].

Now, we consider *thick configuration*, which differs from the basic one in that $L_1 = L_2$ = 2.5mm and, thus, S = 8.5mm, see **Figure 6**. Compared to **Figures 3** and **5**, the mesh-containing structures seem now less appropriate for high-efficiency polarization conversion. Indeed, conversion is typically far from the perfect one, whereas reflections are quite strong. For

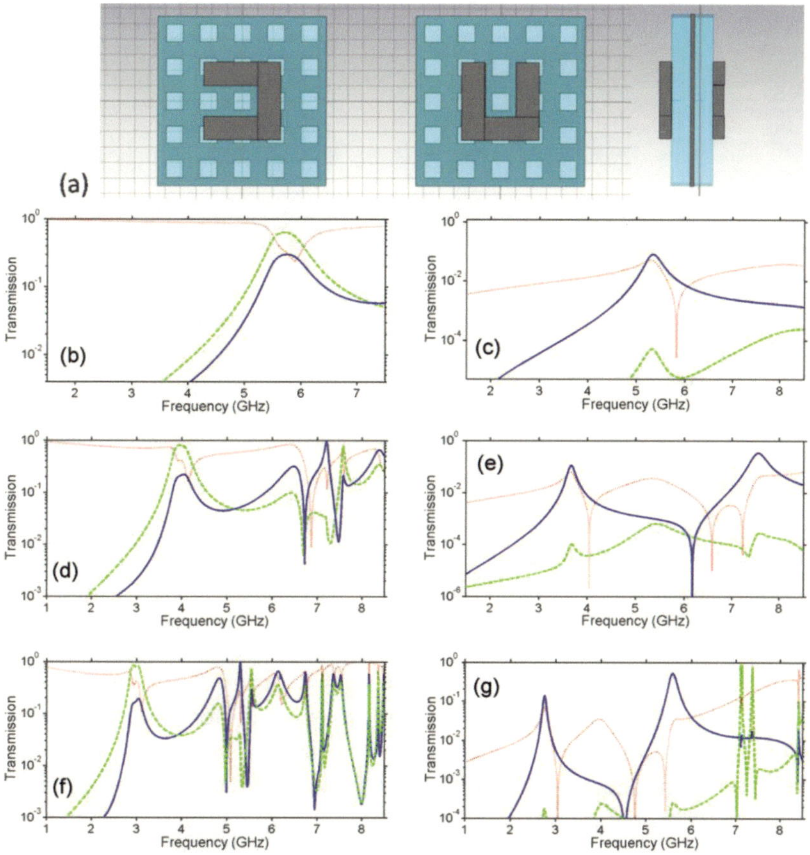

Figure 6. (a) Front, back, and side view of a unit cell, and transmission for the thick configuration (b, d, f) without and (c, e, g) with metallic mesh; (b, c) ε = 2.1, (d, e) ε = 5.8, (f, g) ε = 11.4; solid blue line – $|T^f_{xy}| = |T^b_{yx}|$, dashed green line – $|T^f_{yx}| = |T^b_{xy}|$, and dotted red line – $|T^f_{xx}| = |T^b_{xx}| = |T^f_{yy}| = |T^b_{yy}|$.

instance, in **Figure 6(e)** we obtain $|T^f_{xy}|\approx 0.11$ and $|T^f_{xy}|\approx 0.33$ at f = 3.65 and f = 7.55 GHz. In **Figure 6(g)**, $|T^f_{xy}|\approx 0.14$ and $|T^f_{xy}|\approx 0.51$ at f = 2.74 and f = 5.57 GHz, respectively, and $|T^f_{yx}| > 0.96$ at the two narrow maxima near f = 7.12 GHz (separation is about 25 MHz). At the same time, the mesh-free structures enable high-efficiency conversion and rather wide bands already at ε = 5.8 and, thus, can be more useful. For example, $|T^f_{yx}|\approx 0.79$ and $|T^f_{xy}| > 0.96$ at f = 3.9 and f = 7.2 GHz in **Figure 6(d)**. To compare, in **Figure 6(f)**, $|T^f_{yx}| > 0.8$ at f = 2.89 GHz and $|T^f_{xy}| > 0.95$ at f = 5.29 GHz, for ε = 11.4.

As follows from the comparison of **Figure 6(b, d, f)** with **Figure 6(c, e, g)**, the case of a large distance between metasurfaces in combination with the evanescent-wave regime of the metallic mesh is less appropriate for the practical use. At the same time, polarization conversion can be obtained in such structures for two different polarization states of the incident wave, see **Figure 6(g)**. Hence, this is not a unique property of the mesh-free structures, compare to

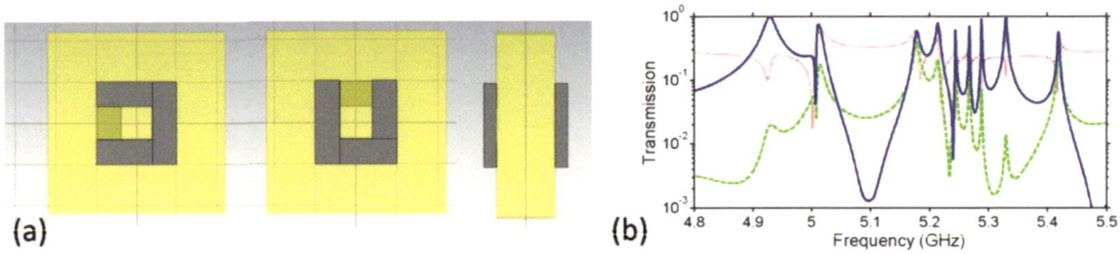

Figure 7. (a) Front, back, and side view of a unit cell, and transmission for the thick configuration with $L = d + L_1 + L_2 = 11mm$ and $S = 14mm$, and (b) transmission in the case without metallic mesh, ε = 35.4; solid blue line – $|T^f_{xy}| = |T^b_{yx}|$, dashed green line – $|T^f_{yx}| = |T^b_{xy}|$, and dotted red line – $|T^f_{xx}| = |T^b_{xx}| = |T^f_{yy}| = |T^b_{yy}|$.

Figures 3 and **5**. Regardless of this, the combination of variations in ε with those in L_1 and L_2 gives a big freedom for design. Also in this case, a scaling rule can be introduced, in the analogy with the rule of $\varepsilon^{-1/2}$ and Eqs. (8) and (9). For instance, for the first maximum of $|T^f_{xy}|$ in **Figure 6(c, e, g)** one can use Eq. (8), in which 0.36 is replaced with 0.23. This approximation is very rough, and the use of other functions than the exponential one may give a much better result in this case. However, it demonstrates the main trend quite correctly: the thicker the dielectric layers, the stronger the deviation from the rule of $\varepsilon^{-1/2}$.

Let us check whether an increase of L up to the values, at which classical Fabry-Perot resonances may appear in the dielectric layers, can improve the performance in terms of polarization conversion and asymmetric transmission. As an example, **Figure 7** presents the results for the structure, which is distinguished from those in **Figures 3(b, d, f, h)** and **6(b, d, f)** only in the values of ε and L. Transmission is shown for a high-frequency range, in which the multiple bands of one-way polarization conversion are located. For instance, the maxima of $|T^f_{xy}| \approx 0.98$ and $|T^f_{xy}| \approx 0.975$ are observed at $f = 4.93$ and $f = 5.32$ GHz, where $S/\lambda = 0.164$ and $\varepsilon^{1/2}L/\lambda = 0.68$, and $S/\lambda = 0.178$ and $\varepsilon^{1/2}L/\lambda = 0.74$, respectively. Hence, Fabry-Perot type interferences may, in principle, contribute to the resulting polarization conversion mechanism, although analytical description of this regime would require taking into account the effective contribution of SRRs. One can see that the maxima of $|T^f_{xy}|$, at which diodelike asymmetric transmission is well pronounced, are connected with the different resonances whose Q-factors may be strongly distinguished from each other. It follows from the obtained results in **Figure 7** that the mesh-free structures with thick dielectric layers may enable a nearly perfect multiband polarization conversion.

3.3. Role of array period

In addition, the effects exerted by variations of the substrate parameters on transmission and polarization conversion are sensitive to the array period, a. Changing the period, we change the distance between the neighboring subwavelength resonators in each metasurface. The same is related to the resonators belonging to the adjacent unit cells and different metasurfaces. Thus, it may be expected that such changes affect coupling and, hence, locations and manifestations of polarization conversion resonances. In **Figure 8**, the results are

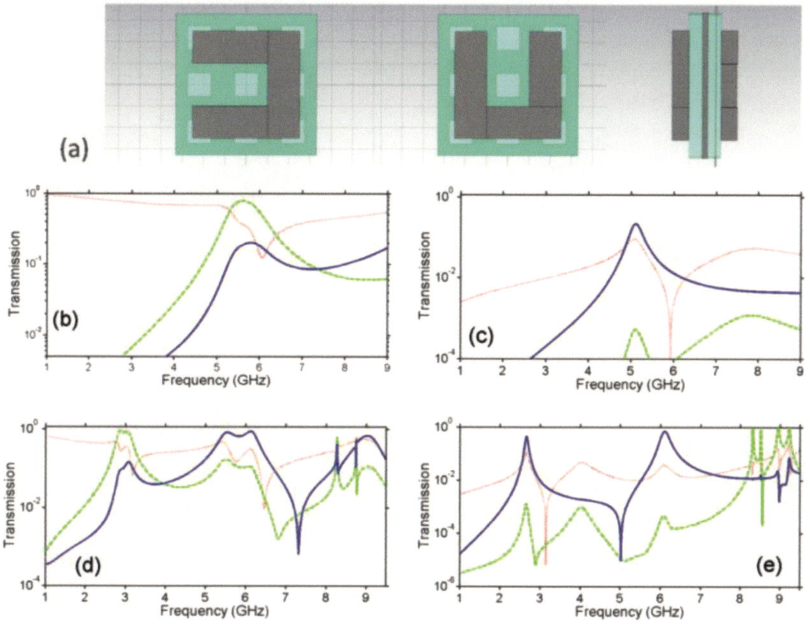

Figure 8. (a) Front, back, and side view of a unit cell, and transmission at a small array period, $a = 3p$, in cases (b, d) without and (c, e) with metallic mesh; (b, c) $\varepsilon = 2.1$, (d, e) $\varepsilon = 11.4$; solid blue line – $|T^f_{xy}|$, dashed green line – $|T^f_{yx}|$, and dotted red line – $|T^f_{xx}| = |T^f_{yy}|$.

presented for four structures, which differ from the basic configuration in that now $a = 3p$, while the remaining parameters are the same. Accordingly, $a/\lambda < 0.5$ at $f < 11.3$ GHz. Surprisingly, the obtained dependences are similar to the case of thick dielectric layers in **Figure 6**. This occurs for both the mesh-free and the mesh-containing structures. Thus, the former allow one obtaining high-efficiency polarization conversion starting from smaller values of ε than the latter. At the same time, as seen in **Figure 8(d, e)**, ε=11.4 is large enough for obtaining efficient polarization conversion and asymmetric transmission in the structures both with and without a mesh.

Finally, we consider the case of a larger array period, $a = 7p$, while the remaining parameters are the same as for the basic configuration in **Figure 3**. The results are presented in **Figure 9**. Now, $a/\lambda < 0.5$ at $f < 4.87$ GHz. Thus, higher diffraction orders may propagate at normal incidence starting from 9.74 GHz. The basic features are not changed, as compared to **Figures 3** and **6**. High-efficiency polarization conversion is observed in the mesh-containing structures with $\varepsilon = 2.1$ and $\varepsilon = 11.4$, see **Figure 9(c, e)**, and in the mesh-free structures at $\varepsilon = 11.4$, see **Figure 9(d)**. In particular, the first and second maxima with $|T^f_{yx}| > 0.8$ can be achieved in the mesh-free structure in **Figure 9(d)**. In the mesh-containing structures, $|T^f_{xy}| > 0.96$ and $|T^f_{xy}| > 0.99$ is obtained for the first maximum in **Figure 9(c)** and **Figure 9(e)**, respectively. Hence, high-efficiency transmission may occur even when the distance between the neighboring SRRs of the same metasurface is rather large. Thus, the energy can be *harvested* from a large

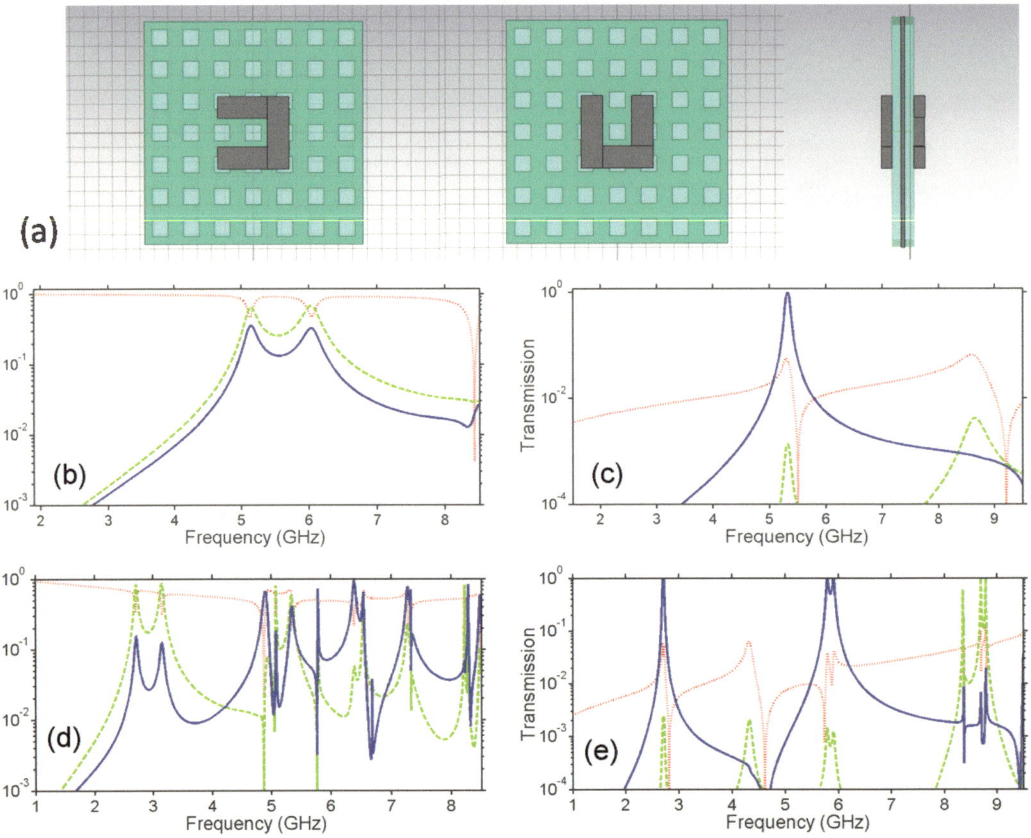

Figure 9. (a) Front, back, and side view of a unit cell, and transmission at a large array period, $a = 7p$, in cases (b, d) without and (c, e) with the metallic mesh; (b, c) $\varepsilon = 2.1$, (d, e) $\varepsilon = 11.4$; solid blue line – $|T^f_{xy}|$, dashed green line – $|T^f_{yx}|$, and dotted red line – $|T^f_{xx}| = |T^f_{yy}|$.

area and then converted into the orthogonal polarization. In other words, the coupled metasurfaces work like an *array of energy concentrators*. It is interesting that the harvesting occurs in the structures with and without a mesh. A detailed study is needed to clarify possible connection of this regime with (spoof) surface plasmons.

The effect of an array period on transmission is illustrated by the data shown in **Table 1**. One can see that spectral locations of the polarization conversion bands/maxima are rather weakly affected by the period. This indicates the dominant role of subwavelength resonances of the complex unit cells and coupling of metasurfaces owing to the resonators located at the front of each other. The array period strongly affects manifestations of the resonances, e.g., due to the difference in the value of Q. The presence of the mesh may lead to strong polarization conversion at the other incident polarization than in the corresponding mesh-free structures for all three values of the period. Nevertheless, the dominant effect of subwavelength resonances of the unit cells on the location of the polarization conversion bands/maxima remains in all of the cases studied.

http://dx.doi.org/10.5772/65950

Small array period, $a = 3p$	**Basic array period, $a = 5p$**	**Large array period, $a = 7p$**
	Structure without mesh, $\varepsilon = 2.1$	
Band of $\lvert T^f_{yx}\rvert$ around 5.58 GHz	Two maxima of $\lvert T^f_{yx}\rvert$ at 5.27 and 6.09 GHz	Two maxima of $\lvert T^f_{yx}\rvert$ at 5.15 and 6.04 GHz
	Structure with mesh, $\varepsilon = 2.1$	
Maximum of $\lvert T^f_{xy}\rvert$ at 5.1 GHz	Maximum of $\lvert T^f_{xy}\rvert$ at 5.31 GHz	Maximum of $\lvert T^f_{xy}\rvert$ at 5.32 GHz
	Structure without mesh, $\varepsilon = 11.4$	
Band of $\lvert T^f_{yx}\rvert$ around $f = 2.95$ GHz	Two maxima of $\lvert T^f_{yx}\rvert$ at 2.76 and 3.17 GHz	Two maxima of $\lvert T^f_{yx}\rvert$ at 2.71 and 3.15 GHz
	Structure with mesh, $\varepsilon = 11.4$	
Two maxima of $\lvert T^f_{xy}\rvert$ at 2.66 and 6.1 GHz	Maximum at 2.66 GHz and twin maximum at 5.72/5.82 GHz for $\lvert T^f_{xy}\rvert$	Twin maximum at 2.701/2.717 GHz and twin maximum at 5.8/5.91 GHz for $\lvert T^f_{xy}\rvert$

Table 1. Comparison of the bands/transmission maxima at which polarization conversion takes place, for $a = 3p$, $a = 5p$, and $a = 7p$; $p = 4.4$ mm.

4. Concluding remarks

To summarize, dielectric substrate parameters can strongly affect polarization conversion and asymmetric transmission regimes achievable in low-profile structures comprising two coupled metasurfaces that represent arrays of subwavelength resonant elements. The comparison has been carried out for the structures with and without metallic mesh and, thus, with and without tunneling. It has been demonstrated that sometimes the former and sometimes the latter can be preferable for the practical use. Perfect polarization conversion usually means that asymmetry in transmission is also perfect, and vice versa. This remains true at least if higher diffraction orders in the hosting air space are evanescent. The obtained results illustrate the main similarities and differences of the cases of a low-ε and high-ε substrate, a thin and thick dielectric layer, and a large and small array period. The role of a dielectric substrate that is directly adjusted to the subwavelength resonators is to modify the resonances that are located in the subwavelength range and shift more resonances to this range. Thus, it is far beyond of being just a mechanical support for metallic elements. In the case of small thickness of dielectric layers, the ability of preserving the secondary electromagnetic characteristics related to transmission and polarization is very general. This means that the required phase-matching conditions that may result in a nearly perfect transmission in the structures with a mesh can be satisfied in a wide range of ε-variation. This enables a scaling rule, which differs from the classical rule of $\varepsilon^{-1/2}$, while the polarization and transmission characteristics are preserved. The same remains true for the impedance-matching conditions in the structures without a mesh. Generally speaking, tunneling as a part of the resulting conversion mechanism is not necessary but desirable in many cases. For the fabrication reasons, structures with rather thick dielectric layers might be preferable. Moreover, thick layers help to obtain wider bands of polarization conversion, which are however not perfect. Unfortunately, conditions of nearly perfect polarization conversion are not satisfied for these structures in a wide ε-range, while thickness is fixed. Hence, more accurate adjustment of geometrical parameters is needed for a given material. In spite of this, variations in ε remain an efficient tool for manipulation by subwavelength resonances and related polarization conversion. The presented results give one useful guidelines, which are hopefully sufficient at least for initial-stage design. It is noteworthy that the resulting structures are electrically thin, e.g., $S/\lambda < 1/10$ and even $S/\lambda < 1/30$ for the selected operation regimes, when a substrate with $\varepsilon > 5$ is utilized. Although the zero-loss approximation has been used here, the basic effects are expected to remain when the actual Ohmic losses are taken into account, at least if Q-factor is not very high (i.e., resonances in zero-loss approximation are not very sharp). In particular, substrate parameters should exert a strong effect on the resulting performance at microwave, terahertz, and, probably, infrared frequencies. Finally, it should be noted that polarization conversion regimes that involve circular polarization can also be efficiently controlled by variations of the substrate parameters, while the individual metasurfaces are kept without change.

Acknowledgements

A.E.S. thanks the National Science Centre of Poland for financial assistance under the project MetaSel DEC-2015/17/B/ST3/00118, and TUBITAK for partial support in the framework of the Visiting Researcher Program. A part of this work is supported by the projects DPT-HAMIT, ESF-EPIGRAT, and NATO-SET-181, and by TUBITAK under the project nos. 107A004, 109A015, and 109E301. E.O. acknowledges partial support from the Turkish Academy of Sciences.

Author details

Andriy E. Serebryannikov[1]*, Mehmet Mutlu[2] and Ekmel Ozbay[2]

*Address all correspondence to: andser@amu.edu.pl

1 Faculty of Physics, Adam Mickiewicz University, Poznań, Poland

2 Nanotechnology Research Center—NANOTAM, Bilkent University, Ankara, Turkey

References

[1] Serebryannikov AE, Mutlu M, Ozbay E. Dielectric inspired scaling of polarization conversion subwavelength resonances in open ultrathin chiral structures. Appl. Phys. Lett. 2015;**107**:221907. DOI: 10.1063/1.4936603

[2] Gan X, Shiue RJ, Gao Y, Assefa S, Hone J, Englund D. Controlled light–matter interaction in graphene electrooptic devices using nanophotonic cavities and waveguides. IEEE Trans. Select. Top. Quant. Electron. 2014;**20**:6000311. DOI: 10.1109/JSTQE.2013.2273412

[3] Song K, Liu Y, Fu Q, Zhao X, Luo C, Zhu W. 90° polarization rotator with rotation angle independent of substrate permittivity and incident angles using a composite chiral metamaterial. Opt. Express. 2013;**21**:7439–7446. DOI: 10.1364/OE.21.007439

[4] Song K, Zhao X, Li Y, Fu Q, Luo C. A frequency-tunable 90°- polarization rotation device using composite chiral metamaterials. Appl. Phys. Lett. 2013;**103**:101908. DOI: 10.1063/1.4820810

[5] Yao Y, Kats MA, Shankar P, Song Y, Kong J, Loncar M, Capasso F. Wide wavelength tuning of optical antennas on graphene with nanosecond response time. Nano Lett. 2014;**14**:214–219. DOI: 10.1021/nl403751p

[6] Pendry JB. Negative refraction makes a perfect lens. Phys. Rev. Lett. 2000;**85**:3966–3969. DOI: 10.1103/PhysRevLett.85.3966

[7] Pendry JB. A chiral route to negative refraction. Science. 2004;**306**:1353–1355. DOI: 10.1126/science.1104467

[8] Kriegler CE, Rill MS, Linden S, Wegener M. Bianisotropic photonic metamaterials. IEEE Trans. Select. Top. Quant. Electron. 2010;**16**:367–375. DOI: 10.1109/JSTQE.2009.2020809

[9] Liu N, Liu H, Zhu S, Giessen H. Stereometamaterials. Nat. Photon. 2009;**3**:157–162. DOI: 10.1038/nphoton.2009.4

[10] Mutlu M, Akosman AE, Ozbay E. Broadband circular polarizer based on high-contrast gratings. Opt. Lett. 2012;**37**:2094–2096. DOI: 10.1364/OL.37.002094

[11] Torres V, Sanchez N, Etayo D, Ortuno D, Navarro-Cia M, et al. Compact dual-band terahertz quarter-wave plate metasurface. IEEE Photon. Technol. Lett. 2014;**26**:1679–1682. DOI: 10.1109/LPT.2014.2330860

[12] Li Z, Liu W, Cheng H, Chen S, Tian J. Realizing broadband and invertible linear-to-circular polarization converter with ultrathin single-layer metasurface. Sci. Rep. 2015;**5**:18106. DOI: 10.1038/srep18106

[13] Cheng H, Chen S, Yu P, Li J, Deng L, Tian J. Mid-infrared tunable optical polarization converter composed of asymmetric graphene nanocrosses. Opt. Lett. 2013;**38**:1567–1569. DOI: 10.1364/OL.38.001567

[14] Pfeiffer C, Zhang C, Ray V, Guo LJ, Grbic A. High-performance bianisotropic metasurfaces: asymmetric transmission of light. Phys. Rev. Lett. 2014;**113**:023902. DOI: 10.1103/PhysRevLett.113.023902

[15] Zhao Y, Belkin MA, Alu A. Twisted optical metamaterials for planarized ultrathin broadband circular polarizers. Nat. Commun. 2012;**3**:870. DOI: 10.1038/ncomms1877

[16] Fedotov VA, Mladyonov PL, Prosvirnin SL, Rogacheva AV, et al. Asymmetric propagation of electromagnetic waves through a planar chiral structure. Phys. Rev. Lett. 2006;**97**:167401. DOI: 10.1103/PhysRevLett.97.167401

[17] Plum E, Fedotov VA, Zheludev NI. Planar metamaterial with transmission and reflection that depend on the direction of incidence. Appl. Phys. Lett. 2009;**94**:131901. DOI: 10.1063/1.3107264

[18] Singh R, Plum E, Menzel C, Rockstuhl C, Azad AK, Cheville RA, Lederer F, Zhang W, Zheludev NI. Terahertz metamaterial with asymmetric transmission. Phys. Rev. B. 2009;**80**:153104. DOI: 10.1103/PhysRevB.80.153104

[19] Mutlu M, Akosman AE, Serebryannikov AE, Ozbay E. Asymmetric chiral metamaterial circular polarizer based on four U-shaped split ring resonators. Opt. Lett. 2011;**36**:1653–1655. DOI: 10.1364/OL.36.001653

[20] Ma X, Huang C, Pu M, Hu C, Feng Q, Luo X. Multi-band circular polarizer using planar spiral metamaterial structure. Opt. Express. 2012;**20**:16050–16058. DOI: 10.1364/OE.20.016050

[21] Pan C, Ren M, Li Q, Fan S, Xu J. Broadband asymmetric transmission of optical waves from spiral plasmonic metamaterials. Appl. Phys. Lett. 2014;**104**:121112. DOI: 10.1063/1.4869297

[22] Menzel C, Helgert C, Rockstuhl C, Kley EB, Tuennermann A, et al. Asymmetric transmission of linearly polarized light at optical metamaterials. Phys. Rev. Lett. 2010;**104**:253902. DOI: 10.1103/PhysRevLett.104.253902

[23] Mutlu M, Akosman AE, Serebryannikov AE, Ozbay E. Diodelike asymmetric transmission of linearly polarized waves using magnetoelectric coupling and electromagnetic wave tunneling. Phys. Rev. Lett. 2012:**108**,213905. DOI: 10.1103/PhysRevLett.108.213905

[24] Serebryannikov AE. One-way diffraction effects in photonic crystal gratings made of isotropic materials. Phys. Rev. B. 2009;**80**:155117. DOI: 10.1103/PhysRevB.80.155117

[25] Li XF, Ni X, Feng LA, Lu MH, He C, Chen YF. Tunable unidirectional sound propagation through a sonic-crystal-based acoustic diode. Phys. Rev. Lett. 2011;**106**:084301. DOI: 10.1103/PhysRevLett.106.084301

[26] Serebryannikov AE, Ozbay E, Nojima S. Asymmetric transmission of terahertz waves using polar dielectrics. Opt. Express. 2014;**22**:3075–3088. DOI: 10.1364/OE.22.003075

[27] Rodriguez-Ulibarri P, Beruete M, Navarro-Cia M, Serebryannikov AE. Wideband unidirectional transmission with tunable sign-switchable refraction and deflection in nonsymmetric structures. Phys. Rev. B. 2013;**88**:165137. DOI: 10.1103/PhysRevB.88.165137

[28] Feng F, Ayache M, Huang J, Xu YL, Lu MH, Chen YF, et al. Nonreciprocal light propagation in a silicon photonic circuit. Science. 2011;**333**:729–733. DOI: 10.1126/science.1206038

[29] Mutlu M, Cakmakyapan S, Serebryannikov AE, Ozbay E. One-way reciprocal spoof surface plasmons and relevant reversible diodelike beaming. Phys. Rev. B. 2013;**87**: 205123. DOI: 10.1103/PhysRevB.87.205123

[30] Serebryannikov AE, Cakmak AO, Ozbay E. Multichannel optical diode with unidirectional diffraction relevant total transmission. Opt. Express. 2012;**20**:14980–14990. DOI: 10.1364/OE.20.014980

[31] Markovich DL, Andriyieuski A, Zalkovskij M, et al. Metamaterial polarization convertor analysis: limits of performance. Appl. Phys. B. 2013;**112**:143–152. DOI: 10.1007/s00340-013-5383-8

[32] Serebryannikov AE, Beruete M, Mutlu M, Ozbay E. Multiband one-way polarization conversion in complementary split-ring resonator based structures by combining chirality and tunneling. Opt. Express. 2015;**23**:13517–13529. DOI: 10.1364/OE.23.013517

[33] Li Z, Chen S, Tang C, Liu W, Cheng H, et al. Broadband diodelike asymmetric transmission of linearly polarized light in ultrathin hybrid metamaterial. Appl. Phys. Lett. 2014;**105**:201103. DOI: 10.1063/1.4902162

[34] Shi J, Liu X, Yu S, Vu T, Zhu Z, Ma HF, Cui TJ. Dual-band asymmetric transmission of linear polarization in bilayered metamaterial. Appl. Phys. Lett. 2013;**102**:191905. DOI: 10.1063/1.4773516

[35] Shi JH, Ma HF, Guan CY, Wang ZP, Cui TJ. Broadband chirality and asymmetric transmission in ultrathin 90°-twisted Babinet-inverted metasurfaces. Phys. Rev. B. 2014;**89**:165128. DOI: 10.1103/PhysRevB.89.165128

[36] Xu HX, Wang GM, Qi MQ, Cai T, Cui TJ. Compact dual-band circular polarizer using twisted Hilbert-shaped planar metamaterials. Opt. Express. 2013;**21**:24912–24921. DOI: 10.1364/OE.21.024912

[37] Zhou L, Wen W, Chan CT, Cheng P. Electromagnetic-wave tunneling through negative-permittivity media with high magnetic fields. Phys. Rev. Lett. 2005;**94**:243905. DOI: 10.1103/PhysRevLett.94.243905

[38] Hou B, Wen H, Leng Y, Wen W. Electromagnetic wave transmission through subwavelength metallic meshes sandwiched between split rings. Appl. Phys. Lett. 2005;**87**:201114. DOI: 10.1063/1.2133915

[39] Sun W, He Q, Hao J, Zhou L. A transparent metamaterial to manipulate electromagnetic wave polarizations. Opt. Lett. 2011;**36**:927–929. DOI: 10.1364/OL.36.000927

[40] Semouchkina E. Development of miniature microwave components by using high contrast dielectrics. In: Minin I, editor. Microwave and Millimeter Wave Technologies from Photonic Bandgap Devices to Antenna and Applications. Intech; Rijeka. 2010. pp. 231–256. DOI: 10.5772/9062

[41] Li TQ, Liu H, Li T, Wang SM, Wang FM, Wu RX, Chen P, et al. Magnetic resonance hybridization and optical activity of microwaves in a chiral metamaterial. Appl. Phys. Lett. 2008;**92**:131111. DOI: 10.1063/1.2905285

[42] Liu N, Liu H, Zhu S, Liu H, Genov DA, Wu DM, et al. Magnetic plasmon hybridization and optical activity at optical frequencies in metallic nanostructures. Phys. Rev. B. 2007;**76**:073101. DOI: 10.1103/PhysRevB.76.073101

[43] Liu H, Cao JX, Zhu SN, Liu N, Ameling R, Giessen H. Lagrange model for the optical properties of stereometamaterials. Phys. Rev. B. 2010;**81**:241403. DOI: 10.1103/PhysRevB.81.241403

Chapter 2

Active and Tunable Metamaterials

Yongjun Huang, Jian Li and Guangjun Wen

Additional information is available at the end of the chapter

http://dx.doi.org/10.5772/67198

Abstract

In this chapter, we will first present the research progress on the active and tunable metamaterials based on different realization methods, such as varactor diodes, liquid crystals, superconductivity, and structural-shifting structures. Then we focus on the achievements in our research group for the tunable metamaterials by using the ferrite as the substrate of metamaterials. We will present the designs and theories of single-, dual-, and triple-band tunable metamaterials based on the ferrite and the design of metamaterial absorbers based on the ferrite. It will indicate that the proposed tunable metamaterials have many advantages compared with other active and tunable metamaterials.

Keywords: tunable metamaterials and absorbers, ferrite

1. Introduction

Metamaterial [1] is a kind of artificial material. When the unit cell of a metamaterial is much smaller than the working wavelength, it can be considered as a homogeneous medium. Metamaterial exhibits many novel electromagnetic properties [2–7] and has important potential applications [8–12], and has attracted wide research attention. However, there are many unsolved problems and bottlenecks, including the large loss, narrow bandwidth, and untunable frequency band. The active tunable metamaterials have shown novel advantages for solving the issues that appeared in the conventional metamaterial.

Here we conclude the recently developed active and tunable metamaterials based on the varactor diodes, liquid crystals, superconductivity, and structural-shifting structures. First, the tunable metamaterial based on the varactor diodes is proposed by Gil et al. [13] and Velez et al. [14]. As shown in **Figure 1(a)**, the diodes are integrated in the regular split ring resonators (SRRs), the resonance of the structure can be adjusted by controlling the DC bias acting on the diodes. After that, various magnetic negative metamaterials and negative index metamaterials based on such varactor diodes are reported and experimentally demonstrated. For instances, researchers experimentally verified that the refractive index of S-shaped metamaterial loaded with diodes (**Figure 1b**) can be changed via altering the magnitude of DC bias [15]. By investigating one SRR unit loaded with a capacitor located at different position or loaded with capacitor with different capacitances (**Figure 1c**), the resonance can also be shifted [16]. Another similar design is achieved by integrating the BST capacitor into the SRR as shown in **Figure 1(d)** [17]. This kind of design can be flexibly integrated in the regular microstrip circuits to achieve the tunable band-pass/band-stop filters, antennas, and nonlinear devices [18, 19].

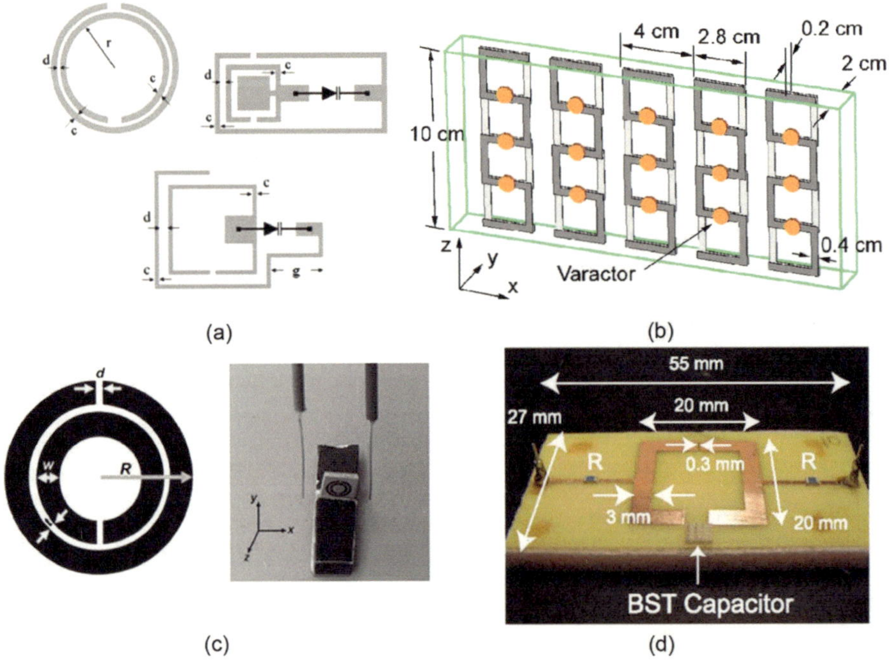

Figure 1. Tunable metamaterials based on varactor diodes. (a) SRR loaded with diodes [13], (b) S-shaped resonator loaded with diodes [15], (c) SRR loaded with capacitor [16], and (d) SRR loaded with BST capacitor [17].

It is well known that the dielectric permittivity of liquid crystal can be adjusted by changing the magnitude and direction of external electric field and magnetic field, and/or the temperature [20, 21]. Therefore, one can integrate the liquid crystal into the conventional metamaterials to realize the tunable designs. For example, researchers experimentally investigated the SRR and Omega-shaped magnetic negative metamaterials infiltrated with the liquid crystals as shown in **Figure 2(a)** and **(b)** [22, 23]. Then the tunability is achieved by controlling the direction of external magnetic field. At the same time, other researcher exhibited a similar SRR magnetic negative metamaterials loaded with the liquid crystals (see **Figure 2c**) and the tunability is obtained by changing the external electric field strengths [24]. At higher operating frequencies, it works as well for the liquid crystal-based tunable metamaterials. Researchers

http://dx.doi.org/10.5772/67198

numerically investigated the reconfigurable index of refraction from negative through zero to positive values by changing equivalently the index of liquid crystal for the kind of tunable metamaterial operating at near-infrared wavelength [25], as shown in **Figure 2(d)**. Another tunable fishnet metamaterials infiltrated by liquid crystals operated at optical ranges was numerically demonstrated as well recently as shown **Figure 2(e)** [26]. It has shown that the liquid crystal can be used to realize the tunable metamaterials and several external control methods can adjust and tune the properties of such metamaterials. However, the tuning range is not wide enough as observed in the previously mentioned references.

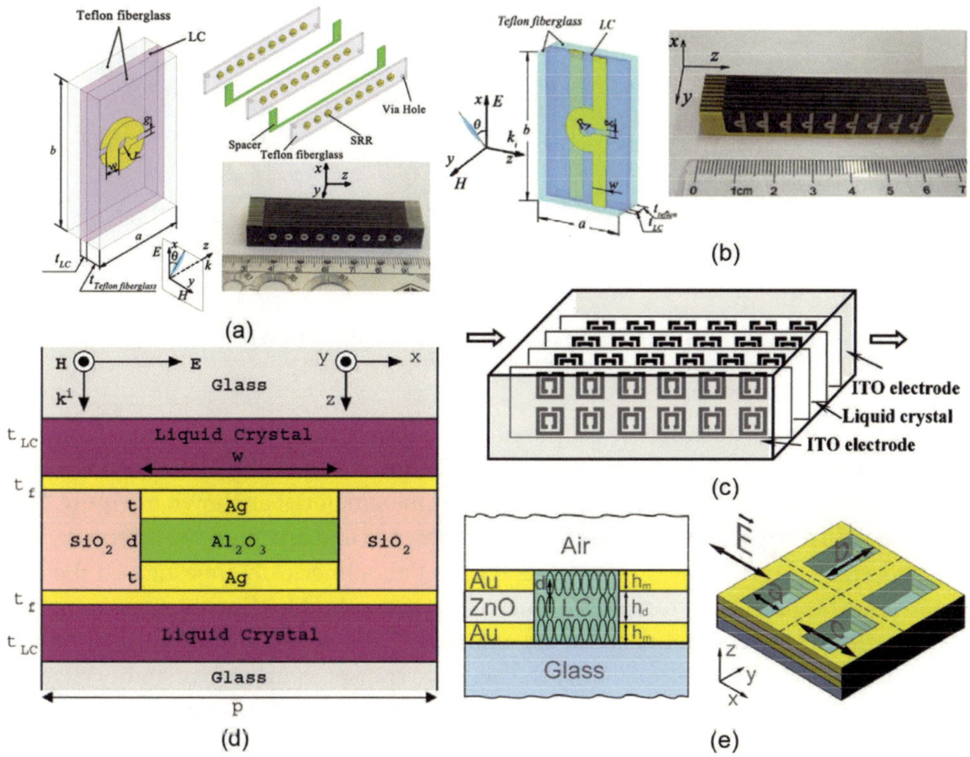

Figure 2. Tunable metamaterials based on liquid crystals. (a) Circular SRR structure [22], (b) Omega-shaped resonator [23], (c) square SRR structure [24], (d) near-infrared structure [25], and (e) fishnet structure [26].

Recently, other kinds of tunable metamaterials were also proposed by using, e.g., superconductivity and MEMS structures [27, 28] as shown in **Figure 3(a)** and **(b)**. For the superconductivity-based metamaterial, the tunability is obtained by changing the external DC or RF magnetic field [27]. For the MEMS metamaterial, the tunability is achieved by heating the nanostructure with rapid thermal annealing method to alter the direction of the SRR [28]. Another two similar structurally shifted metamaterials operated at microwave and THz range are shown in **Figure 3(c)** and as well, and these two designs can alter the electromagnetic responses by mechanically changing the relative positions for part of the SRRs in the array vertically and/or horizontally [29, 30].

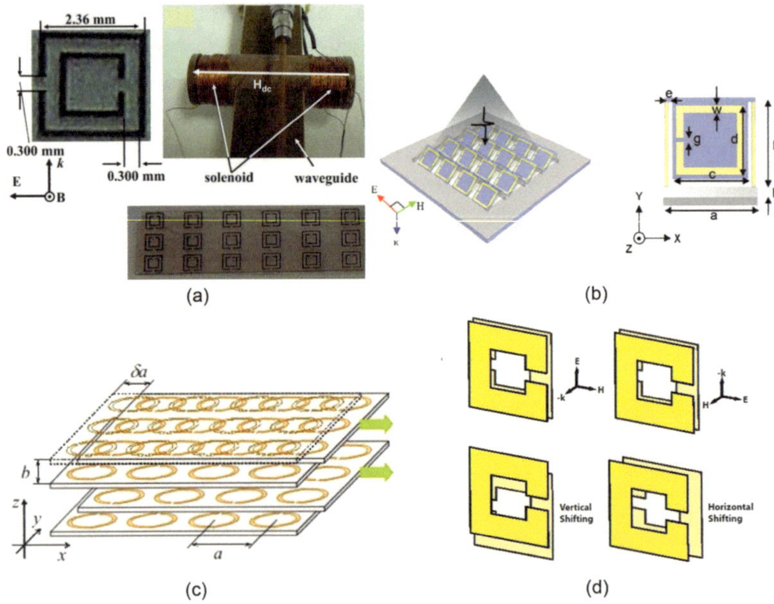

Figure 3. Other tunable metamaterial configurations. (a) Superconductivity [27], (b) MEMS [28], (c) structural shifted structure [29], and (d) THz structural shifted structure [30].

On the other hand, some researchers proposed ferrite-based metamaterials, that is, substituting ferrites such as yttrium iron garnets (YIG) for the SRR structures to obtain the negative permeability [31–38]. Dewar first proposed the ferrite-based metamaterial as shown in **Figure 4(a)** and gave the theoretical expressions of the effective parameters and analyzed the interactions of the ferrites and wires [31, 32]. Rachford et al. and Cai et al. numerically investigated the electromagnetic wave propagation properties of such metamaterials [33, 34] with minor modification (see **Figure 4b**) for the easy fabrication. He et al. and Zhao et al. fabricated the metamaterial samples and investigated experimentally the negative transmission and tunability characteristics [35, 36] at different frequencies as shown in **Figure 4(c)** and **(d)**. In our previous works, we also analyzed the theoretical parameters, investigated the negative refractive index characteristics, and discussed the different refractive characteristics in a broad frequency band [37, 38].

Similarly, due to the magnetic properties of the ferrites, it can be controlled with the applied DC field. The ferrites can be embedded into the conventional resonant structure metamaterials in order to realize the tunability. The experimental studies on the SRR metamaterial loaded with ferrite was carried out. The tunable characteristics of working frequency band for the single magnetic negative metamaterial and the double negative metamaterials are demonstrated as shown in **Figure 5(a)** and **(b)** [39, 40]. Our research has also shown that the ferrite embedded into the conventional metamaterials can not only realize the tunable frequency band, but also can realize the dual or even tri-band tunable characteristics (as shown in **Figure 5c**) [41]. This will be discussed in details in the next section. At the same time, researches used two different kinds of ferrites to realize the tunable dual-band metamaterials (as shown in **Figure 5d**) [42]. This realization process is simple, and its working frequency band can be adjusted in a wide frequency band with the applied DC magnetic field.

http://dx.doi.org/10.5772/67198

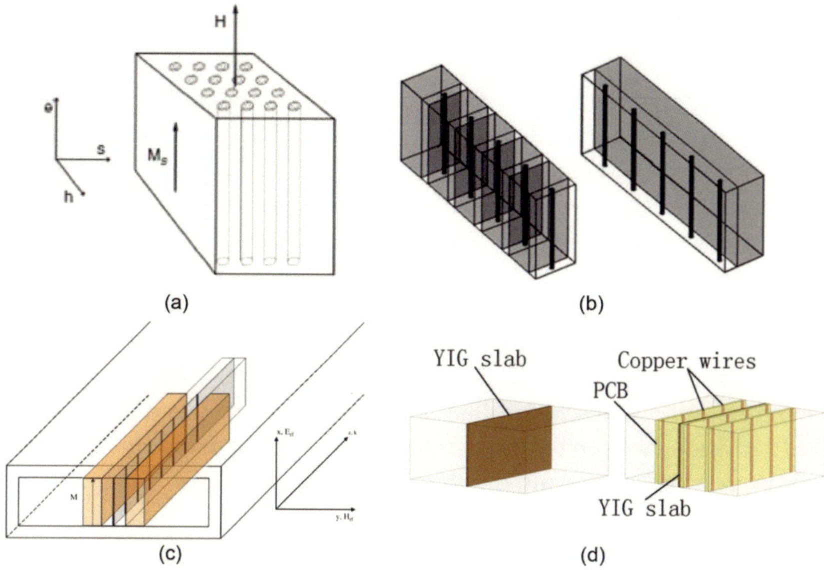

Figure 4. Tunable metamaterials based on the ferrite. (a) Original design [31, 32], (b) modification achievements [34], and (c and d) experimental demonstrations [35, 36].

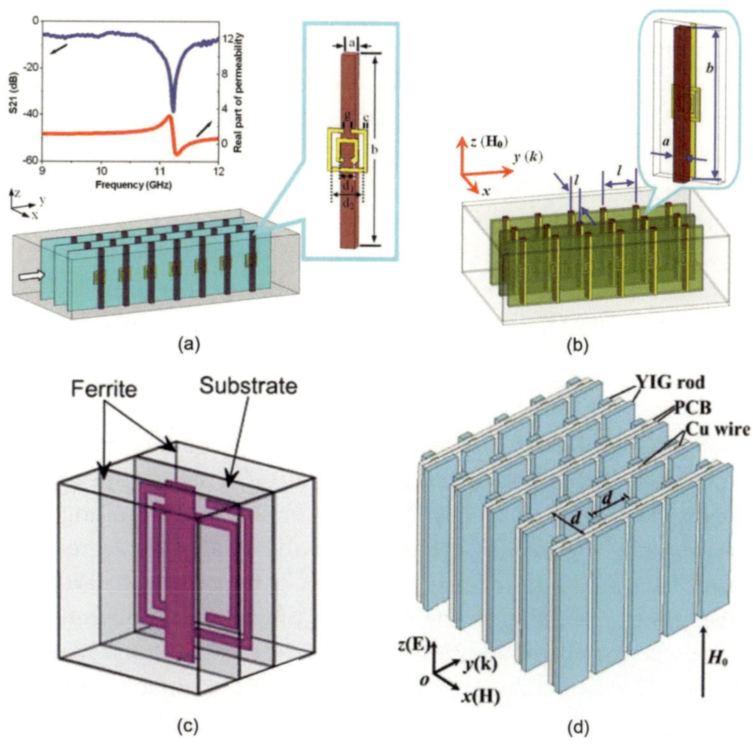

Figure 5. Tunable metamaterials loaded with ferrite as the substrate. (a) Magnetic negative structure [39], (b) double negative structure [40], and (c and d) dual-band structures [41, 42].

2. Theoretical analysis

Here the theoretical analysis of the proposed ferrite-inspired metamaterials is presented. As shown in **Figure 6(a)**, when a transverse electromagnetic (TEM) wave acts on the ferrite slabs along the *y*-axis (the electric field along the *z*-axis), and a DC magnetic bias acts on the ferrite slabs along the *z*-axis, the effective permeability of ferrite has the following property: $\mu_{eff} = (\mu^2 - \kappa^2)/\mu$, where $\mu = 1 + \omega_m\omega_0/(\omega_0^2 - \omega^2)$, $\kappa = \omega_m\omega/(\omega_0^2 - \omega^2)$ [43], and $\omega_0 = \gamma H_0$ stands for ferromagnetic resonance (FMR) frequency, γ denotes the gyromagnetic ratio, $\omega_m = 4\pi Ms\gamma$ represents the characteristic frequency of the ferrite slabs, ω is the angular frequency, and H_0 indicates the DC magnetic bias. For the ferrite slabs used here (TT1-3000 series of Trans-Tech Ltd., the saturation magnetization $4\pi Ms = 3000$ G, $\varepsilon_r = 12.9$, $\tan\delta = 0.0005$, and resonance beamwidth $\Delta H = 228$ Oe), the damping describing loss of ferrite should be taken into consideration. In that case, the FMR frequency can be revised to $\omega_0 + j\alpha\omega$, where α stands for the damping of the ferromagnetic precession. As a result, the effective permeability has changed to

$$\mu_{eff} = 1 - \frac{\omega_m}{\frac{\omega^2}{\omega_0 + \omega_m} - \omega_0 - j\alpha\omega\frac{\omega^2}{(\omega_0 + \omega_m)^2 + 1}} \tag{1}$$

The higher order infinitesimal α is usually ignored due to the very low loss in ferrite slabs.

The effective permeability is calculated under different magnetic biases based on Eq. (1), as shown in **Figure 6(b)**. It can be seen that the effective permeability of ferrite slabs exhibits all typical resonant characteristics with different DC magnetic biases. The negative $\mathrm{Re}(\mu_{eff})$ is achieved above the resonant frequencies. For one condition as an example, the $\mathrm{Re}(\mu_{eff})$ is negative in the frequency range of 9.5–14 GHz and changes to positive after 15 GHz when $H_0 = 2.2$ kOe. So, when combining the ferrite slabs into the classic magnetic metamaterial structures which exhibit negative permeability frequency band below or above the negative band of ferrite slabs, one can expect to achieve two negative $\mathrm{Re}(\mu_{eff})$ frequency band. On the other hand, for the continuous wires of the classic SRR-wires [2] and short wire pairs [44] structures (the Ω-like structures [45] can be also considered as continuous wires), the negative permittivity frequency band is wide enough which can cover the two negative permeability frequency bands. Therefore, the ferrite-based metamaterials have the ability to achieve the dual-band negative refractive index properties.

Moreover, as shown in **Figure 6(b)**, the magnetic resonant frequency of ferrite slabs is shifted from 9.5 to 12 GHz when H_0 increases from 2.2 to 3 kOe. So the negative permeability frequency band of ferrite slabs can be shifted arbitrarily and reversibly by controlling the magnetic bias. Furthermore, as shown in **Figure 6(b)**, the positive part of $\mathrm{Re}(\mu_{eff})$ is decreased when increasing of the magnetic bias. So the resonance frequencies of the SRR, Ω-like resonators and short wire pairs can be altered, resulting in the changing of the second operating band. Then we numerically investigate the above-mentioned three types of dual-band metamaterial to demonstrate the theoretical expectations. We also investigate the tunable triple-band realization and a single-band THz ferrite-based tunable metamaterials. Lastly, by using the ferrite, we propose two kinds of tunable and broadband metamaterial absorbers (MAs) in details.

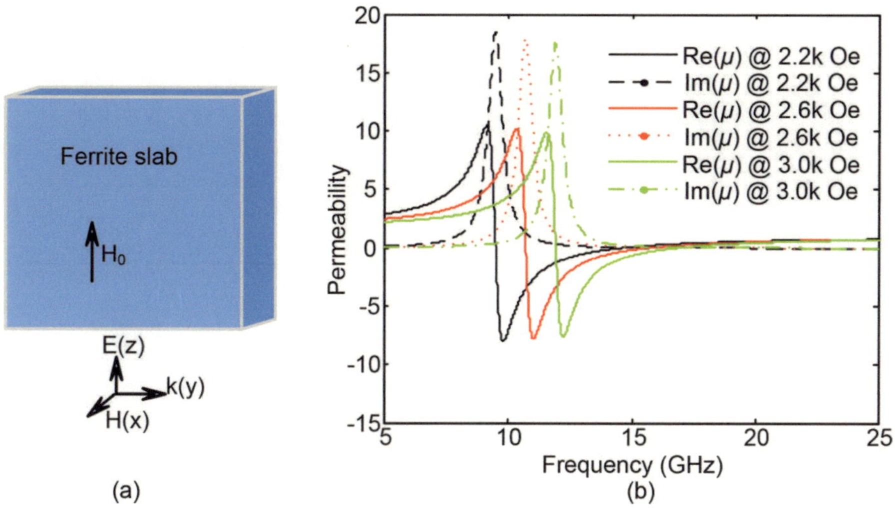

Figure 6. (a) A sample presentation of the ferrite under a DC magnetic bias, and (b) theoretical effective permeability of ferrite slab under different DC magnetic biases [41].

3. Numerical demonstrations for ferrite-based tunable metamaterials

3.1. Tunable dual-band configurations

The tunable dual-band metamaterial composed of ferrite slabs and SRR-wires is numerically discussed first. The schematic view of such structure with unit cell dimension of 2 mm is designed and shown in **Figure 7(a)**. As can be seen, the SRR and wire are placed in a 0.6-mm Rogers RT/duroid 5880 ($\varepsilon_r = 2.2$, tan $\delta = 0.0009$) substrate and separated by 0.25 mm, and then two 0.7-mm TT1-3000 ferrite slabs are placed on the two sides of such substrate. In simulations, the structure is excited by a TEM wave with propagation vector (k) along the y-axis, and electric field vector (E) along the z-axis, as shown in **Figure 7(a)**. A DC magnetic bias acts on the ferrite slabs along the z-axis to excite the negative permeability property.

First, the single-band SRR-wires metamaterial and ferrites-wires metamaterial are demonstrated, respectively, to show the negative operating bands. The dimensional values of the two single-band metamaterials are optimized. For the SRR-wires structure, we have $w_1 = 0.4$ mm, $d_1 = 1.6$ mm, $d_2 = 1.2$ mm, $w_2 = 0.1$ mm, and $w_3 = 0.3$ mm. In this structure, a fictitious media ($\varepsilon_r = 12.9$, tan $\delta = 0.0005$) is used to compare the ferrite media. The results shown in **Figure 7(b-i)** indicate that the single-band SRR-wires metamaterial has a negative operating band centered at 14.1 GHz. The transmission peak is −0.44 dB at 13 GHz. For the ferrite-wire structure, the fictitious media is replaced by ferrite media and the SRR is deleted. A 2-kOe magnetic bias acts on the ferrite slabs along the z-axis. The results shown in **Figure 7(b-ii)** indicate that the single-band ferrite-wire metamaterial has a negative operating band centered at 8.5 GHz. The transmission peak is −6.6 dB at 8.2 GHz.

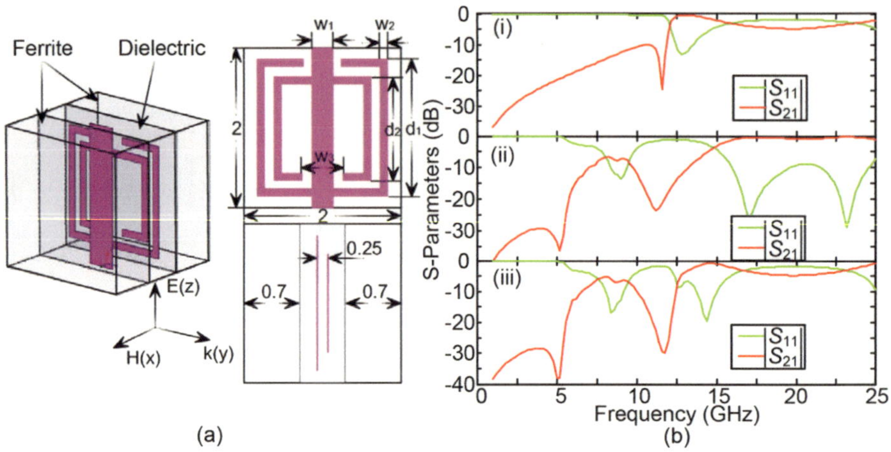

Figure 7. (a) The schematic of dual-band metamaterial composed of ferrite slabs and SRR-wires structure and (b) simulated transmission and reflection characteristics for the discussed configurations [41].

Then the combined dual-band metamaterial is investigated numerically. When the applied DC magnetic bias is set as 2 kOe, as shown in **Figure 7(b-iii)**, it clearly shows that there are two passbands centered at 8.4 and 15.3 GHz, respectively. The two transmission peaks are −5.6 and −1 dB at 8 and 14.6 GHz. Comparing the two single-band metamaterials and the dual-band metamaterial, the first passband contributed by ferrite slabs is not changed, while the second passband contributed by SRR in the combined dual-band metamaterial has a few blueshift. This is because that the positive part of the effective permeability is less than 1, as shown in **Figure 6**, so the resonance frequency of the SRR within the ferrite media is larger than the condition in the dielectric fictitious media.

To demonstrate the dual negative refractive index properties, the effective impedance and refractive index are obtained from the simulated S-parameters by using the retrieval method [46], and shown in **Figure 8(a)**. In **Figure 8(a)**, we see that the EM waves can be transmitted in the metamaterial in the above-mentioned two passbands. The real part of effective refractive index is below zero in such two passbands with very low loss due to the small imaginary part of effective refractive index.

To show the tunable property of the ferrite-based dual-band metamaterial, the transmission characteristics of such structure at different DC magnetic biases ranging from 2.2 to 3 kOe are numerically investigated and shown in **Figure 8(b)**. It can be seen that both the passbands are shifted accordingly. Specifically, the first passband is shifted from 8.5 to 10 GHz with tuning rate of about 1.875 MHz/Oe and the second passband is shifted from 14.6 to 15.8 GHz with tuning rate of about 1.5 MHz/Oe. The transmission peaks and bandwidths for the first operating band are not changed and the second operating band is narrowed when increasing the magnetic bias. As shown in **Figure 6**, this is because when increasing the magnetic bias, the negative permeability near the magnetic resonance is shifted rapidly; however, the positive permeability of the upper negative region is shifted slowly. Therefore, the resonance frequency of SRR within the ferrite media is shifted slowly.

http://dx.doi.org/10.5772/67198

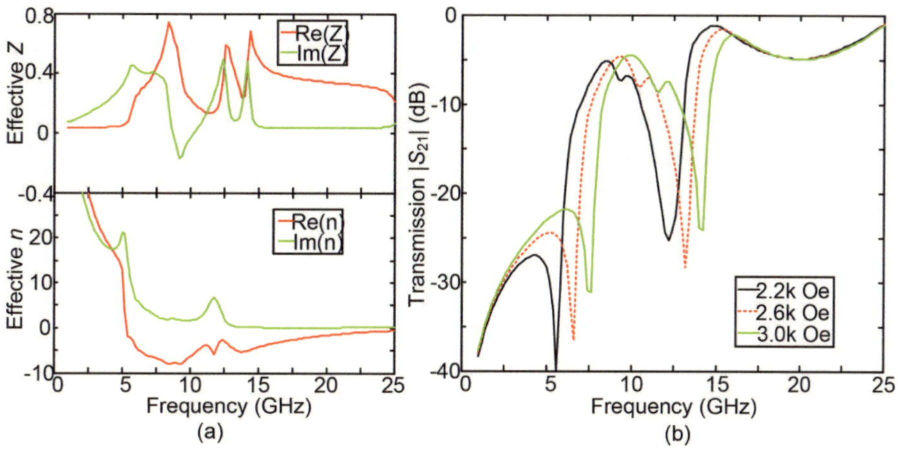

Figure 8. (a) Retrieved effective impedance and refraction, and (b) simulated tunability of the ferrite-based SRR-wire metamaterial [41].

Then the second kind of ferrite-based tunable dual-band metamaterial is discussed. We add the ferrite slabs to the Ω-like resonator [45] to achieve dual-band property. As shown in **Figure 9(a)**, the Ω-like resonators are placed in a 0.6-mm Rogers RT/duroid 6010 ($\varepsilon_r = 10.2$, $\tan\delta = 0.0023$) substrate and separated by 0.25 mm, and then two 0.7-mm TT1-3000 ferrite slabs are placed at the two sides of the substrate. The dimension of this dual-band metamaterial unit is $2 \times 2 \times 3$ mm^3. The other structural values are optimized numerically: $w_1 = 0.4$ mm, $r = 1.6$ mm, and $w_2 = 0.2$ mm.

As shown in **Figure 9(b)**, the results indicate that the single-band Ω-like metamaterial has a negative operating band centered at 12.1 GHz, and the transmission peak is −0.2 dB at 12.6 GHz. The ferrite-based metamaterial under a magnetic bias of 2 kOe has two passbands, centered at 7 and 13.9 GHz, respectively. The transmission peaks are −6 and −1.6 dB at 8 and 13.5 GHz, accordingly. Also, the second passband contributed by Ω-like resonators in the combined dual-band metamaterial is larger than the condition in the single-band Ω-like metamaterial.

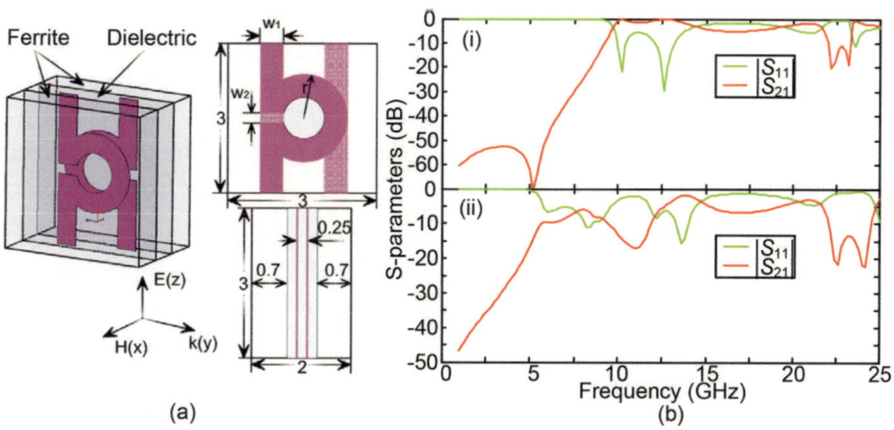

Figure 9. (a) The schematic of dual-band metamaterial composed of ferrites and Ω-like resonators, and (b) simulated transmission and reflection characteristics of the Ω-like metamaterial alone and loaded with ferrite [41].

The retrieved effective impedance and refractive index of such dual-band metamaterial, shown in **Figure 10(a)**, indicate that the EM waves can be transmitted in such metamaterial in the above-mentioned two passbands. The transmission properties of the structure are further simulated in different magnetic biases ranging from 2.2 to 3 kOe and shown in **Figure 10(b)**. It is seen that the two operating bands are shifted as well. Specifically, the first passband is shifted from 8.4 to 9.8 GHz with tuning rate of about 1.75 MHz/Oe, and the second passband is shifted from 13.7 to 15.3 GHz with tuning rate of about 2 MHz/Oe.

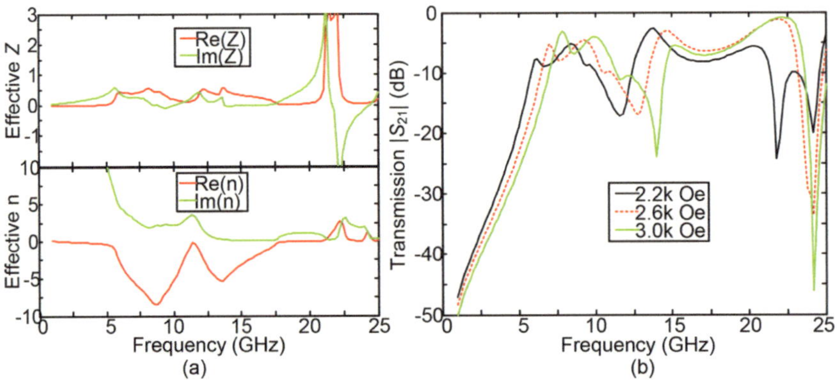

Figure 10. (a) Obtain effective impedance and refraction, and (b) numerical demonstrated tunability for the ferrite-based Ω-like metamaterial [41].

We now discuss the last kind of tunable dual-band metamaterial composed of ferrites and short wire pairs. As presented in **Figure 11(a)**, the short wire pairs are placed on a 0.25-mm Rogers RT/duroid 6010 substrate. Two TT1-3000 ferrite slabs are placed on the two sides of the substrate with air gaps, and then two other ferrite slabs are put on the other side of the substrate. The dimension of this dual-band metamaterial unit is 3 × 2 × 4 mm^3. The other structure values are optimized numerically: w_1 = 0.2 mm, w_2 = 0.4 mm, d = 3.4 mm, and w_3 = 0.6 mm.

As shown in **Figure 11(b)**, the results indicate that the single-band metamaterial has a negative passband centered at 16.5 GHz. The transmission peak is −0.2 dB at 15.9 GHz. The ferrite-based dual-band metamaterial under a magnetic bias of 2 kOe has two passbands centered at 7.9 and 16.5 GHz, respectively. The two transmission peaks are −7.3 and −0.5 dB at 8 and 16.2 GHz, respectively. The second passband contributed by short wire pairs in the combined dual-band metamaterial is larger than the condition in the single-band short wire pairs metamaterial.

As shown in **Figure 12(a)**, the retrieved effective impedance and refractive index of such metamaterial indicate that the EM waves can be transmitted in the metamaterial in the above-mentioned two passbands. The transmission properties of the dual-band metamaterial are further simulated in different magnetic biases ranging from 2.2 to 3 kOe, and are shown in **Figure 12(b)**. It can be seen that such two operating bands are shifted as well. Specifically, the first passband is shifted from 8.5 to 10.7 GHz with tuning rate of about 2.75 MHz/Oe. However, the second passband is shifted slightly, from 16.2 to 16.5 GHz with tuning rate of about 0.375 MHz/Oe. The second passband is narrowed when increasing the magnetic bias.

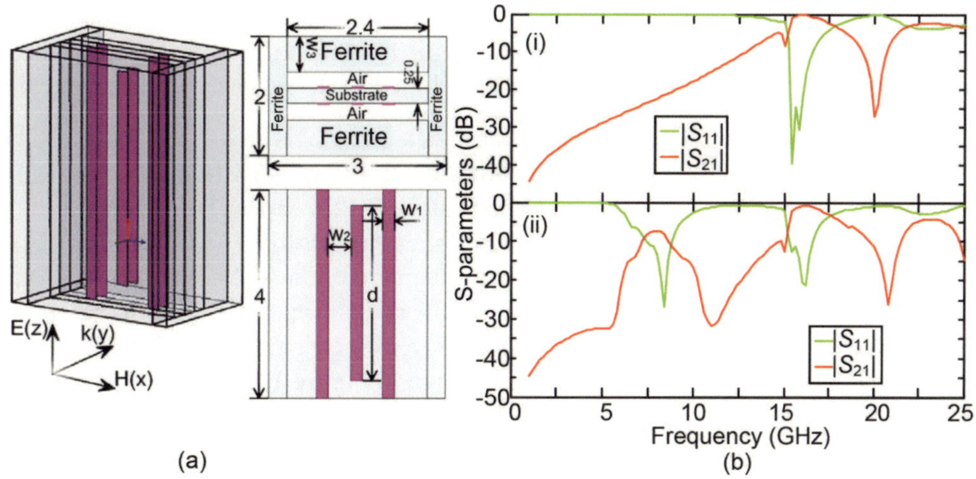

Figure 11. (a) The schematic of dual-band metamaterial composed of ferrites and short wire pairs, and (b) simulated transmission and reflection characteristics for the single-band short wire pair metamaterial and combined dual-band ferrite-based short wire pair metamaterial [41].

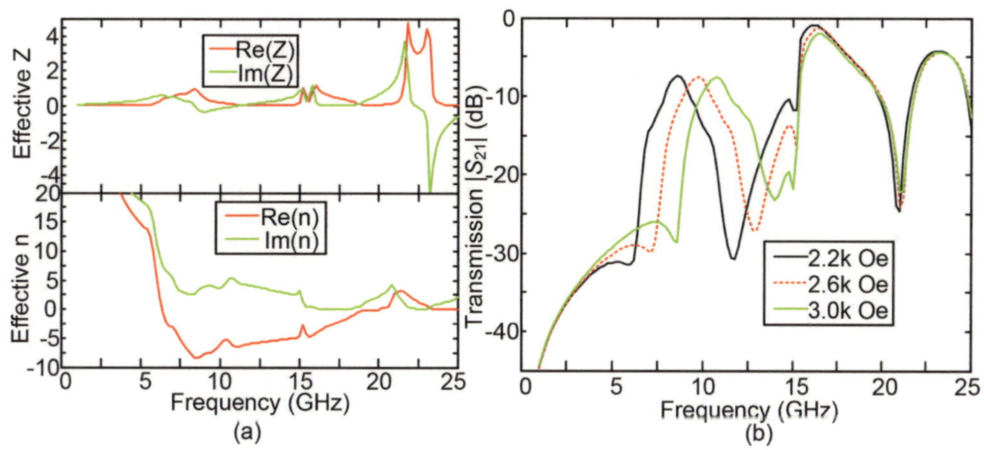

Figure 12. (a) Obtained effective impedance and refraction, and (b) the numerically demonstrated tunability for the ferrite-based short wire pair metamaterial [41].

From the above theoretical analysis and numerical investigations, the three dual-band metamaterials with different structures exhibit the same transmission characteristics and negative refraction. The two passbands of all the three metamaterials can be shifted arbitrarily and reversibly by easily changing the magnetic bias. Therefore, we can conclude that embedding the ferrites in the conventional resonance structure can fabricate the dual-band and tunable metamaterials.

3.2. Tunable triple-band configuration

The tunable triple-band negative permeability metamaterial is proposed by merging the ferrite slabs into the previously published dual-band negative permeability metamaterial based on single-loop resonator (SLR) [47]. The SLRs provide two separate magnetic resonances, and the ferrite slabs under an applied DC magnetic bias exhibit the third magnetic resonance. All the three magnetic resonances can be controlled by adjusting the intensity of the magnetic bias.

The unit cell of the proposed tunable triple-band metamaterial is shown in **Figure 13(a)** [48]. A 0.03-mm-thick copper SLR is printed on one side of a Rogers TMM 4 substrate ($\varepsilon_r = 4.5$, loss tangent tan $\delta = 0.002$). A commercial ferrite slab (G-4256 series, Trans-Tech Ltd.), with saturation magnetization $4\pi Ms = 1600$ Gs, resonant beamwidth $\Delta H = 84$ Oe, $\varepsilon_r = 15.1$, and loss tangent tan $\delta = 0.0002$ is deposited on the other side of the TMM 4 substrate. As mentioned in Refs. [31, 32], the ferrite would damage the resonances of the metallic resonators if these two elements are directly touched. Therefore, a dielectric substrate should be inserted between the ferrite and the SLR structure (as shown in **Figure 13a**) to decouple this strong interaction.

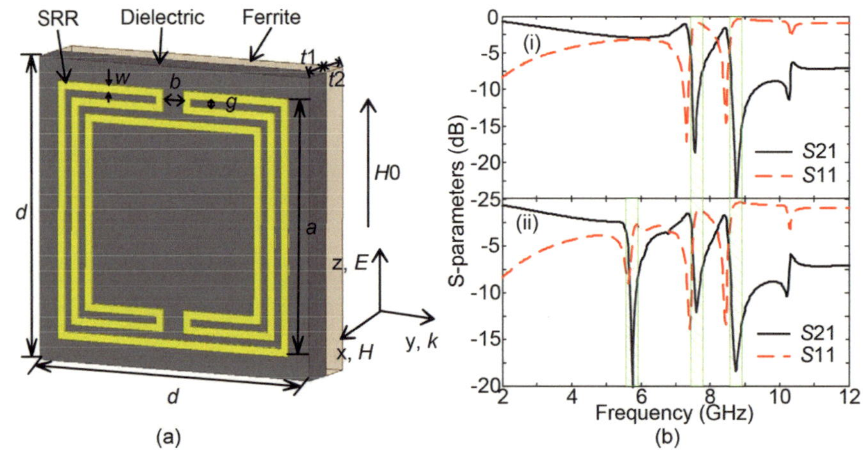

Figure 13. (a) Schematic representation of the tunable triple-band metamaterial unit cell and (b) numerical results of the transmission/reflection of the dual-band SLR structure metamaterial and the triple-band metamaterial [48].

In our simulations, the unit cell of the proposed ferrite-based metamaterial is placed in the center of a waveguide (with cross section of 4 mm × 4 mm) with perfect electric boundaries along the *z*-axis and perfect magnetic boundaries along the *x*-axis. The applied DC magnetic bias H_0 is set on the ferrite slab along the *z*-axis and the two wave ports placed in the front and back of the metamaterial along the *y*-axis. The geometric parameters of the metamaterial, obtained by numerical optimization, are shown as follows: $a = 3.4$ mm, $b = 0.3$ mm, $d = 4$ mm, $g = w = 0.1$ mm, and $t_1 = t_2 = 0.5$ mm.

The transmission and reflections of the SLR array with and without ferrite under the DC magnetic bias of 1.4 kOe are computed, and the results are shown in **Figure 13(b)**. In order to make a fair comparison, in the case without ferrite slab, a same sized fictitious medium which has the relative dielectric constant of 15.1 and a loss tangent of 0.0002 is used. **Figure 13(b-i)** shows the case for dual-band SLR structure. There are three transmission deeps appearing at 7.5, 8.7, and 10.2 GHz, respectively. The first two transmission deeps

http://dx.doi.org/10.5772/67198

correspond to the magnetic resonances and the last one corresponds to the electric resonance, which have been numerically demonstrated in [47]. As shown in **Figure 13(b-ii)**, when a ferrite slab is merged, four transmission deeps appear at 5.7, 7.6, 8.7, and 10.2 GHz, respectively. Intuitively, the lowest deep results from the ferrite slab and the other three deeps come from the SLR structure. As we can see, the presence of ferrite has little influence on the three transmission deeps of the SLRs (only very slight frequency shifts), and the resonant frequency of the ferrite slab in simulation also agrees well with the theoretical result in **Figure 6**.

For better understanding the resonant behaviors of the ferrite-based metamaterial, in **Figure 14** we investigate the surface current distribution on the metallic SLR at each resonant frequency. It is clearly found that for the first transmission deep at 5.7 GHz, almost no surface currents are observed and so we can conclude that this resonance is contributed by the ferrite slab. For the other three transmission deeps—at 7.6, 8.7, and 10.2 GHz—the characteristics of surface currents are similar to that analyzed in [47].

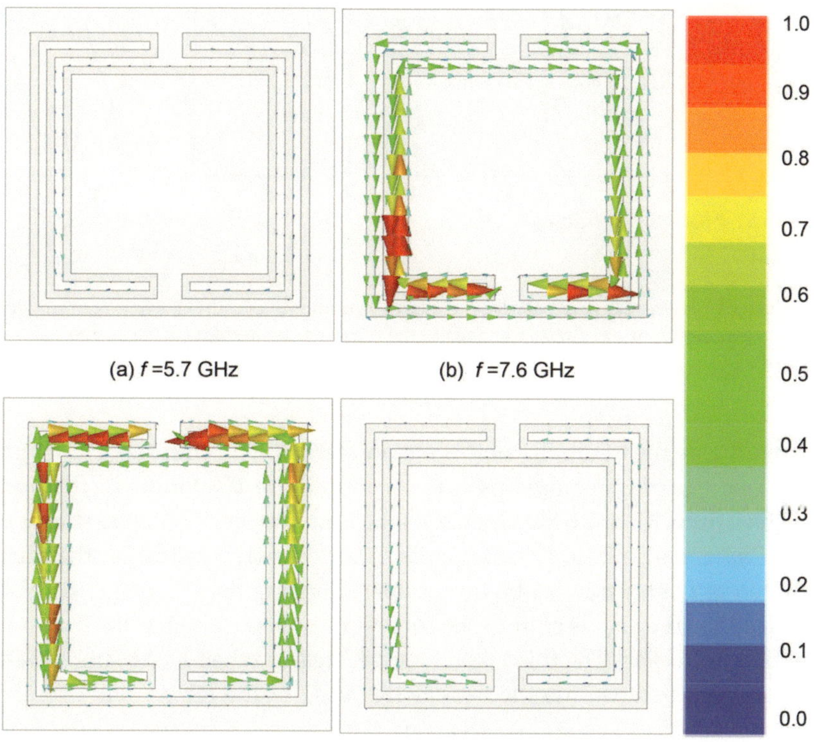

Figure 14. The simulated surface current distributions of the SLR structure at four resonant frequencies: (a) 5.7 GHz, (b) 7.6 GHz, (c) 8.7 GHz, and (d) 10.2 GHz. All the subplots use the same scale as shown in the right [48].

The effective constitutive parameters of the ferrite-based metamaterial are further determined from the numerically computed S-parameters by using the retrieval method [46]. As shown in **Figure 15**, the effective permeability curve exhibits three negative magnetic resonances around 5.7, 7.5, and 8.5 GHz, where three μ-negative bands are obtained in 5.7–5.9, 7.5–7.7, and 8.5–9.0 GHz, respectively. It is also shown that a weak electric resonance appears at 10.2 GHz, which is not strong enough to obtain a ε-negative band. Comparison of theoretical permeability of ferrite medium (**Figure 6**) and the numerically retrieved effective permeability of the ferrite slab (inset of **Figure 15b**) shows that the intrinsic resonant frequencies of the ferrite in these two conditions agree very well. However, the μ-negative band in the simulation is much narrower than that of the theoretical analysis, because the theoretical result corresponds to a bulk ferrite medium, while the numerical one corresponds to a finite ferrite slab. Through this comparison, we can conclude that the intrinsic resonant frequency of the ferrite slab is not affected significantly by the magnetic resonances of the metallic SLR structure.

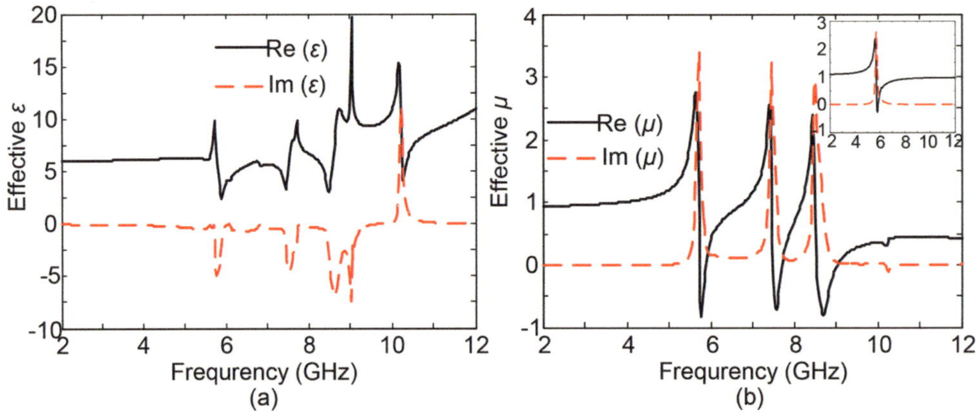

Figure 15. Retrieved effective media parameters: (a) effective permittivity and (b) effective permeability of the triple-band metamaterial. The inset figure in (b) corresponds to the retrieved permeability of the ferrite slab simulated at the same condition [48].

Now, let us investigate the tunability of this metamaterial under different DC magnetic biases. The tunable properties of transmissions as shown in **Figure 16(a)** indicate that increasing the DC magnetic bias from 1.2 to 1.6 kOe increases the first transmission deep sharply from 5.2 to 6.3 GHz; the second transmission deep increases very slowly; contrarily, the third transmission deep decreases from 8.8 to 8.6 GHz, because adjusting the DC magnetic bias in a small range cannot lead to an obvious shift of the second resonance which is very closely above the resonance of the ferrite slab. The third resonance decreases when increasing the DC magnetic bias in a small range because the theoretical $|\mathrm{Re}(\mu_{eff})|$ of the ferrite slab changes to bigger values as shown in **Figure 6**.

The effective permeability curves under different DC magnetic biases are also determined from the simulated S-parameters, and the results are shown in **Figure 16(b)**. It is found that when increasing the DC magnetic bias, the first resonance increases sharply; the second resonance increases very slowly; and contrarily, the third resonance decreases. In all the cases under different DC magnetic biases, three μ-negative bands are obtained. We have also numerically found that when further increasing the intensity of the DC magnetic bias, first the frequencies of all the three resonances increase, then the resonance of the ferrite slab will overlap with the two resonances of the SLR, and lastly the resonance of the ferrite slab will be above the two resonances of the SLR.

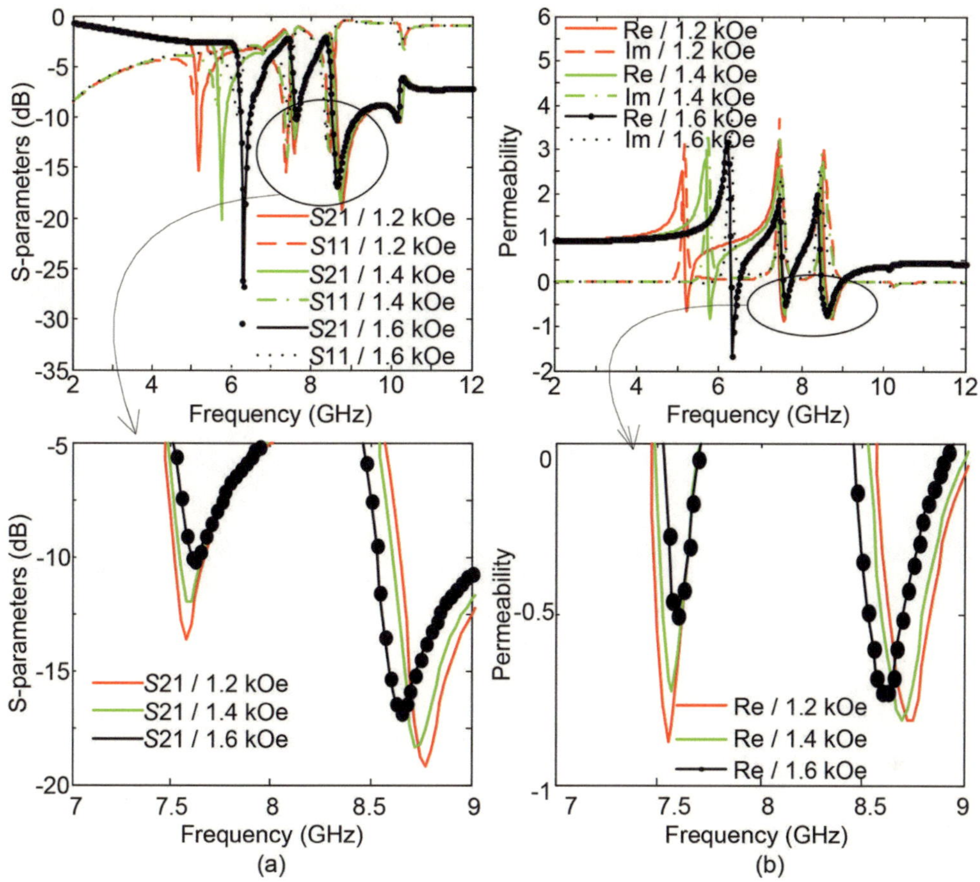

Figure 16. (a) Simulated transmissions and (b) retrieved effective permeability of the triple-band metamaterial under different dc magnetic biases ranging from 1.2 to 1.6 kOe [48].

3.3. Tunable THz ferrite-based metamaterial

Now we discuss the tunable THz ferrite-based metamaterial by using the well-developed ferrite film and metallic films technique [49]. Special ferrite films [such as the Lu Bi Fe O (LuBiIG) garnet film (see **Figure 17a**) [50] prepared by liquid phase epitaxy (LPE) method on a gadolinium gallium garnet (GGG) substrate] are used to achieve negative permeability. For the ferrite film used (saturation magnetization $4\pi Ms = 1750$ Gs, resonant beamwidth $\Delta H = 1$ Oe, $\varepsilon_r = 3$), the negative permeability band with very low magnetic loss can be obtained at the THz region. The silver films deposited on polyimide (Kapton 500HN, Krempel, Vaihingen/Enz, Germany) substrates using inkjet printing techniques (see **Figure 17b**) [51] are utilized to achieve negative permittivity. The substrate polyimide, with a relative permittivity of $\varepsilon_r = 2.2$, and loss tangent of tan δ = 0.0002, decouples the interactions between the ferrite films and the silver films.

Figure 17(c) shows the schematic view of the LuBiIG–silver–LuBiIG tunable metamaterial. The silver strip is deposited first on one of the two polyimide substrates, and such two substrates are placed close to the two LuBiIG films. The TEM wave propagates along the *y*-direction with the electric field along the *z*-direction. The applied DC magnetic field acts on the LuBiIG film along the *z*-direction. **Figure 17(d)** shows the theoretically calculated effective permeability and it exhibits typical resonant characteristics as well. When the applied DC magnetic field is set as 50 kOe, the real part of effective permeability is below zero in the frequency range of 0.142–0.145 THz. The imaginary part is much smaller than the real part and it means very low magnetic loss and thus can be ignored. Noticeably, the resonant frequencies of the ferrite film increase from 0.1397 to 0.1453 THz as the magnetic bias increases from 49 to 51 kOe.

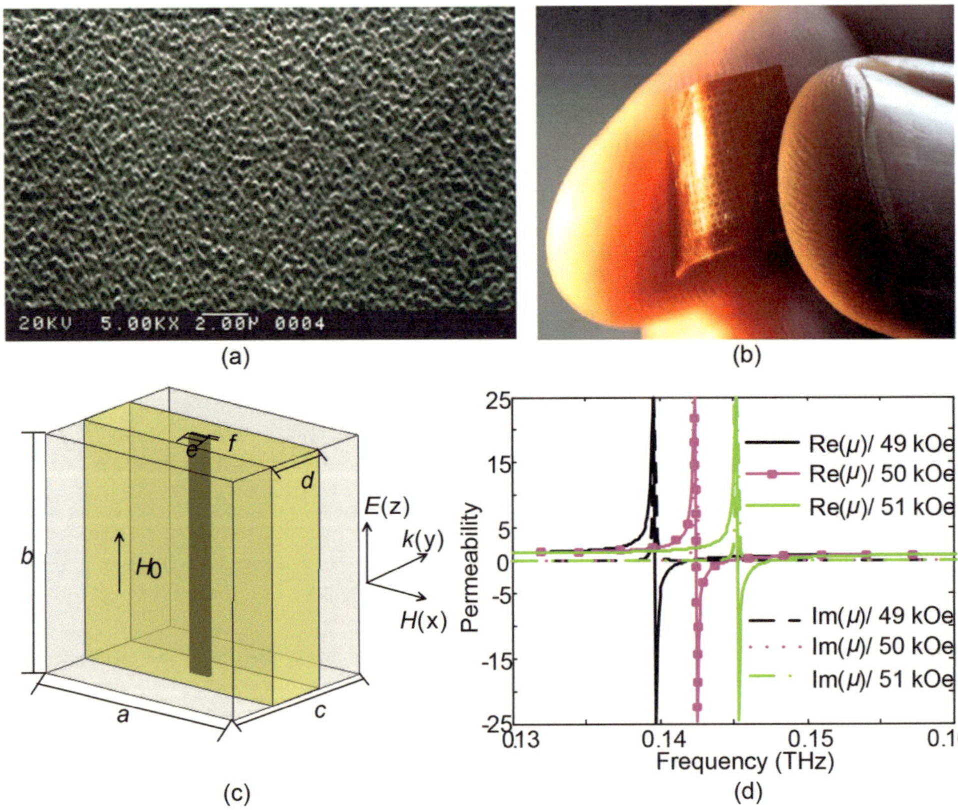

Figure 17. (a) Surface pattern images of the LuBiIG thin film [50], (b) photo of the silver SRR film [51], and (c and d) the tunable THz ferrite metamaterial unit and theoretical permeability of the ferrite film [49].

After the physical parameters of LuBiIG film, substrate, and silver are determined above, we now optimize the size of each film. The previous theoretical analysis results indicate that the effective permeability of LuBiIG film under 50 kOe is below zero in the frequency range of 0.142–0.145 THz, and the plasma frequency of silver strip array is about 1.5 THz when the periodicity *d* is about 100 μm. Based on these results, the optimized transmission results are depicted in **Figure 18**. The optimized parameters are found to be the following: a = 140 μm,

$b = c = 100$ μm, $d = 18.4$ μm, $e = 16$ μm, and $f = 2$ μm. From the simulation results as shown in **Figure 18**, the THz metamaterials have a passband centered at 0.1413 THz. The operating bandwidth (S21 > 8 dB) is about 2.4 GHz.

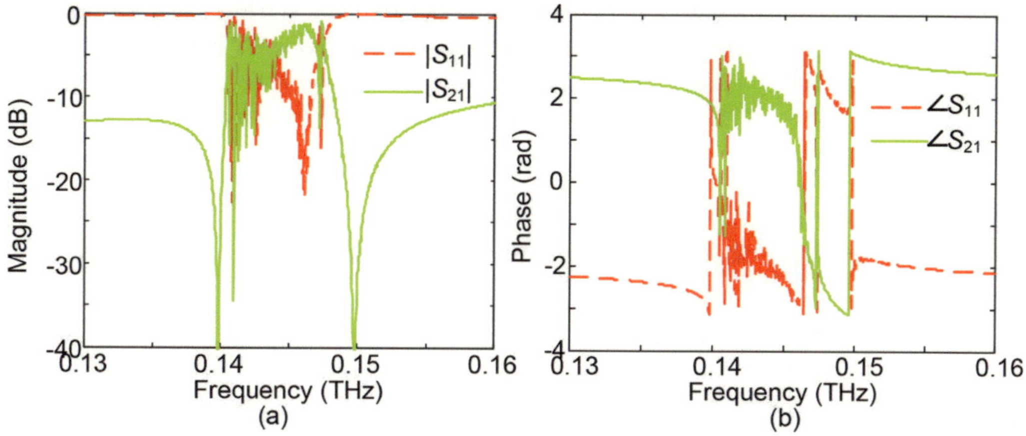

Figure 18. Simulated transmission and reflection characteristics LuBiIG–silver–LuBiIG structure metamaterials. (a) Magnitude. (b) Phase [49].

The effective refractive index values of the metamaterial are retrieved from the simulated parameters by using the retrieval method [46] and are shown in **Figure 19(a)**. It is seen that the metamaterial has negative refractive index in the range of 0.1401–0.1434 THz. Comparing the frequency bands of negative permeability (**Figure 17d**) and negative refractive index (**Figure 19a**) at same applied DC magnetic bias, a few frequency shifts appear because of the integrated dielectric substrate. Moreover, the real part of effective refractive index has a fixed value close to 10 at the frequency below 0.1401 THz and above 0.1434 THz.

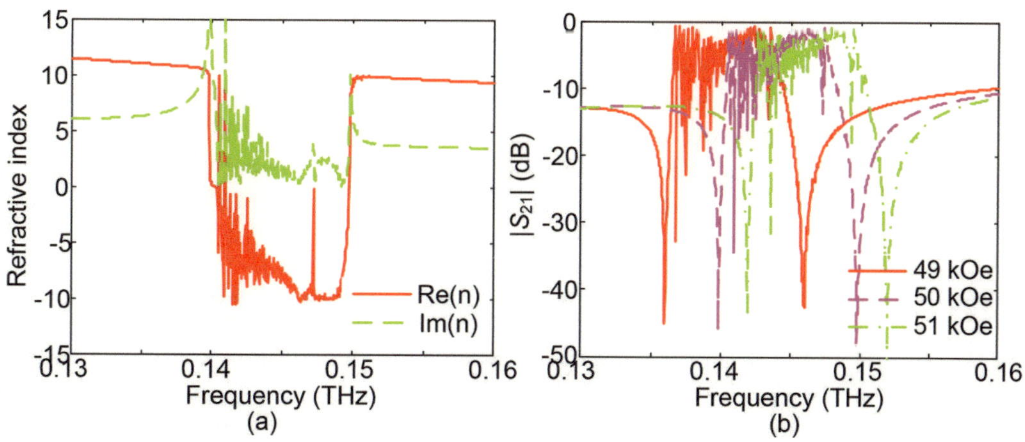

Figure 19. (a) Effective refractive index retrieved from S-parameters, and (b) tunability characteristics of metamaterials numerically demonstrated [49].

The tunability property is demonstrated earlier by tuning the applied DC magnetic field. As shown in **Figure 19(b)**, the transmission band shifts from 0.1385 to 0.1442 THz, when the applied DC magnetic bias is increased from 49 to 51 kOe. The tunability of such metamaterial depends on the tunability of the ferrite film. And the silver strip has a negative permittivity within a wide frequency region just below the plasma frequency. Therefore, the operating band of the proposed metamaterials can be tuned in a wide frequency band. Moreover, the shifted operating bandwidth and transmission peak do not vary strongly when the applied magnetic field is changed. This is mainly because of the fix saturation magnetization of the ferrite film.

4. Demonstrations for ferrite-based tunable metamaterial absorbers

Metamaterial absorbers (MAs) [52] have received most research attention due to the flexible design, subwavelength thickness, and high absorbing performance. MAs can be achieved at most of the frequency region by selecting the unit cell dimensions accordingly. The existence of such a huge operating spectral range enables the applications of MAs in various fields such as energy harvesting and thermal emitting and sensing [53–55]. Recently, various designs have been reported for MAs [56–60] and the corresponding numerical analysis and theory methods can be referred to, e.g., in [61–64].

However, the unique properties of MAs can be attributed to associated resonances, and thus MAs typically have narrow operating bandwidths. Researchers have proposed different methods to expand the operating bandwidths of MAs [65–69], while the flexible control of the MAs is still an intractable issue. Recent reports have shown that the absorbing frequencies of MAs can be dynamically adjusted by integrating tunable media (e.g., varactor diodes [70], liquid crystals [71], graphene [72], or phase-change materials [73]) to the traditional passive MAs. In some specific designs, researchers have advised to use oxide films [74] or film-coupled colloidal nanoantennas [75] as superstrate/substrate to adjust the absorbing frequencies. At the same time, we also designed a mechanic-tunable MA with moveable dielectric cover layer [76]. However, most of the reported designs have limited tuning range (e.g., [71–76]). Here we will first theoretically demonstrate that a properly geometrically configured ferrite-based metamaterial [77] has the ability to achieve high absorptivity with the added possibility to magnetically tune the absorption band, and then further demonstrate that by integrating a ferrite (the resonance bandwidth characteristic is not crucial any more) as the substrate or superstrate into a conventional planar MA, magnetic biased frequency-tunable absorption can also be achieved [78].

4.1. Tunable broadband metamaterial absorber based on ferrite and wire

The ferrite-wire MA unit cell is shown in **Figure 20(a)**. A 0.017-mm continuous copper wire is placed in the center of FR4 ($\varepsilon_r = 4.4$, $\tan\delta = 0.02$) substrate. Then two ferrite slabs (TT1-390, $4\pi Ms = 2150$ Gs, $\varepsilon_r = 12.7$, $\Delta H = 648$ Oe) are placed on the two sides of the FR4 substrate. The applied DC magnetic bias is applied on the ferrite slabs along the z-axis.

http://dx.doi.org/10.5772/67198

When an incident TEM wave propagates along the *y*-axis with electric field along the *z*-axis (shown in **Figure 20(a)**), the theoretical effective permeability of the ferrite slabs has been already analyzed as represented in Eq. (1).

The effective permeability of the ferrite slab used here under a magnetic bias of 3.2 kOe is calculated based on Eq. (1) and the results are shown in **Figure 20(b)**. It is seen that there is a magnetic resonance at 12 GHz and the real part of effective permeability is negative when frequency is larger than the magnetic resonance frequency. With the enlarged ΔH, the effective magnetic loss tangent $\tan\delta_\mu = \mathrm{Im}(\mu)/|\mathrm{Re}(\mu)|$ increases strongly in the negative permeability frequency band. Specifically, the $\tan\delta_\mu$ here has a value larger than 2.5 in the frequency region of 12.3–14.3 GHz. It indicates that the ferrite slabs can absorb most of the incident magnetic field component. Moreover, as shown in **Figure 20(c)**, the magnetic resonance of the ferrite slab can be shifted by adjusting the applied DC magnetic bias. When the magnetic bias increases from 2.6 to 3.4 kOe, the obtained magnetic resonance shifts from 10 to 12.5 GHz. And the big $\tan\delta_\mu$ is kept very well in the process of altering the applied DC magnetic bias.

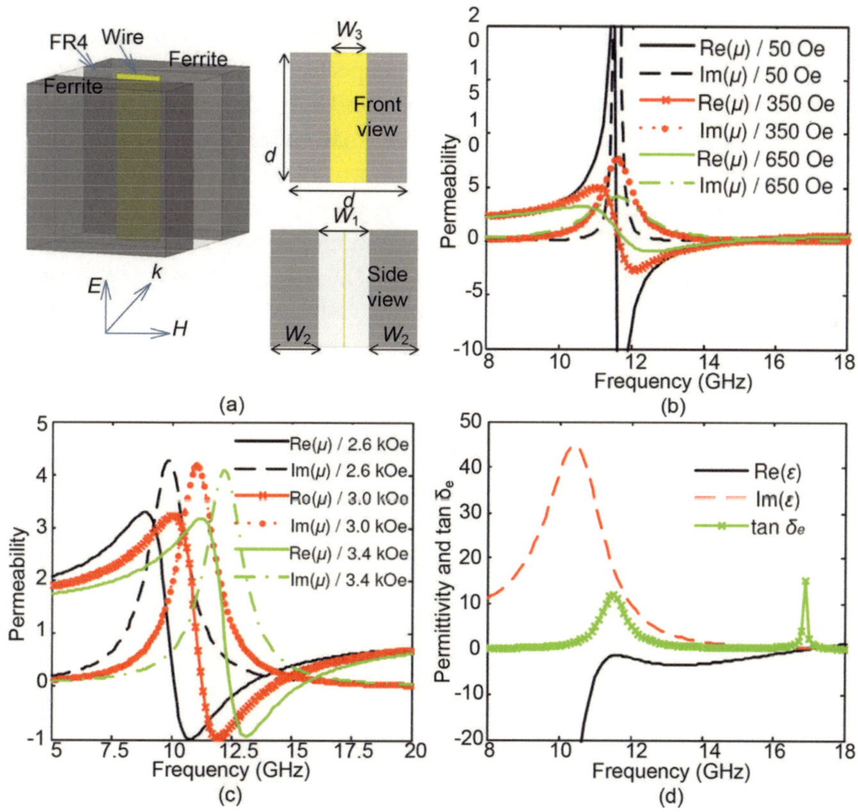

Figure 20. (a) Schematic diagram of the ferrite-wire MMA unit cell, (b) calculated theoretical effective permeability of the ferrite slabs with different resonance beamwidth ΔH and under a magnetic bias of 3.2 kOe, (c) calculated theoretical effective permeability of the ferrite slabs under different magnetic biases, and (d) calculated theoretical effective permittivity and effective dielectric loss tangent of the wire embedded in the ferrite slabs [77].

On the other hand, the effective permittivity of wires embedded in the ferrite slabs can be obtained as [31, 32]

$$\varepsilon_{\mathrm{eff}} = \varepsilon_f - \frac{\frac{\sigma_{\mathrm{eff}}}{\omega}}{j + \left(\frac{\omega \mu_0 a^2 \sigma_{\mathrm{eff}}}{2\pi}\right)\left[ln\frac{r_2}{r_1} + \frac{\mu_{\mathrm{eff}}}{\mu_0}\left(ln\frac{a}{r_2} - 1.06\right)\right]} \tag{2}$$

where ε_f is the permittivity of the ferrite slab, σ_{eff} is the effective conductivity of the wire, r_1 and r_2 are the effective radii of the wire and the substrate layer, respectively, and a is the periodic lattice constant. As a result, the effective permittivity of wires can be calculated as shown in **Figure 20(d)**. It can be known that the real part of the effective permittivity is negative in a wide frequency region from DC to 17 GHz. And the effective dielectric loss tangent $\tan\delta_\varepsilon = \mathrm{Im}(\varepsilon)/|\mathrm{Re}(\varepsilon)|$ has a very large value ($\tan\delta_\varepsilon > 1$) in the range of 10–13 GHz, also shown in **Figure 20(d)**. Therefore, the continuous wires can absorb most of the incident electric field component. Lastly, we can expect that the ferrite-wire MA can absorb the incident EM waves with a very high absorptivity.

Next, based on the above theoretical analysis, we numerically discuss the absorption properties by optimizing the structural parameters to achieve the perfect impedance match and the highest absorption. As shown in **Figure 21(a)**, the powers of transmission $T(\omega)$, reflectance $R(\omega)$, and absorptivity $A(\omega) = 1 - T(\omega) - R(\omega)$ are presented respectively. The final structure parameters are obtained: $d = 2$ mm, $w_1 = 0.959$ mm, $w_2 = 0.909$ mm, and $w_3 = 0.585$ mm. It can be seen that both the transmission $T(\omega)$ and reflectance $R(\omega)$ have very small values in the X-band (8–12 GHz) and as a result the absorptivity $A(\omega)$ can achieve near-uniform value. Specifically, the smallest reflectance $R(\omega)$ is about 0 at 9.9 GHz. The peak absorptivity $A(\omega)$ is about 98.2% at 9.9 GHz, and the frequency band of $A(\omega)$ over 90% is about 2.3 GHz. It should be noted that the high performance absorbing frequency band in simulation is lower than the frequency band of theory. This is due to the fact that the theoretical results discussed previously are under ideal conditions [31, 32]. That means the theoretical expressions in this chapter only provide the design guidance, and the numerical optimizations are more accurate and useful for the proposed ferrite-based MA.

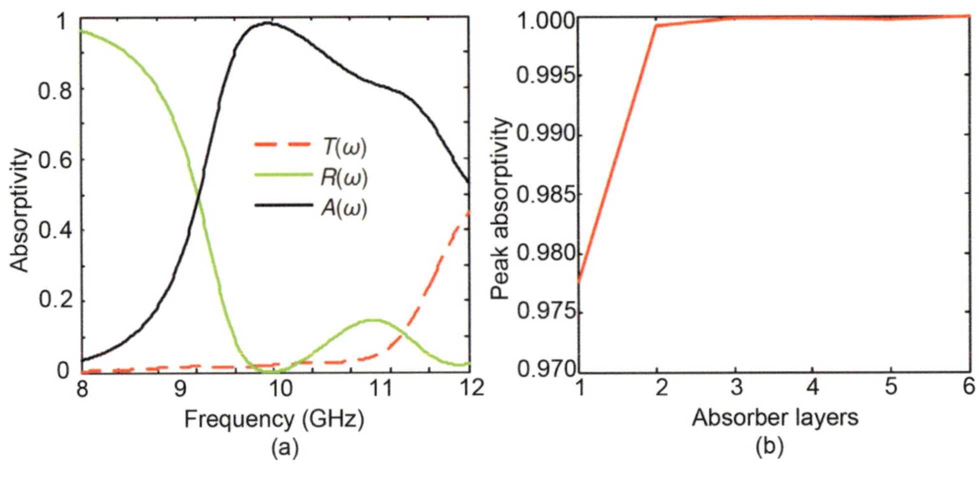

Figure 21. (a) Simulated transmission, reflectance and absorptivity characteristics of the ferrite-wire MMA under a magnetic bias of 3.2 kOe, and (b) absorptivity characteristic of ferrite-wire MA as a function of the number of layers [77].

Moreover, the absorptivity can be further enhanced by adding more layers. For instance, **Figure 21(b)** shows the absorptivity property as a function of the layers. As can be seen, $A(\omega)$ rises quickly with the increased layers and approaches asymptotically to 1. Two MA layer can achieve an absorptivity of about 99.92%. At this condition, the entire MA thickness is only $\lambda/5$ for the center operating frequency of 9.9 GHz.

In order to demonstrate the tunability of the operating frequency, **Figure 22(a)** shows the simulated shift property of the MA in different magnetic biases. As can be seen, the peak absorptivity and the bandwidth are not changed, the frequency of the peak absorptivity $A(\omega)$ shifts from 6.75 to 11.75 GHz as the applied DC magnetic bias changes from 2 to 4 kOe with a near-linear tuning rate of 2.625 MHz/Oe. We can further expect that the proposed ferrite-wire MA can operate at higher frequencies, such as terahertz region based on the discussed techniques [49].

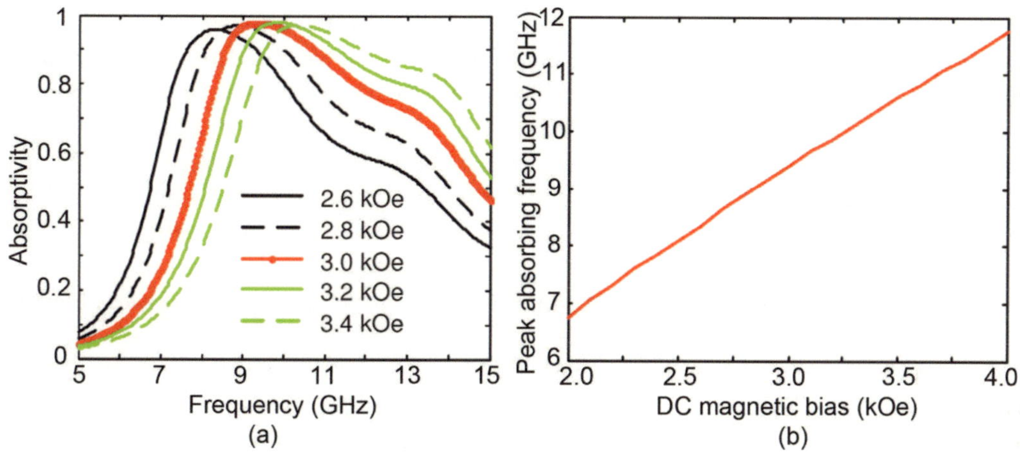

Figure 22. Peak shift properties of the absorptivity of the ferrite-wire MMA under different magnetic biases. (a) The absorption band characteristics under different magnetic biases, and (b) the frequency of the peak absorptivity as a function of the magnetic bias [77].

4.2. Tunable ferrite-based metamaterial absorbers

The new ferrite inspired MAs are schematically shown in **Figure 23(a)**. Similar to [76], we use the electric-LC (ELC) resonators with a metallic plate to achieve a conventional MA. The ELC array is printed on one side of the FR4 substrate and a full-sized metallic plate is covered on the other side of the FR4. For the purpose of achieving tunable property, we design two strategies by integrating ferrite into the passive MA. In the first case (MA1), the ferrite slab is inserted between the FR4 and ground plane [see the left plot of **Figure 23(b)**]. Both the ferrite and the FR4 layers service together as a substrate. In the second case (MA2), the ferrite layer is employed as a superstrate [see **Figure 23(b)**, right plot] and another FR4 layer is placed between the ferrite and the passive MA. The two MAs are analyzed and measured in an X-band rectangular waveguide closed measurement system as shown in **Figure 23(c)**.

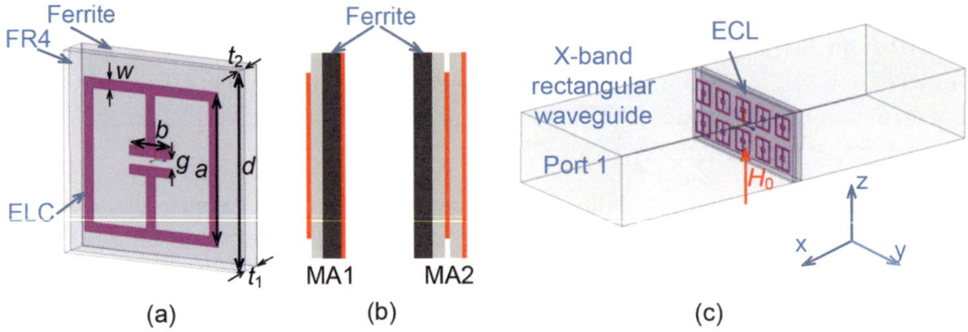

Figure 23. (a) Unit cell of the ferrite-based tunable MA, (b) side views of two MA models, and (c) schematic representation of measurements [78].

The relative permittivity of ferrite used in this chapter is 13.8 (loss tangent is 0.0002), and $4\pi Ms$ = 1830 Gs, ΔH = 22 Oe. As shown in **Figure 24**, when the applied DC magnetic bias increases from 1.1 to 1.9 kOe, the permeability of ferrite shows blueshift with shift rate of about 20 MHz/Oe. At the frequency band of interest (X-band), it can be seen that the effective permeability changes from positive to negative when the magnetic field increases. This changing feature can be used for dynamical tunability of the host media of ELC. Different from our previous work [77] where the ferrite's tunable permeability is employed for synthesizing magnetically controllable metamaterials, the frequency band we are interested in is outside the intrinsic resonant band of ferrite. Therefore, the required ferrite is not limited by its narrow resonant bandwidth and can be generally substituted by any other types of ferrites.

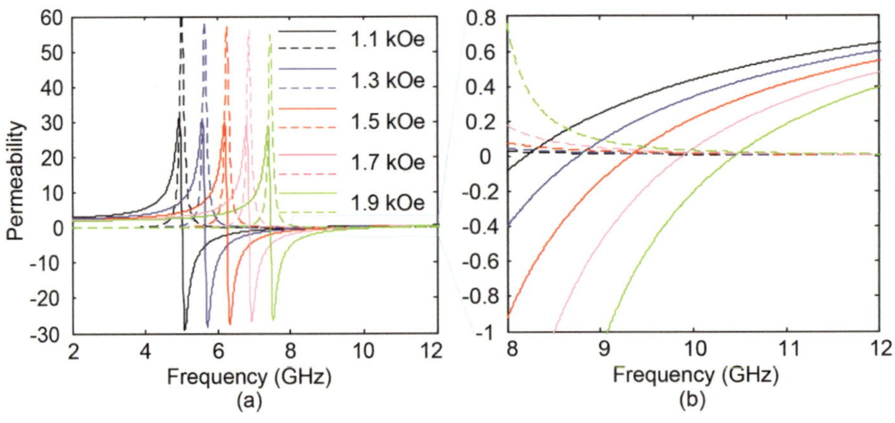

Figure 24. Permeability dispersion of the ferrite under different magnetic fields, where solid and dashed curves denote real and imagery parts, respectively. Zoomed plot shows the permeabilities of ferrite in the frequency range of interest (8–12 GHz) [78].

Now we optimize the ELC geometrical parameter by employing numerical simulations (CST). Due to the lab-experimental restriction, the thicknesses of the metallic patterns, FR4, and ferrite slab are fixed to be 0.034, 0.5, and 1 mm, respectively. The geometrical parameters for

MA1 are optimized to be a = 2.94 mm, b = 1 mm, w = g = 0:2 mm, and d = 4.5 mm. For the case of MA2, the additional FR4 layer would result in the red-shift of the resonance outside of 8–12 GHz. To avoid this problem, it should slightly reduce the size of the ELC resonator to a = 2.82 mm. The applied DC magnetic bias is preset as 1.55 and 1.6 kOe in simulations for both structures. The two MA samples are fabricated and measured in an X-band rectangular waveguide measurement system by using Agilent N5230A vector network analyzer (VNA). Specifically, the FR4 with ELC resonators and ferrite slab are put into the waveguide sequentially for MA1; the ferrite slab, FR4, and passive MA are staked one over the other within MA2. For these two structures, a metallic plate tightly covers the waveguide port to prevent any energy leakage. A tunable z-directed DC magnetic field generator is applied on the MA samples. Finally, the rectangular waveguide filled with MA samples is connected to the coaxial cable of VNA by using a waveguide-to-coaxial converter.

As shown in **Figure 25(a)** and, **(b)** each MA indicates a clear near-uniform absorption peak. The measured peak frequencies are higher than the simulated ones slightly. This is because the ferrite slabs are difficult to be placed seamlessly to the FR4 inside the waveguide. However, it should be noticed that this does not affect the tunability property of the ferrite-based MAs.

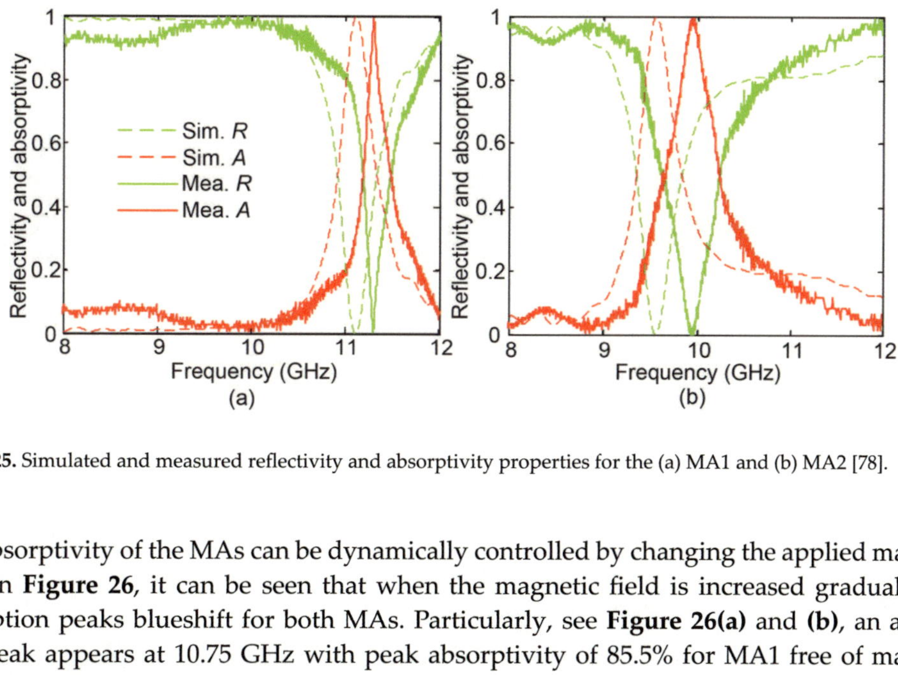

Figure 25. Simulated and measured reflectivity and absorptivity properties for the (a) MA1 and (b) MA2 [78].

The absorptivity of the MAs can be dynamically controlled by changing the applied magnetic bias. In **Figure 26**, it can be seen that when the magnetic field is increased gradually, the absorption peaks blueshift for both MAs. Particularly, see **Figure 26(a)** and **(b)**, an absorption peak appears at 10.75 GHz with peak absorptivity of 85.5% for MA1 free of magnetic field in simulation. When the magnetic field is gradually increased from 1.15 to 1.95 kOe, the absorbing frequency moves from 10.99 to 11.29 GHz in simulation (11.17 to 11.45 GHz in measurement). Meanwhile, the absorptivity increases from 97.8% to nearly 100% and then drops to 93.3% in simulation (increases from 85% to nearly 100% and then descends to 80% in measurement). Similar frequency shift and absorptivity change performance can be found in **Figure 4(c)** and **(d)** for MA2, as well. It is easy to conclude that the frequency shift rates of MA1 and MA2 are about 0.36 and 0.18 MHz/Oe, respectively.

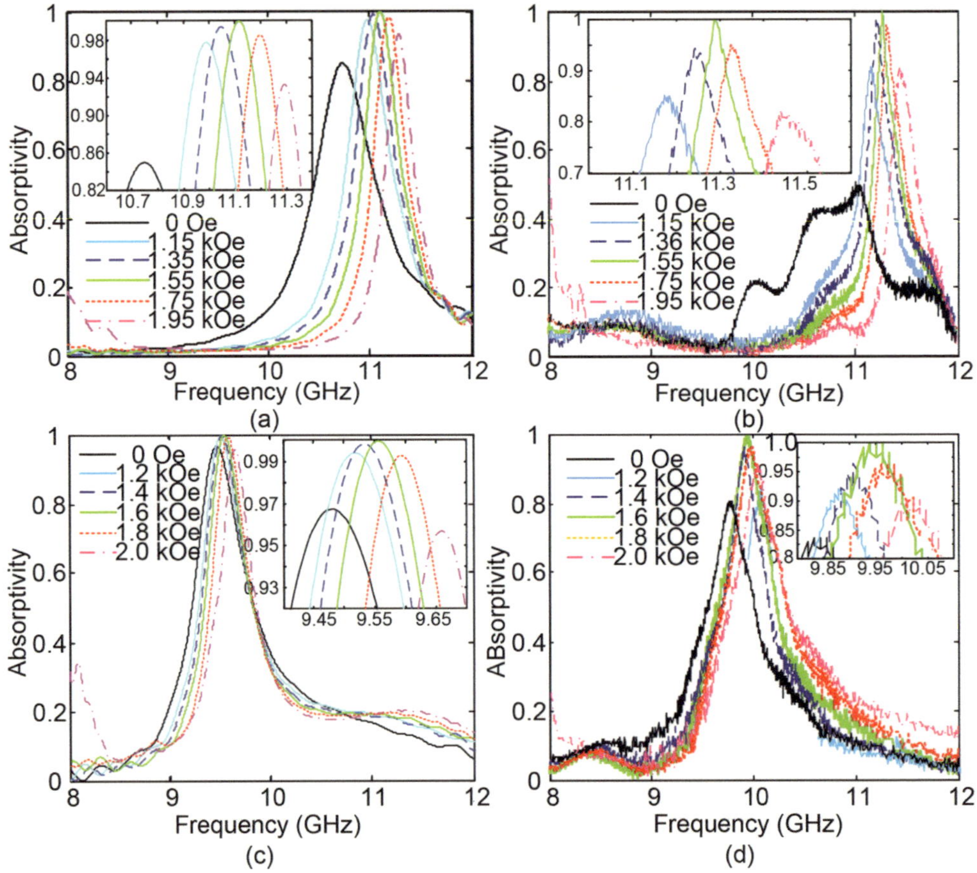

Figure 26. (a and c) Simulated and (b and d) measured absorptivity properties under different magnetic fields for MA1 and MA2, respectively. Insets show the zoom-in plots [78].

When closely looked at the measured results and the corresponding simulations at the same magnetic bias, the absorbing frequencies for both MAs have higher measured values compared with simulations. Moreover, compared to the two MA configurations, the shift rate of MA1 is larger than MA2. This is because the resonance-induced localized electromagnetic fields are mainly located in the substrates rather in the superstrates, and therefore the changing features of the MAs are more sensitive on the substrates.

In the above discussions, both thicknesses of the ferrite slab and the FR4 are fixed at 1 and 0.5 mm, respectively. Now we further discuss the effects of the thicknesses of the ferrite and the FR4 on the absorbing performances of the tunable MAs. For examples, we fix the thickness of one layer (either ferrite or FR4) and change the other layer to see the tunable absorptions of the two MAs. For MA1, the FR4 (ferrite) slab is fixed to be 0.5 mm (1 mm) and the ferrite (FR4) layer is changed from 0.6 to 1.4 mm (0.1 to 0.7 mm). The corresponding numerical results are concluded in **Figure 27(a)** and **(b)**. For MA2, the FR4 (ferrite) layer is fixed to be 0.5 mm (1 mm) and the ferrite (FR4) layer is tuned from 0.8 to 1.1 mm (0.3 to 0.6 mm), and the results

http://dx.doi.org/10.5772/67198

are shown at **Figure 27(c)** and **(d)**. For MA2, it should be noted that we only change one FR4 layer which acts as a superstrate and the FR4 substrate between the ELC and ground plane is kept unchanged.

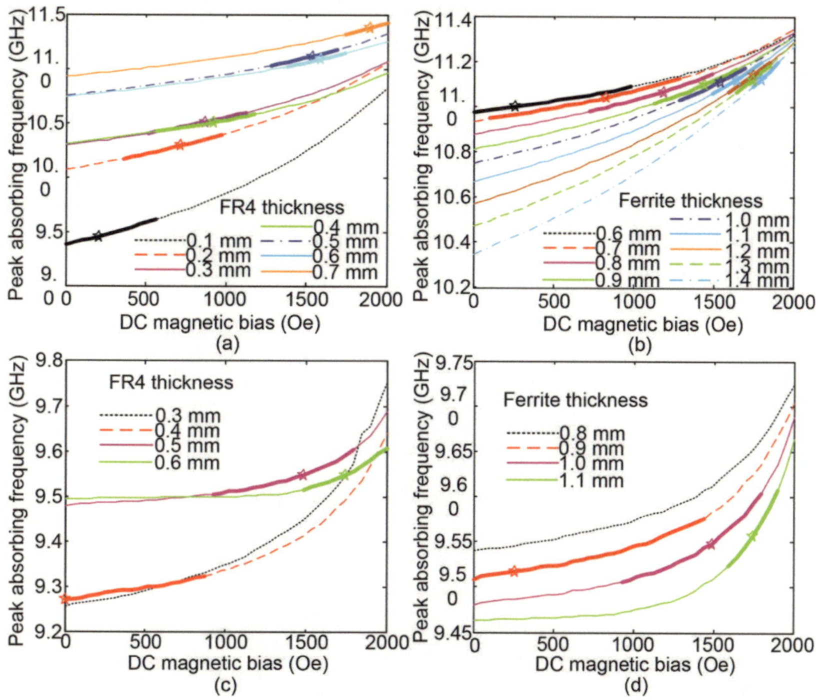

Figure 27. Simulated absorbing frequencies of MAs for MA1 when the thicknesses of (a) FR4 and (b) ferrite layers are changed, and for MA2 when the thicknesses of (c) FR4 and (d) ferrite layers are changed. The star and bold part at each curve correspond to the nearly uniform absorptivity and the band of absorptivity larger than 98.5%.

More specifically, when the magnetic field is increased, the resonant frequencies of the MAs blueshift. This is in accordance to the theoretical results shown in **Figure 24**. For a larger thickness of the FR4 or ferrite slab, a larger magnetic bias is required to ensure a nearly uniform absorptivity (represented as stars in **Figure 27**) and the magnetic field for a high absorptivity (>98.5%) is reduced (see the bold part of each curve). Fortunately, one can also increase the magnetic bias to reduce the effective permeability of the ferrite. The impedance matching can be regained at a certain magnetic field at that case. Moreover, the shift rate of the absorbing frequency is nearly linear when the magnetic field is weak, and becomes strongly nonlinear when the magnetic field increases, especially for the MA2 [see **Figure 27(c)** and **(d)**]. On the other hand, some unfavorable properties can also be obtained when the thickness of the FR4 or the ferrite slab is increased. Specifically, for thicker FR4, the initial peak absorptivity frequencies increase, and meanwhile the frequency shift rates decrease for both MAs. Contrarily, as shown in **Figure 27(b)** and **(d)**, inverse trends are found when the thicknesses of the ferrite layers are increased.

From previous discussions, we know that the resonant frequency of the MA becomes more sensitive to the magnetic field for thinner FR4, but the high absorptivity band is reduced as well. As a result, it is important to get an equilibrium such that the frequency shift range is relatively large and at the same time the MA has a uniform absorptivity in the tuning range. As an example, we choose FR4 between the ELC and ground plane to be 0.65-mm thick for MA2, and for the other FR4 to be 0.1-mm thick. For this condition, the ferrite slab is close enough to the ELC and so can give more contributions on the tunability, and at the same time, the other enlarged FR4 can offset the impedance mismatching between MA and background medium. In **Figure 28** we see that the resonant frequency of the MA can be shifted from 8.6 to 9.65 GHz and the absorptivity at each resonance can be kept higher than 98.5%, when the magnetic field increases from 0 to 1.9 kOe. The obtained shift range is as large as 11.5%. This is a very wide frequency tuning range (more than three times larger than those reported in previous works [71, 72, 74]).

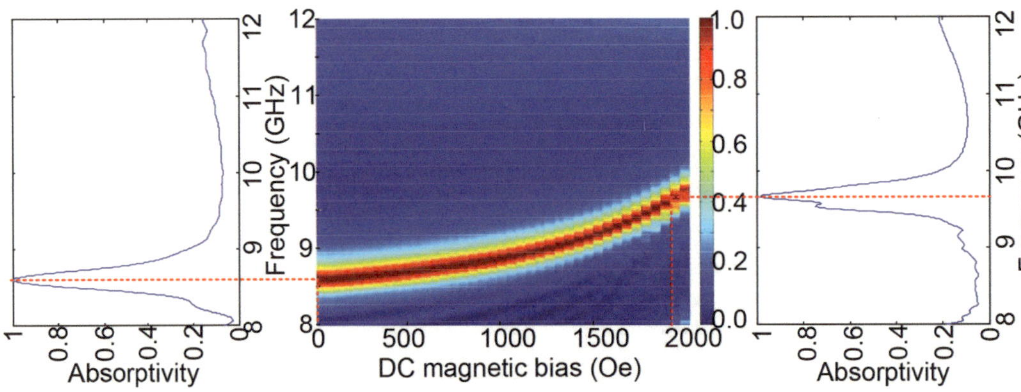

Figure 28. Simulated absorptivity of MA2 under different magnetic biases with optimized dimensional parameters. The middle plot presents the absorptivity spectra as a function of magnetic field and frequency. The left and right plots show the absorptivities at magnetic field of 0 and 1900 Oe, respectively [78].

5. Conclusions

In this chapter, we presented the recent achievements in our research group for the tunable metamaterials by using the ferrite as the substrate of metamaterials. We presented the designs and theories of single-, dual-, and triple-band tunable metamaterials based on the ferrite, and the design of metamaterial absorbers based on the ferrite. It indicates that the proposed tunable metamaterials have many advantages compared with other active and tunable metamaterials.

Acknowledgements

This work was supported by the National Natural Science Foundation of China (Grant No. 61371047, 61601093), and the Science and Technology Planning Project of Guangdong Province

http://dx.doi.org/10.5772/67198

and Sichuan Province of China (Grant No. 2016A010101036, 2016GZ0116, 2016GZ0061). Y. Huang also gratefully acknowledges the Scholarship Award for Excellent Doctoral Student granted by Ministry of Education of China (Grant No. A03003023901006).

Author details

Yongjun Huang*, Jian Li and Guangjun Wen

*Address all correspondence to: yongjunh@uestc.edu.cn

Centre for RFIC and System Technology, School of Communication and Information Engineering, University of Electronic Science and Technology of China, Chengdu, China

References

[1] N. Engheta and R. W. Ziolkowski. Metamaterials: Physics and Engineering Explorations. New Jersey: John Wiley & Sons; 2006.

[2] R. A. Shelby, D. R. Smith, and S. Schultz. Experimental verification of a negative index of refraction. Science. 2001;**292**:77–79.

[3] S. H. Lee, C. M. Park, Y. M. Seo, and C. K. Kim. Reversed Doppler effect in double negative metamaterials. Phys. Rev. B. 2010;**81**:241102(R).

[4] Z. Duan, C. Guo, and M. Chen. Enhanced reversed Cherenkov radiation in a waveguide with double-negative metamaterials. Opt. Express. 2011;**19**:13825–13830.

[5] T. J. Cui, X. Q. Lin, Q. Cheng, H. F. Ma, and X. M. Yang. Experiments on evanescent-wave amplification and transmission using metamaterial structures. Phys. Rev. B. 2006;**73**:245119.

[6] C. Argyropoulos, N. M. Estakhri, F. Monticone, and A. Al`u. Negative refraction, gain and nonlinear effects in hyperbolic metamaterial. Opt. Express. 2013;**21**:15037–15047.

[7] H. Y. Dong, J. Wang, K. H. Fung, and T. J. Cui. Super-resolution image transfer by a vortex-like metamaterial. Opt. Express. 2013;**21**:9407–9413.

[8] X. Zhang and Z. Liu. Superlenses to overcome the diffraction limit. Nat. Mater. 2008;**7**:435–441.

[9] D. Schurig, J. J. Mock, B. J. Justice, S. A. Cummer, J. B. Pendry, A. F. Starr, and D. R. Smith. Metamaterial electromagnetic cloak at microwave frequencies. Science. 2006;**314**:977–980.

[10] C. M. Watts, X. Liu, and W. J. Padilla. Metamaterial electromagnetic wave absorbers. Adv. Mater. 2012;**24**:OP98–OP120.

[11] R. Marqu´es, F. Mart´ın, and M. Sorolla. Metamaterials With Negative Parameters: Theory, Design and Microwave Applications. New Jersey: John Wiley & Sons; 2007.

[12] N. K. Grady, J. E. Heyes, D. R. Chowdhury, Y. Zeng, M. T. Reiten, A. K. Azad, A. J. Taylor, D. A. R. Dalvit, and H. T. Chen. Terahertz metamaterials for linear polarization conversion and anomalous refraction. Science. 2013;**340**:1304–1307.

[13] I. Gil, J. Bonache, J. García-García, et al. Tunable metamaterial transmission lines based on varactor-loaded split-ring resonators. IEEE Transactions on Microwave Theory and Techniques. 2006;**54**:2665-2674.

[14] A. Vélez, J. Bonache, and F. Martin. Varactor-loaded complementary split ring resonators (VLCSRR) and their application to tunable metamaterial transmission lines. IEEE Microwave Wireless Components Lett. 2008;**18**:28–30.

[15] D. Wang, L. Ran, H. Chen, et al. Active left-handed material collaborated with microwave varactors. Appl. Phys. Lett. 2007;**91**:164101.

[16] K. Aydin and E. Ozbay. Capacitor-loaded split ring resonators as tunable metamaterial components. J. Appl. Phys.. 2007;**101**:024911.

[17] T. H. Hand and S. A. Cummer. Frequency tunable electromagnetic metamaterial using ferroelectric loaded split rings. J. Appl. Phys. 2008;**103**:066105.

[18] H. Chen, B. I. Wu, L. Ran, et al. Controllable left-handed metamaterial and its application to a steerable antenna. Appl. Phys. Lett. 2006;**89**:053509.

[19] I. V. Shadrivov, S. K. Morrison, Y. S. Kivshar. Tunable split-ring resonators for nonlinear negative-index metamaterials. Opt. Express. 2006;**14**:9344–9349.

[20] Q. Zhao, L. Kang, B. Li, et al. Tunable negative refraction in nematic liquid crystals. Appl. Phys. Lett. 2006;**89**:221918.

[21] L. Kang, Q. Zhao, B. Li, et al. Experimental verification of a tunable optical negative refraction in nematic liquid crystals. Appl. Phys. Lett. 2007;**90**:181931.

[22] F. Zhang, Q. Zhao, L. Kang, et al. Magnetic control of negative permeability metamaterials based on liquid crystals. Appl. Phys. Lett. 2008;**90**:193104.

[23] F. Zhang, L. Kang, Q. Zhao, et al. Magnetically tunable left handed metamaterials by liquid crystal orientation. Opt. Express. 2009;**17**:4360–4366.

[24] Q. Zhao, L. Kang, B. Du, et al. Electrically tunable negative permeability metamaterials based on nematic liquid crystals. Appl. Phys. Lett. 2007;**90**:011112.

[25] D. H. Werner, D. H. Kwon, I. C. Khoo, et al. Liquid crystal clad near-infrared metamaterials with tunable negative-zero-positive refractive indices. Opt. Express. 2007;**15**:3342–3347.

[26] A. Minovich, D. N. Neshev, D. A. Powell, et al. Tunable fishnet metamaterials infiltrated by liquid crystals. Appl. Phys. Lett. 2010;**96**:193103.

[27] M. C. Ricci, H. Xu, R. Prozorov, et al. Tunability of superconducting metamaterials. IEEE Trans. Appl. Superconductivity. 2007;**17**:918–921.

[28] H. Tao, A. C. Strikwerda, K. Fan, et al. MEMS based structurally tunable metamaterials at terahertz frequencies. J. Infrared Milli. Terahz. Waves. 2011;**32**:580–595.

[29] M. Lapine, D. Powell, M. Gorkunov, et al. Structural tunability in metamaterials. Appl. Phys. Lett. 2009;**95**:084105.

[30] E. Ekmekci, A. C. Strikwerda, K. Fan, et al. Frequency tunable terahertz metamaterials using broadside coupled split-ring resonators. Phys. Rev. B. 2011;**83**:193103.

[31] G. Dewar. A thin wire array and magnetic host structure with n < 0. J. Appl. Phys. 2005;**97**:10Q101.

[32] G. Dewar. Minimization of losses in a structure having a negative index of refraction. New J. Phys. 2005;**7**:161.

[33] F. J. Rachford, D. N. Armstead, V. G. Harris, et al. Simulations of ferrite-dielectric-wire composite negative index materials. Phys. Rev. Lett. 2007;**99**:057202.

[34] X. B. Cai, X. M. Zhou, and G. K. Hu. Numerical study on left-handed materials made of ferrite and metallic wires. Chin. Phys. Lett. 2005;**23**:348–351.

[35] Y. He, P. He, S. D. Yoon, et al. Tunable negative index metamaterial using yttrium iron garnet. J. Magnetism Magnetic Mater. 2007;**313**:187–191.

[36] H. Zhao, J. Zhou, Q. Zhao, et al. Magnetotunable left-handed material consisting of yttrium iron garnet slab and metallic wires. Appl. Phys. Lett. 2007;**91**:131107.

[37] Y. J. Cao, G. J. Wen, K. M. Wu, et al. A novel approach to design microwave medium of negative refractive index and simulation verification. Chin. Sci. Bull. 2007;**52**:433–439.

[38] Y. Huang, G. Wen, T. Li, et al. Low-loss, broadband and tunable negative refractive index metamaterial. J. Electromag. Ana. Appl. 2010;**2**:104–110.

[39] L. Kang, Q. Zhao, H. Zhao, et al. Magnetically tunable negative permeability metamaterial composed by split ring resonators and ferrite rods. Opt. Express. 2008;**16**:8825–8834.

[40] L. Kang, Q. Zhao, H. Zhao, et al. Ferrite-based magnetically tunable left-handed metamaterial composed of SRRs and wires. Opt. Express. 2008;**16**:17269–17275.

[41] Y. J. Huang, G. J. Wen, Y. J. Yang, et al. Tunable dual-band ferrite-based metamaterials with dual negative refractions. Appl. Phys. A. 2012;**106**:79–86.

[42] K. Bi, J. Zhou, H. Zhao, et al. Tunable dual-band negative refractive index in ferrite-based metamaterials. Optics Express. 2013;**21**:10746–10752.

[43] B. Lax and K. J. Button. Microwave Ferrites and Ferrimagnetics. New York: McGraw-Hill; 1962.

[44] J. Zhou, L. Zhang, G. Tuttle, et al. Negative index materials using simple short wire pairs. Phys. Rev. B. 2006;**73**:041101.

[45] J. Huangfu, L. Ran, H. Chen, et al. Experimental confirmation of negative refractive index of a metamaterial composed of Ω-like metallic patterns. Appl. Phys. Lett. 2004;**84**:1537–1539.

[46] D. R. Smith, D. C. Vier, T. Koschny, et al. Electromagnetic parameter retrieval from inhomogeneous metamaterials. Phys. Rev. E. 2005;**71**:036617.

[47] E. Ekmekci and G. Turhan-Sayan. Single-loop resonator: dual-band magnetic metamaterial structure. Electron. Lett.. 2010;**46**:324–325.

[48] J. Zhong, F. Wang, G. Wen, et al. Tunable triple-band negative permeability metamaterial consisting of single-loop resonators and ferrite. J. Electromag. Waves Applicat. 2013;**27**:267–275.

[49] Y. J. Huang, G. J. Wen, and T. Q. Li, et al. Design and characterization of tunable terahertz metamaterials with broad bandwidth and low loss. IEEE Antennas Wireless Propagat. Lett. 2012;**11**:264–267.

[50] Q. H. Yang, H. W. Zhang, Y. L. Liu, Q. Y. Wen, and J. Zha. An artificially garnet crystal materials using in terahertz waveguide. Chin. Phys. Lett.. 2008;**25**:3957–3960.

[51] M. Walther, A. Ortner, H. Meier, U. Löffelmann, P. J. Smith, and J. G. Korvink. Terahertz metamaterials fabricated by inkjet printing. Appl. Phys. Lett.. 2008;**95**:251107.

[52] N. I. Landy, S. Sajuyigbe, J. J. Mock, D. R. Smith, and W. J. Padilla. Perfect metamaterial absorber. Phys. Rev. Lett. 2008;**100**:207402.

[53] X. Liu, T. Tyler, T. Starr, A. F. Starr, N. M. Jokerst, and W. J. Padilla. Taming the blackbody with infrared metamaterials as selective thermal emitters. Phys. Rev. Lett. 2011;**107**:045901.

[54] M. Yin, X. Y. Tian, L. L. Wu, and D. C. Li. A broadband and omnidirectional electromagnetic wave concentrator with gradient woodpile structure. Opt. Express. 2013;**21**:19082–19090.

[55] W. Withayachumnankul, H. Lin, K. Serita, C. M. Shah, S. Sriram, M. Bhaskaran, M. Tonouchi, and C. Fumeaux. Sub-diffraction thin-film sensing with planar terahertz metamaterials. Opt. Express. 2012;**20**:3345–3352.

[56] W. Zhu and X. Zhao. Metamaterial absorber with random dendritic cells. Eur. Phys. J. Appl. Phys. 2010;**50**:21101.

[57] J. Grant, Y. Ma, S. Saha, A. Khalid, and D. R. S. Cumming. Polarization insensitive, broadband terahertz metamaterial absorber. Opt. Lett. 2012;**36**:3476–3478.

[58] L. Huang, D. R. Chowdhury, S. Ramani, M. T. Reiten, S.-N. Luo, A. J. Taylor, and H.-T. Chen. Experimental demonstration of terahertz metamaterial absorbers with a broad and flat high absorption band. Opt. Lett. 2012;**37**:154–156.

[59] W. Zhu and X. Zhao. Metamaterial absorber with dendritic cells at infrared frequencies. J. Opt. Soc. Am. B. 2009;**26**:2382–2385.

[60] K. B. Alici, A. B. Turhan, C. M. Soukoulis, and E. Ozbay. Optically thin composite resonant absorber at the near-infrared band: A polarization independent and spectrally broadband configuration. Opt. Express. 2011;**19**:14260–14267.

[61] Q.-Y. Wen, Y.-S. Xie, H.-W. Zhang, Q.-H. Yang, Y.-X. Li, and Y.-L. Liu. Transmission line model and fields analysis of metamaterial absorber in the terahertz band. Opt. Express. 2009;**17**:20256–20265.

[62] H.-T. Chen. Interference theory of metamaterial perfect absorbers. Opt. Express. 2012; **20**:7165–7172.

[63] T. Wanghuang, W. Chen, Y. Huang, and G. Wen. Analysis of metamaterial absorber in normal and oblique incidence by using interference theory. AIP Adv. 2013;**3**:102118.

[64] Y. Pang, H. Cheng, Y. Zhou, and J. Wang. Analysis and design of wire-based metamaterial absorbers using equivalent circuit approach. J. Appl. Phys. 2013;**113**:114902.

[65] J. Zhong, Y. Huang, G. Wen, H. Sun, P. Wang, and O. Gordon. Single-/dual-band metamaterial absorber based on cross-circular-loop resonator with shorted stubs. Appl. Phys. A. 2012;**108**:329–335.

[66] Y. Huang, Y. Tian, G. Wen, and W. Zhu. Experimental study of absorption band controllable planar metamaterial absorber using asymmetrical snowflake-shaped configuration. J. Opt. 2013;**15**:055104.

[67] H. Li, L. H. Yuan, B. Zhou, X. P. Shen, Q. Cheng, and T. J. Cui. Ultrathin multiband gigahertz metamaterial absorbers. J. Appl. Phys. 2011;**110**:014909.

[68] J. Sun, L. Liu, G. Dong, and J. Zhou. An extremely broad band metamaterial absorber based on destructive interference. Opt. Express. 2011;**19**:21155–21162.

[69] Y. Cui, K. H. Fung, J. Xu, H. Ma, Y. Jin, S. He, and N. X. Fang. Ultrabroadband light absorption by a sawtooth anisotropic metamaterial slab. Nano Lett. 2012;**12**:1443–1447.

[70] J. Zhao, Q. Cheng, J. Chen, M. Q. Qi, W. X. Jiang, and T. J. Cui. A tunable metamaterial absorber using varactor diodes. New J. Phys. 2013;**15**:043049.

[71] D. Shrekenhamer, W.-C. Chen, and W. J. Padilla. Liquid crystal tunable metamaterial absorber. Phys. Rev. Lett. 2013;**110**:177403.

[72] A. Andryieuski and A. V. Lavrinenko. Graphene metamaterials based tunable terahertz absorber: Effective surface conductivity approach. Opt. Express. 2013;**21**:9144–9155.

[73] T. Cao, L. Zhang, R. E. Simpson, and M. J. Cryan. Mid-infrared tunable polarization-independent perfect absorber using a phase-change metamaterial. J. Opt. Soc. Am. B. 2013;**30**:1580–1585.

[74] Q.-Y. Wen, H.-W. Zhang, Q.-H. Yang, Z. Chen, Y. Long, Y.-L. Jing, Y. Lin, and P.-X. Zhang. A tunable hybrid metamaterial absorber based on vanadium oxide films. J. Phys. D: Appl. Phys. 2012;**45**:235106.

[75] A. Moreau, C. Cirac'ı, J. J. Mock, R. T. Hill, Q. Wang, B. J. Wiley, A. Chilkoti, and D. R. Smith. Controlled-reflectance surfaces with film-coupled colloidal nanoantennas. Nature (London). 2012;**492**:86–90.

[76] W. Zhu, Y. Huang, I. D. Rukhlenko, G. Wen, and M. Premaratne. Configurable metamaterial absorber with pseudo wideband spectrum. Opt. Express. 2012;**20**:6616–6621.

[77] Y. J. Yang, Y. J. Huang, G. J. Wen, J. P. Zhong, H. B. Sun, and O. Gordon. Tunable broadband metamaterial absorber consisting of ferrite and wire. Chin. Phys. B. 2012;**21**:038501.

[78] Y. Huang, G. Wen, W. Zhu, J. Li, L.-M. Si, and M. Premaratne. Experimental demonstration of a magnetically tunable ferrite based metamaterial absorber. Opt. Express. 2014;**22**:16408–16417.

Chapter 3

Metasurfaces for Spatial Light Manipulation

Jian Wang and Jing Du

Additional information is available at the end of the chapter

http://dx.doi.org/10.5772/66319

Abstract

Light manipulation means that one can take advantages of different physical dimensions of lightwaves to realize flexible light control. Light manipulation over subwavelength propagation distances can be realized using metasurfaces. There are two categories of metasurfaces divided by the material type of unit structure, that is, plasmonic and dielectric metasurfaces. For plasmonic metasurfaces, they are made on the basis of metallic meta-atom whose optical responses are driven by the plasmon resonances supported by metallic particles. For dielectric metasurfaces, unit structure is constructed with high refractive index dielectric resonators such as silicon, germanium, or tellurium, which can support electric and magnetic dipole responses based on Mie resonances. The responses of plasmonic and dielectric metasurfaces are all relevant to the characteristics of unit structure. One can manipulate the electromagnetic field of lightwave scattered by the metasurfaces through designing the dimension parameters of each unit structure in the metasurfaces. In this chapter, we review our recent research progress in light manipulation using plasmonic and dielectric metasurfaces. It is believed that metasurfaces based nanophotonic devices are one of the most potential devices applied in various fields such as beam steering, spatial light modulator, nanoscale-resolution imaging, sensing, quantum optics devices, and even optical communication networks.

Keywords: light manipulation, metamaterials, metasurfaces, plasmonic metasurfaces, dielectric metasurfaces, orbital angular momentum, vector beams, Bessel beams, space-division multiplexing

1. Introduction

The fundamental properties of lightwaves can be described from the basic electrical field expression of electromagnetic waves, $\overrightarrow{E(x,y,z,t)} = \overrightarrow{e(z,t)} \cdot F(x,y,t) \cdot |A(z,t)| \cdot \Phi(z,t) \cdot \exp$

$[j(\omega t - kz) + \varphi]$, as displayed in **Figure 1**, where $\overrightarrow{e(z,t)}$ is the polarization, $F(x,y,t)$ is related to the spatial structure, $|A(z,t)|$ is the amplitude, $\Phi(z,t)$ is the phase, $A(z,t) = |A(z,t)| \cdot \Phi(z,t)$ is the complex amplitude, ω is the frequency of the lightwave, k is the wave vector, φ is the initial phase, z is the propagation distance and t is the time [1]. Light manipulation means that one can take advantages of these physical dimensions (wavelength/frequency, time, complex amplitude, polarization, spatial structure) of lightwaves to realize flexible control of electromagnetic waves. Utilizing different physical dimensions of lightwaves, light manipulation has been widely applied in various areas such as wavelength-division multiplexing (WDM) [2], polarization-division multiplexing (PDM) [3], time-division multiplexing (TDM) [4], space-division multiplexing (SDM) [5], advanced modulation format [6], sensing [7], imaging [8], optical tweezers [9], optical knots [10], optical trapping [11], and so on. In order to realize light manipulation on the physical dimensions of lightwaves, there have been lots of approaches such as modulators (e.g. intensity, phase, I/Q modulators, spatial light modulators, etc.), conventional optical elements (e.g. lens, waveplate, polarizer, optical coupler, etc.), and metamaterials and metasurfaces, as listed in **Figure 2** [4–11]. The spatial structure physical dimension has recently attracted increasing interest in wide research areas. In this chapter, we mainly focus on the recent progress in light manipulation on spatial structure physical dimension using metasurfaces.

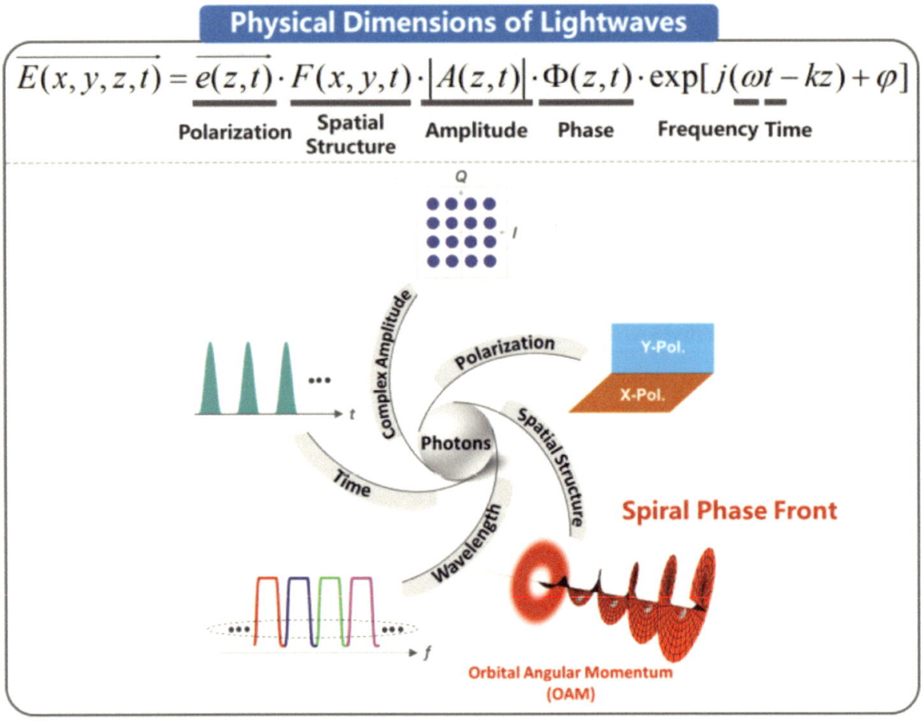

Figure 1. Schematic illustration of physical dimensions of lightwaves.

http://dx.doi.org/10.5772/66319

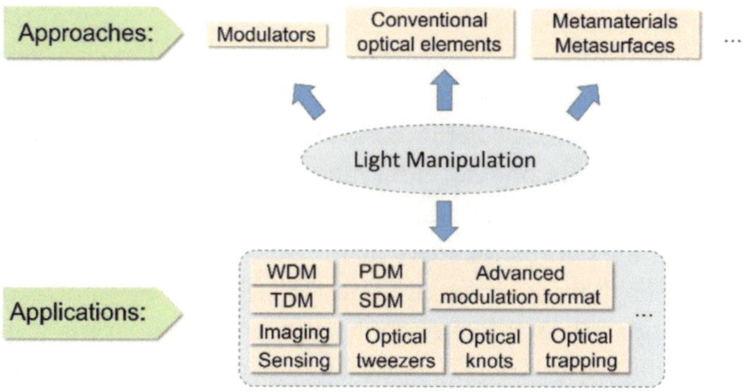

Figure 2. Schematic illustration of approaches and applications of light manipulation.

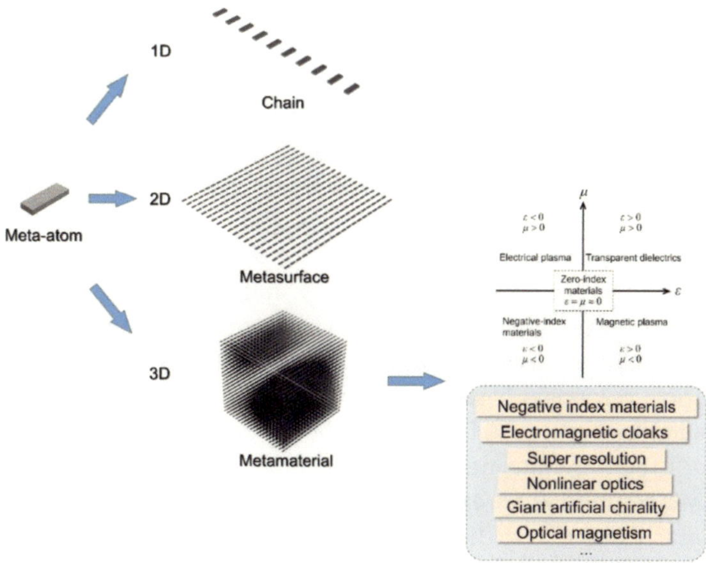

Figure 3. Schematic illustration of meta-atom, 1D chain, 2D metasurface, and 3D metamaterial. Inserts are the representation of the parameter space for permittivity ε and permeability μ and the typical examples of applications of metamaterials.

"Metamaterial" can be defined as a man-made material whose properties can be attained by designing the unit structure. It has been widely focused for a dozen years since it was firstly reported by Smith et al. in 2000 [12]. The unit structure in metamaterial called as meta-atoms or meta-molecules, must be considerably smaller than the operating wavelength, and the distance between neighboring meta-atoms also has subwavelength scale [13]. This subwavelength scale in homogeneity makes the whole metamaterial uniform on the macro of performance, and this phenomenon makes this artificial structure essentially a "material" rather than a device. Therefore, one can arbitrarily arrange the so-called meta-atoms into periodic arrays to build one-dimensional (1D) materials (chains), two-dimensional (2D) materials (metasurfaces), and three-dimensional (3D) materials (metamaterials), as shown in **Figure 3** [14]. For metamaterial, one can even manipulate its macroscopical properties such as permittivity ε and permeability μ. The researches of metamaterial are essentially related to the exploitation of the electromagnetic parameter space for better control of electromagnetic waves, as depicted in **Figure 3**. Utilizing metamaterials, one can enter regions of the electromagnetic parameter space that are not observed in conventional material but are not forbidden by Maxwell's equations such as third quadrant of the parameter space (negative-index materials), zero-index materials and space far away from the non-magnetic line [15]. Taking advantage of these attractive features, as shown in **Figure 1**, metamaterials have been widely applied in optical negative index materials [16], nonlinear optics [17], optical magnetism [18], super resolution [19], giant artificial chirality [20], electromagnetic cloaks [21], and so on.

Besides 3D materials, 2D metasurfaces are also widely focused for their ability of flexible light manipulation (phase, amplitude, polarization) over subwavelength propagation distances [22]. Most of the metasurfaces can be divided into two categories by the material type of unit structure: plasmonic and dielectric, as displayed in **Figure 4**. For plasmonic metasurfaces, they are made on the basis of metallic meta-atom whose optical responses are driven by the plasmon resonances supported by metallic particles. When a metallic particle is placed in an electric field, the conduction electrons are escaped from their equilibrium positions with respect to the core ions, resulting in a polarization of the particle and generating a depolarizing field. In a time-varying external field, this collective motion in the metallic particle can be seen as a Lorentzian oscillator whose characteristic peak in the displacement amplitude is around the resonance frequency and accompanied by a phase shift of π over the spectral width of the resonance [14]. For dielectric metasurfaces, unit structure is constructed with high refractive index dielectric resonators such as silicon, germanium or tellurium, which can support electric and magnetic dipole responses based on Mie resonances [23]. When a dielectric particle is illuminated by a light wave whose frequency is below or near the bandgap frequency of the material of particle, both the magnetic dipole (first Mie resonance) and electric dipole resonances (second Mie resonance) are excited. The magnetic and electric Mie resonance can then enhance the magnetic and electric field at the particle's center at optical frequencies, respectively, and this enhancement is related to the intrinsic properties of dielectric particles [24]. Therefore, the responses of plasmonic and dielectric metasurfaces are all relevant to the characteristics of unit structure such as dimensions and materials. One can manipulate the electromagnetic field of light wave scattered by the metasurfaces through designing the dimension parameters of each unit structure in the metasurfaces [25].

To produce the designed metasurfaces, many attractive previous works have reported the fabrication techniques such as electron-beam lithography [16], focused-ion beam [26], interference lithography [27], and nanoimprint lithography [28]. Electron-beam lithography is a standard method that can fabricate the metasurface with high nanoscale resolution. Focused-ion beam is a more appropriate method for rapid prototyping. Interference lithography can provide a solution for realizing large-scaled metasurface fabrication. Nanoimprint lithography is a more promising technique for combining many advantages of the formers such as high-resolution, large-scaled producing, and low processing cost [28, 29]. The unique properties and the development of fabrication techniques have made metasurfaces applied in various fields such as flat optics [30–41], nonlinear effects [42, 43], photonic Hall effects [44], electromagnetically induced transparency [45], cloaking [46], and so on as shown in **Figure 4**. For instance, the applications of metasurfaces in the field of flat optics mainly include anomalous wave plates [30], reflection/refraction [31, 32], flats lens/axicons [33], mirrors [35], hologram [37], filters [38], optical vortex generation [39, 40], polarization beam splitter [41], and so on. The applications of metasurfaces in the field of nonlinear effects mainly include second/third harmonic generation [42] and enhanced absorbing [43].

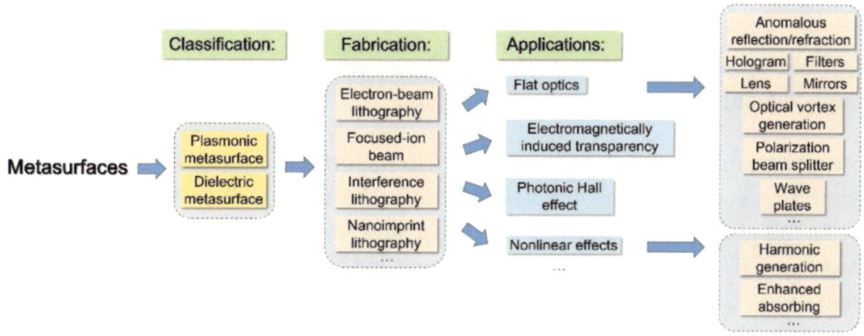

Figure 4. Schematic illustration of classification, fabrication and applications of metasurfaces.

In this review article, we go over our recent progress in light manipulation using plasmonic and dielectric metasurfaces, including metasurfaces-based broadband and selective generation of orbital angular momentum (OAM)-carrying vector beams [47, 48], N-fold OAM multicasting using V-shaped antenna array [49], metasurface on conventional optical fiber facet for linearly polarized mode (LP11) generation [50], OAM beams generation using nanophotonic dielectric metasurface array [51], and Bessel beams generation and OAM multicasting using dielectric metasurface array [52].

2. Light manipulation using plasmonic metasurfaces

Plasmonic metasurfaces are composed of metal nanoparticles based on the plasmon resonances. Several typical types of plasmonic metasurfaces and previous works are reported: (1) metal V-shaped nanoantennas are designed and fabricated on silicon substrate using electron-beam lithography method to manipulate the phase of scattering light and generate OAM beams [22]; (2) Au patch antennas separated from a metal back plane by a MgF2 spacer are designed and fabricated using electron-beam evaporation electron-beam lithography technique to realize anomalous reflections [31]; (3) metal H-shaped nanoantenna arrays are designed and fabricated using printed circuit boards to realize high efficiency conversion from propagating waves to surface waves in microwave regime [32]; (4) metal nanorods are deposited on the glass substrate to generate hologram [37]; (5) freestanding nanofabricated fishnet metasurfaces are designed and fabricated using electron-beam lithography method to realize broadband band pass filter [38]; (6) metal rectangular apertures are arranged into an array with rotational symmetry in metal film to generate OAM beams [39]; (7) metal nanorods are designed and fabricated on glass substrate with a standard electron-beam lithography and lift-off process to generate broadband OAM beams [40]; (8) metal split-ring resonators are designed and fabricated on glass substrate with a standard electron-beam lithography to enhance second harmonic generation [42]. All these works are based on manipulating the plasmon responses of nanoresonators by altering the dimensions, directions and even arrangement of unit structures. Recently, we have also made some progresses in plasmonic metasurfaces on the basis of previous works.

2.1. Metasurfaces-based broadband and selective generation of OAM-carrying vector beams [47, 48]

We propose and design compact metal-assisted metasurfaces to enable broadband generation of OAM-carrying vector beams. **Figure 5(a)** and **(b)** depicts the structure and geometric parameters of metal-assisted metasurfaces. We design two concentric rings in a gold film with a thickness of h = 200 nm. Each ring is composed of 42 rectangular apertures with gradually varied orientation. The rectangular aperture array in the gold film can enhance the transmission of linearly polarized light (perpendicular to the aperture direction), which might be explained as follows: (1) the localized waveguide resonance (each air aperture can be regarded as a truncated rectangular waveguide with four metal walls and two sides open to air); (2) the property of the surface plasmon resonance due to the aperture array. Hence, each rectangular aperture can be regarded as a localized linear polarizer. By controlling the orientation angle of rectangular apertures, we can construct desired spatially variant polarizers to generate OAM-carrying vector beams with right or left circularly polarized input light beam. **Figure 5(c)** shows an example of phasefront and spatial polarization after passing though the metasurfaces, which indicates the generation of OAM-carrying vector beam from circularly polarized beam.

Figure 6 plots spatial phase, power and polarization distributions of output beams under the excitation of input left circularly polarized light ($\boldsymbol{E}_{in} = [1\ i]^T$). We set the wavelength at 1550 nm to characterize the properties of generated OAM-carrying vector beams. The employed metasurfaces has an orientation angle of $\alpha(\varphi) = l\varphi + \alpha_0$, where l varies from +3 to −3 and $\alpha_0 = 0$. We use $\boldsymbol{E}_1$ and $\boldsymbol{E}_2$ to represent the electric field components along directions of $\boldsymbol{e}_1(\varphi)$ and $\boldsymbol{e}_2(\varphi)$ in **Figure 5(b)**, respectively. The first row in **Figure 6** shows spatial phase distribution of $\boldsymbol{E}_1$, indicating that output beams carry OAM with a charge number of l from +3 to −3. The second and third rows in **Figure 6** show spatial power distributions ($P_1 \infty |E_1|^2$, $P_2 \infty |E_2|^2$) along directions of $\boldsymbol{e}_1(\varphi)$ and $\boldsymbol{e}_2(\varphi)$, respectively. It is found that the power component P_1 is much larger than P_2. The extinction ratio (ER), defined by $10 \times \log_{10}(P_1/P_2)$, exceeds 20 dB. Hence, the electric field component $\boldsymbol{E}_2$ along the direction of $\boldsymbol{e}_2(\varphi)$ can be ignored. The fourth and fifth rows in **Figure 6** show spatial power distributions (P_x, P_y) along x and y axes, respectively. The alternative bright and dark power distribution implies that the polarization state rotates with the azimuthal angle φ. Distributions of spatial polarization are shown in **Figure 6** (sixth row), indicating the vector beams producing with a polarization order of l changing from +3 to −3. Generation of vector beams carrying OAM utilizing the proposed metal metasurfaces is successfully confirmed in the simulated results.

The operation bandwidth is also studied. We considered the rectangular apertures in the metal film with orientation angle expression of $\alpha(\varphi) = l\varphi+\alpha_0$ ($l = 1, 2, 3$). We choose input optical beam

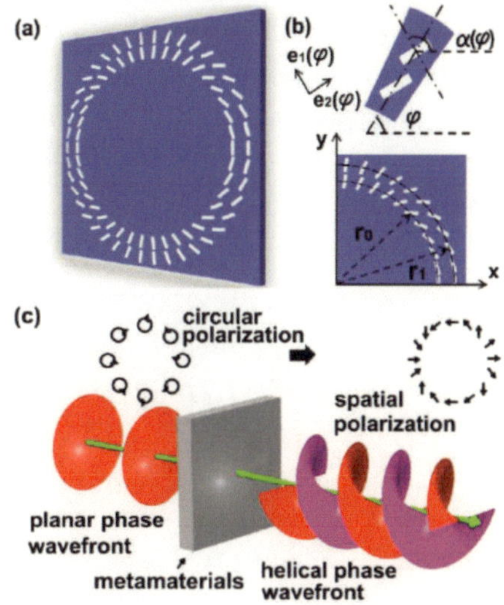

Figure 5. (a) Schematic diagram of metasurfaces for generating vector beams carrying OAM. (b) The radii of two concentric rings are $r_i = (i+6.3) \times 700$ nm ($I = 0, 1$) and the orientation angle respect to the x axis is $\alpha(\varphi) = l\varphi + \alpha_0$ (e.g., $l = 2$, $\alpha_0 = 0$). The dimension of rectangular aperture is 600×140 nm. (c) Concept of generating vector beam carrying OAM (OAM charge number is 2; polarization order is 2) [47].

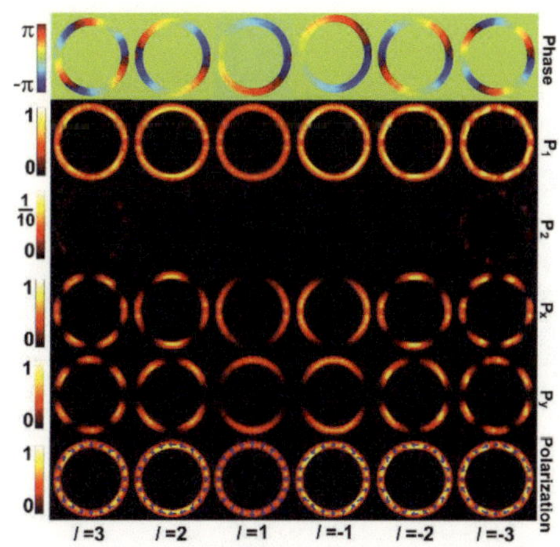

Figure 6. Spatial distributions of phase, power and polarization of generated OAM-carrying vector beams (σ = 1:left circularly polarized input beam, α_0 = 0:along the direction of $e_1(\varphi)$) [47].

of left circular polarization light as the stimulating source. ER and purity are used to analyze the quality of produced vector beams carrying OAM. **Figure 7(a)** shows the relationship between ER and wavelength. One can see that the produced vector beams carrying OAM have high quality in broadband regime of wavelength from 1000 to 2500 nm, covering parts of near-infrared band and mid-infrared band. For topological charge number l of 1 and 2, the ER is maintained above 20 dB covering the wavelength of 1000–2500 nm. For topological charge number l of 3, we achieve ER > 16 dB covering the wavelength of 1000–2500 nm and ER > 20 dB covering the wavelength of 1000–1800 nm. The purity as a function of wavelength of vector beam carrying OAM of $l = 3$ is plotted in **Figure 7(b)**, achieving values above 0.85 covering the wavelength of 1000–2500 nm. The inset pictures show the relationship between weight spectra and topological charge numbers of OAM and polarization order at 1550 nm. The results show that we achieve high-values purities.

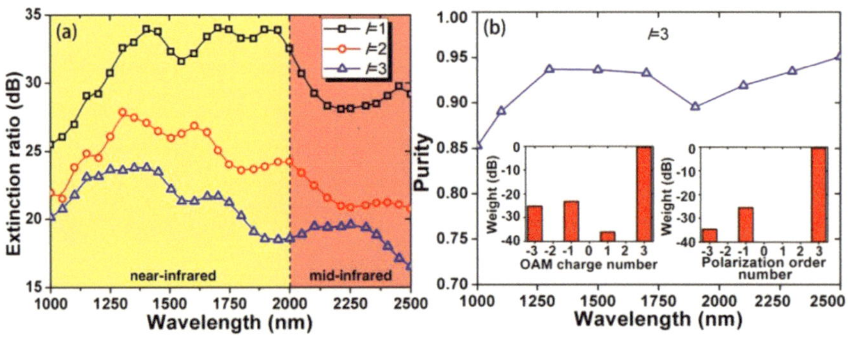

Figure 7. Wavelength-dependent (a) extinction ratio (ER) and (b) purity for the generation of OAM-carrying vector beams. Insets in (b) show weight as functions of OAM charge number (left) and polarization order number (right) at 1550 nm [47].

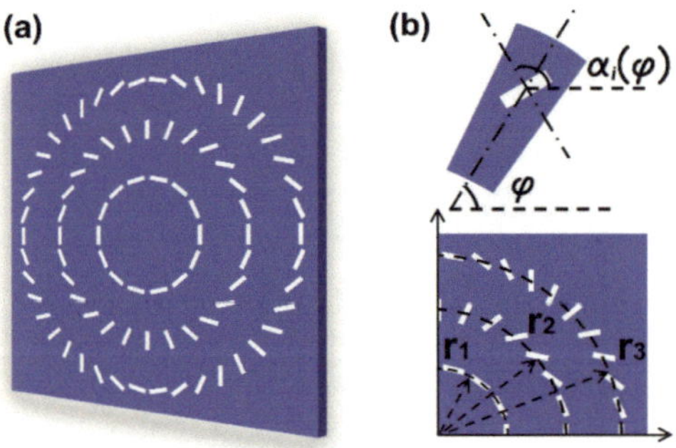

Figure 8. (a) Schematic diagram of proposed metasurfaces. (b) Details of rectangular aperture in the metasurface [48].

In **Figure 5(a)**, two concentric rings are used to excite a single OAM-carrying vector beam. Actually, one ring can also enable the generation of one OAM-carrying vector beam. Hence, it is possible to generate multiple OAM-carrying vector beams by forming multiple concentric

http://dx.doi.org/10.5772/66319

rings in a gold film as shown in **Figure 8**, where each ring contains a rectangular aperture array with each aperture acting as a localized polarizer. In particular, it is expected to selectively generate different orders of OAM-carrying vector beams by shining the circularly polarized light source onto different concentric rings.

Figure 9(a) shows ER as a function of wavelength. One can see broadband operation from 1000 to 1550 nm with an extinction ratio above 20 dB, which indicates high-quality broadband (~500 nm) generation of OAM-carrying vector beams. The fabrication tolerance of the proposed metasurface is also studied by analyzing the relationship between characteristic dependence and the initial orientation angle α_{i0} of each rectangular aperture. The relationship between the extinction ratio and the initial orientation angle α_{20} of each rectangular aperture in the metal film of the second ring is shown in **Figure 9(b)**. When the value of α_{20} is changed from −1 to 7 degrees, the extinction ratio is maintained over 20 dB. The initial orientation angle range is obtained near 8 degrees, indicating a favorable fabrication tolerance of the proposed metasurfaces. We use $\boldsymbol{E}_{1i}$ and $\boldsymbol{E}_{2i}$ to denote the electric field components on the directions of $\boldsymbol{e}_{1i}\,(\varphi)$ and $\boldsymbol{e}_{2i}\,(\varphi)$ (similar to $\boldsymbol{e}_1(\varphi)$ and $\boldsymbol{e}_2(\varphi)$ in **Figure 5(b)**), respectively. We calculate the phase of $\boldsymbol{E}_{1i}$ and the power components on directions of $\boldsymbol{e}_{1i}\,(\varphi)$, $\boldsymbol{e}_{2i}\,(\varphi)$, x and y axes, which are presented by P_1, P_2, P_x, and P_y, respectively. We also calculated the state of spatially variant polarization of the vector beams carrying OAM. All these simulated results are displayed in **Figure 9(c)**. One can clearly indicate that the components of power on the direction of $\boldsymbol{e}_{1i}\,(\varphi)$ has a larger value than that of $\boldsymbol{e}_{2i}(\varphi)$. In **Figure 9(c)**, the phase distributions indicate that the component of electric field $\boldsymbol{E}_{1i}$ has OAM with a topological charge number of $m = i$ (i =1, 2, 3). Spatial distribution of P_x, P_y and the polarization state confirm that the vector beams carrying OAM have a number of polarization order of $l = i$ (i =1, 2, 3) can be confirmed by the spatial distributions of the polarization and power.

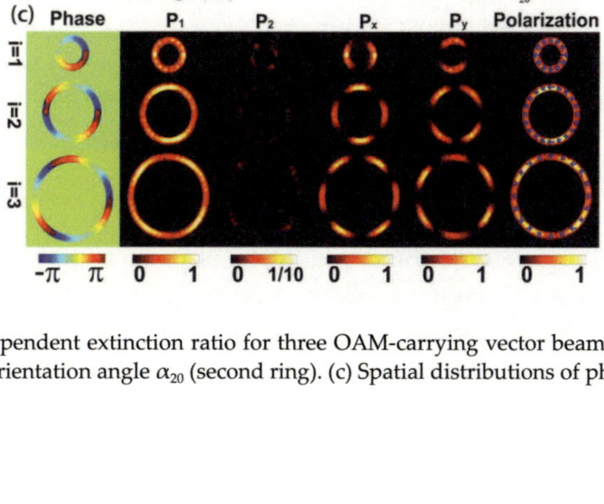

Figure 9. (a) Wavelength-dependent extinction ratio for three OAM-carrying vector beams. (b) The dependence of extinction ratio on the initial orientation angle α_{20} (second ring). (c) Spatial distributions of phase, power components and polarization [48].

2.2. N-fold OAM multicasting using V-shaped antenna array [49]

We design a V-shaped antenna array to realize on-chip multicasting from a single Gaussian beam to four OAM beams. The concept and schematic diagram of N-fold optical multicasting of beams carrying OAM utilizing V-shaped antennas is displayed in **Figure 10**. At the side of multiple OAM multicasting, a specially designed V-shaped antenna array can be seen as a complex phase pattern that can manipulate the wavefront of the input Gaussian beam and then produce multiple OAM beams of collinearly superimposed propagation direction. Then the V-shaped antenna array can duplicate the messages carried by the input Gaussian beam and deliver to multiple optical beams carrying OAM of different topological charge number. Due to the phase singularity of optical beams carrying OAM, the intensity of superimposed multiple optical beams carrying OAM has a dark center. At the side of optical beams carrying OAM demultiplexing, the superimposed multiple optical beams carrying OAM are respectively distributed to multiple end users. At the side of each end user, the helical phase front of the desired optical beam carrying OAM is removed using an inverse phase pattern, resulting in a bright center in the intensity distribution, which can be conveniently separated from other optical beams carrying OAM using spatial filtering.

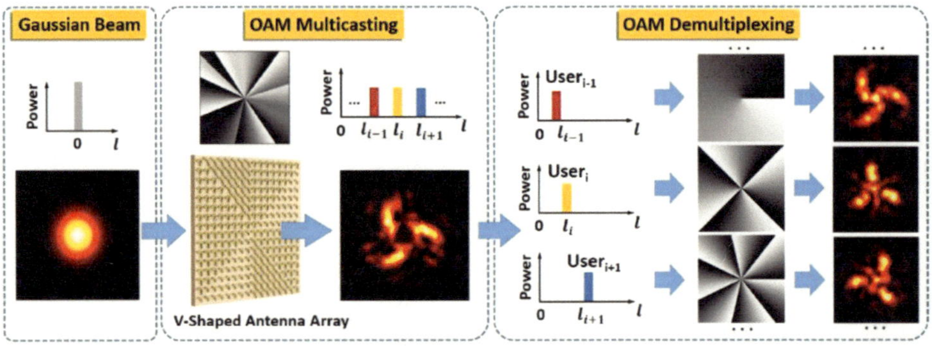

Figure 10. Concept and principle of N-fold multicasting of OAM beams using V-shaped antenna phase array [49].

To enable on-chip OAM multicasting, a pattern search assisted iterative (PSI) algorithm is employed to prepare a specific phase pattern for the simultaneous generation of multiple collinearly superimposed OAM beams. **Figure 11(a)** displays the calculated phase pattern using PSI algorithm to generate four collinearly superimposed OAM beams with charge values of l = 1, 4, 7, 10, respectively. Considering discrete unit of V-shaped antenna array, replace the simulated continuous complex phase pattern with a discrete one is necessary. The right picture of **Figure 11(a)** displays the simulated discrete phase pattern that discretizes the previous continuous complex phase pattern into 32 different values along the azimuthal direction. We then evaluate the degradation of performance of produced collinearly superimposed optical beams carrying OAM caused by the simulated discrete phase pattern by calculating and comparing power distributions of channels of optical beams carrying OAM produced by both continuous and discrete complex phase patterns, as depicted in **Figure 11(b)**. Here the crosstalk for all channels of optical beams carrying OAM is defined by the power ratio of the desired channel of OAM beam (e.g. $l = 4$) to its neighboring ones (e.g. $l = 3$ and $l = 5$). The crosstalk degradation of less than 2 dB is obtained for all four channels of OAM beam after passing through the discrete complex phase pattern. Then we design and simulate the V-shaped antenna array to achieve multicasting from a Gaussian beam to four optical beams carrying OAM by replacing the designed discrete complex phase pattern with corresponding V-shaped antennas, as plotted in **Figure 11(c)–(d)**.

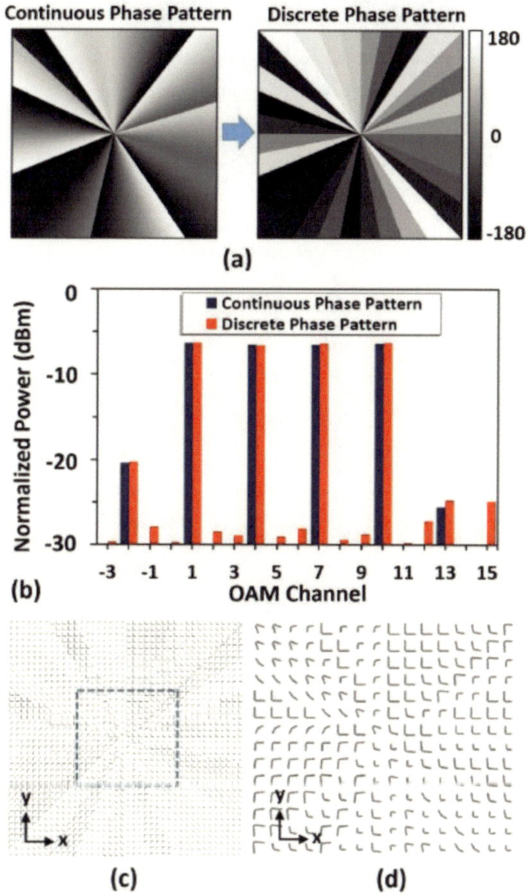

Figure 11. (a) Simulated continuous and discrete phase patterns. (b) Power distributions of channels with different topological charge number of OAM produced by continuous discrete phase patterns. (c) and (d) Top view of designed metal V-shaped antenna array to generate multicasting of optical beams carrying OAM [49].

The simulated intensity distribution of the multicasting optical beams carrying OAM produced by the designed V-shaped antenna array has a dark center, as plotted in **Figure 12(a)**. **Figure 12(b)** and **(c)** show the intensity distributions of channels of optical beams carrying OAM after demultiplexed by the inverse phase pattern. One can see that multicasting channels have bright spots in the center of the beams (**Figure 12(c)**), while undesired channels have dark centers in the intensity distributions (**Figure 12(b)**). We also simulate the power distribution of channels with different topological charge number produced by the designed V-shaped antenna array, as displayed in **Figure 12(d)**. The cross talks of less than −15 dB for all multicast channels are obtained, as plotted in **Figure 12(d)**.

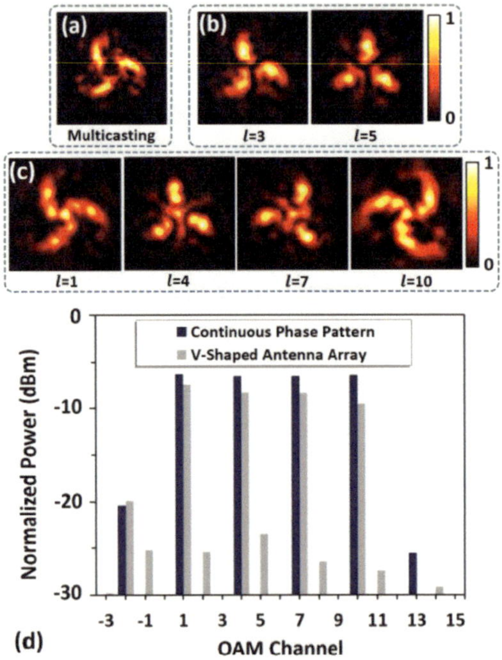

Figure 12. (a) Simulated intensity profiles of produced multicast four optical beams carrying OAM. (b) Simulated intensity profiles of undesired channels after demultiplexed by inverse phase pattern. (c) Simulated intensity profiles of multicasting channels after demultiplexed by inverse phase pattern. (d) Power distributions of channels with different topological charge number of OAM produced by designed V-shaped antenna array compared with that of continuous phase pattern [49].

2.3. Metasurface on conventional optical fiber facet for linearly polarized mode (LP_{11}) generation

We design and fabricate metasurface on the facet of conventional G.652 single mode fiber (SMF) [50]. Here we choose metal rectangle resonator as a unit structure of metasurface. The working principle of a metal rectangle resonator can be explained as follows: the polarization of incident light is along x polarized, which can be decomposed into two perpendicular components corresponding to the long and short edge of the resonator, respectively; the transmission amplitudes of response for a resonator in both components are almost the same while the relative phase retardation is around π, where the linear polarization conversion occurs, resulting in a y-polarized transmitted light. Finite-difference time-domain (FDTD) method is also used to simulate the amplitude and phase response of proposed metal rectangle resonator and its mirror image. The amplitude responses of y-polarized transmitted light of this resonator and its mirror image are almost the same while the phase responses of them have a difference of π. Therefore, this feature can be used to generate higher-order LP mode in the optical fiber. When an x polarized Gaussian beam is irradiating on the metasurface on the facet of G.652 SMF at wavelength of 632.8 nm, it can generate two kinds of transmitted beams with orthogonal polarization. Transmitted beam with x polarization called normal refractive beam that is not influenced by the metasurface can generate LP_{01} mode. Meanwhile, transmitted beam with y polarization called abnormal refractive beam that is influenced by the metasurface can generate LP_{11} mode. Owing to the orthogonal polarization of generated LP_{01} and LP_{11} modes, it is significantly convenient to separate these two modes in optical fiber. **Figure 13(a)** and **(b)** shows the fabricated metasurface on the facet of G.652 SMF to generate LP_{11} mode. The experimental results can be seen in **Figure 13(d)–(g)**. When the polarization direction of incident light is set along x-axis, the LP_{01} mode is x-polarized and LP_{11} mode is y-polarized, as plotted in **Figure 13(d)** and **(e)**. When the polarization direction of incident light is changed to y-axis, the polarization directions of LP_{01} and LP_{11} mode exchange, as seen in **Figure 13(f)** and **(g)**.

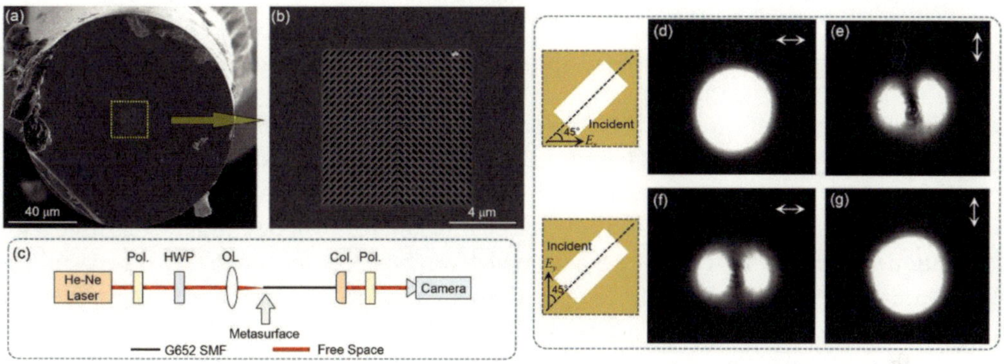

Figure 13. (a) and (b) SEM images of top view of a fabricated metasurface on the facet of SMF to generate LP_{11} mode: (b) details of the metasurface array. (c) Experimental setup for LP_{11} mode generation using the fabricated metasurface. Pol.: polarizer; HWP: half-wave plate; OL: objective lens; Col.: collimator. (d)–(g) Measured intensity profiles of generated LP_{11} mode and LP_{01} mode using the fabricated metasurface on the facet of SMF [50].

3. Light manipulation using dielectric metasurfaces

Dielectric metasurfaces are composed of high refractive index dielectric nanoparticles based on the Mie resonances. Several typical types of dielectric metasurfaces and previous relevant researches have been reported: (1) Er-doped Si-rich silicon nitride nanopillar array is designed and fabricated using RF magnetron sputtering, electron beam lithography, and reactive ion etching methods for enhanced omnidirectional light extraction and OAM beams generation [34]; (2) Si-based metasurfaces are designed and fabricated using electron-beam lithography and reactive ion etching techniques to possess sharp electromagnetically induced–transparency-like resonances in the near-infrared regime [41]; (3) silicon nanobeams antennas are designed and fabricated using low-pressure chemical vapor deposition (LPCVD), electron-beam lithography, and reactive ion etching techniques to generate Bessel beams [29]; (4) amorphous-silicon nanoridges are designed and fabricated using standard electron-beam lithography, and reactive ion etching techniques to realize polarization beam splitting in pixel-level [36]; (5) silicon nanodisks are designed and fabricated using chemical vapor deposition, electron-beam lithography, and reactive ion etching methods to achieve high transmission and full-phase control in visible wavelength [53]; (6) silicon cut-wires array in combination with a silver ground plane are designed and fabricated using chemical vapor deposition, electron-beam lithography, and reactive ion etching techniques to achieve high linear polarization conversion efficiency in the near-infrared band [54]; (7) silicon nanopillar array is designed and fabricated using electron-beam lithography and reactive ion etching techniques to realize low loss micro-lenses in the near-infrared band [55]; (8) dielectric metasurface with a tailored-phase gradient is designed to achieve carpet cloaking at microwave frequencies [56]. All these previous works are based on manipulating the Mie responses of dielectric nanoparticles by altering the dimensions, directions, arrangement of unit structures and even combining multiple resonators with different shapes. Recently, we have also made some progresses in dielectric metasurfaces on the basis of previous works.

3.1. OAM beams generation using nanophotonic dielectric metasurface array

We design and fabricate a kind of chip-scale dielectric metasurface array on silicon-on-insulator (SOI) platform. **Figure 14** shows the schematic drawing of the proposed dielectric metasurface units, consisting of rectangle or ellipse silicon resonators with various dimensions. The working principle of these proposed dielectric resonators can be explained as follows: the polarization of incident light is along x-axis, which can be decomposed into two perpendicular components corresponding to the long and short axis of the resonator, respectively; the reflection amplitudes of response for a resonator in both components are almost the same while the relative phase retardation is around π, where the linear polarization conversion occurs, resulting in a y-polarized reflected light [51].

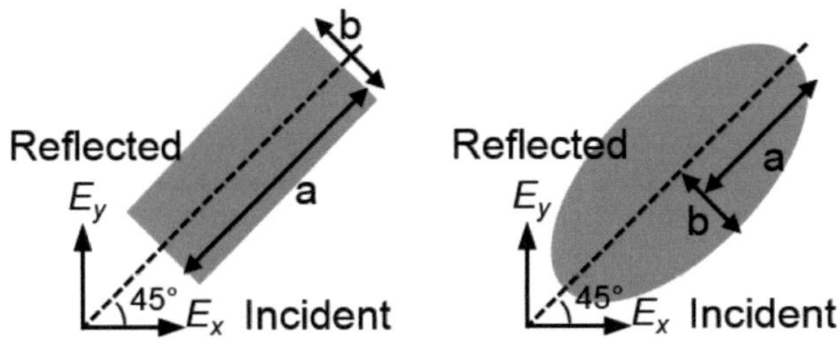

Figure 14. Schematic diagram of top view of designed reflective dielectric rectangle/ellipse metasurface units based on silicon-on-insulator (SOI) platform.

Then we simulate the amplitude and phase responses of the near-field of reflected light of a rectangle dielectric metasurface unit with various dimensions using FDTD method at the wavelength of 1064 nm. The simulated results indicate that the response of the proposed resonator can cover a full phase varying over 0–2π and wide shift of amplitude. It is relatively

http://dx.doi.org/10.5772/66319

easy to find out a series of resonators covering full-phase control while maintaining approximately constant amplitude, which is considerably important for OAM beams generation. Then eight kinds of resonators with different geometric dimensions are chosen to provide an equal-spacing phase shift from 0 to 2π and nearly constant amplitude, as plotted in **Figure 15**. **Figure 16** plots the simulated intensity and phase distributions of OAM beams of topological charge number $l = 1$–4 generated using the rectangle dielectric metasurface array consisted of the eight chosen resonators. The results indicate that the generated OAM beams match the theoretical ones well.

Moreover, we also design the dielectric ellipse metasurface arrays at the wavelength of 632.8nm. **Figure 17** shows the concept of OAM beams generation through dielectric ellipse

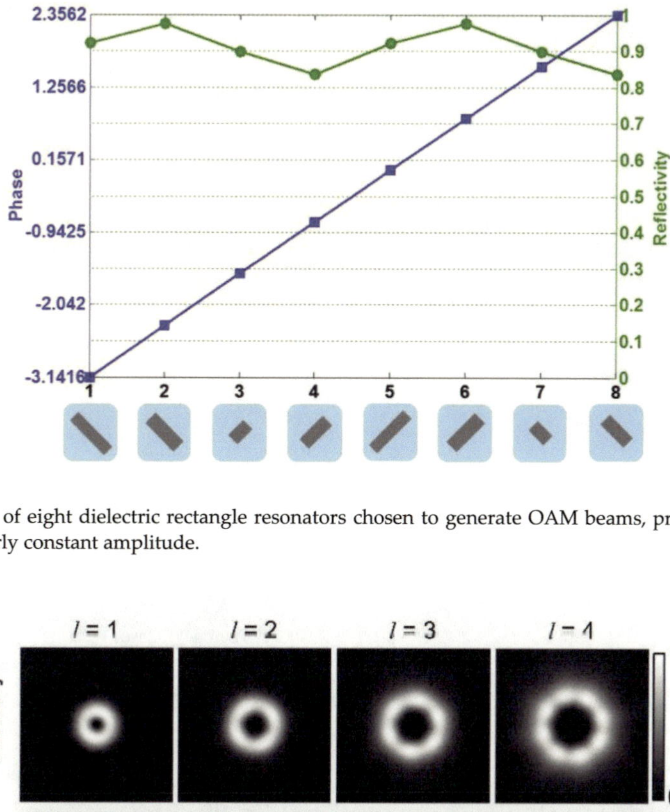

Figure 15. Schematic of eight dielectric rectangle resonators chosen to generate OAM beams, providing a phase shift from 0 to 2π and nearly constant amplitude.

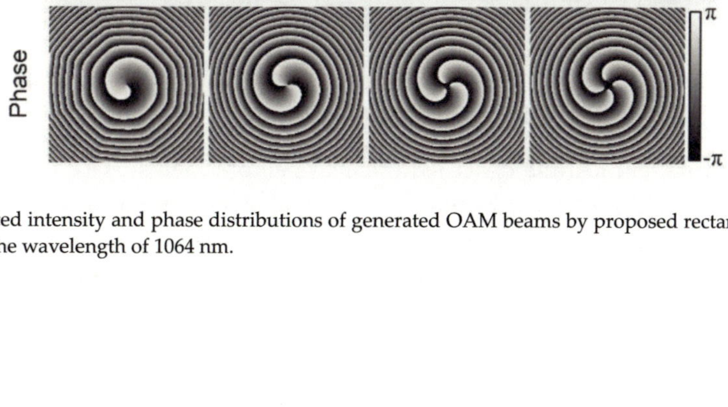

Figure 16. Simulated intensity and phase distributions of generated OAM beams by proposed rectangle dielectric metasurface array at the wavelength of 1064 nm.

metasurface array consisting of a series of resonators covering a full-phase modulation. When a Gaussian beam of vertical polarization is incident to the metasurface, the reflected beam can be transformed into OAM beams of horizontal polarization. In view of polarization conversation happened on the proposed metasurface, one can easily separate the incident and reflected lights. **Figure 18** plots the simulated intensity and phase distributions of OAM beams of topological charge number $l = -4$ to 4 generated using the dielectric ellipse metasurface array consisted of the eight chosen resonators at the wavelength of 632.8 nm. The results indicate that the generated OAM beams match the theoretical ones well.

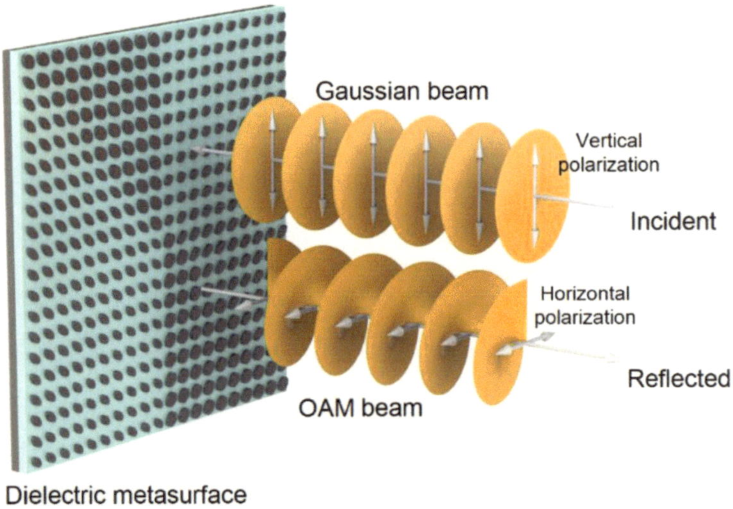

Figure 17. Concept of OAM beams generation using dielectric ellipse metasurface array.

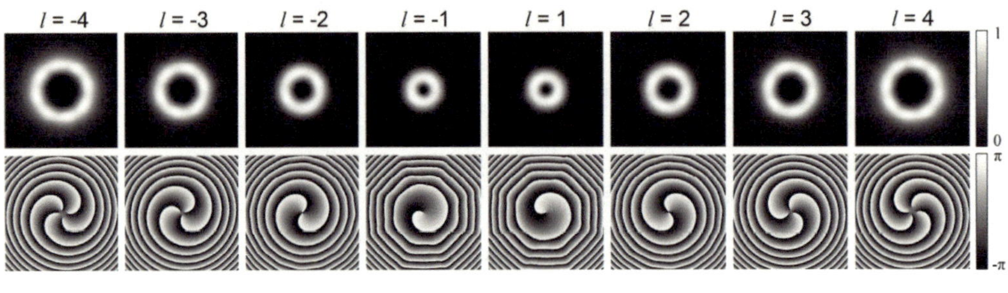

Figure 18. Simulated intensity and phase distributions of generated OAM beams by proposed dielectric ellipse metasurface array at the wavelength of 632.8 nm [51].

We further fabricate and test the designed dielectric ellipse metasurface arrays at the wavelength of 632.8 nm. **Figure 19** shows the experimental results of fabricated dielectric ellipse metasurface arrays. The first line in **Figure 19** plots the measured intensity profiles of generated OAM beams of $l = -4$ to 4 using the fabricated metasurface arrays. The second line in **Figure 19** exhibits the measured intensity profiles of generated OAM beams after interfered by Gaussian beam. One can clearly see from **Figure 19** that the number of the helical arms in the distribution of interference intensity is equal to the topological number l of OAM beam, indicating the favorable results.

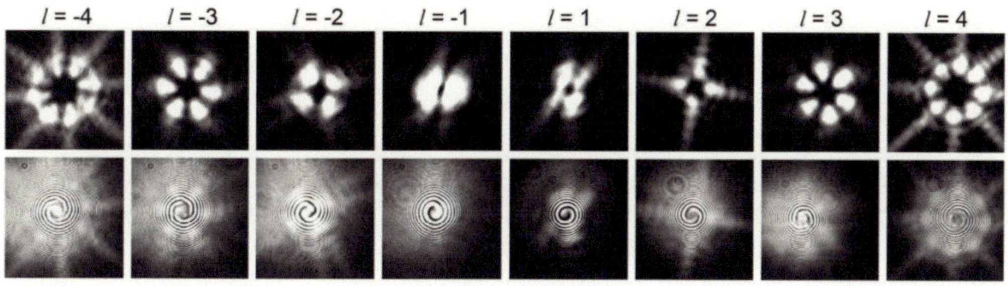

Figure 19. Experimental intensities of generated OAM beams of $l = -4$ to 4 by proposed dielectric metasurface array at the wavelength of 632.8 nm (first line). Experimental intensities of generated OAM beams after interfered by Gaussian beam with $l = 0$ (second line) [51].

3.2. Bessel beams generation and OAM multicasting using dielectric metasurface array

We propose and design dielectric ellipse metasurfaces to realize Bessel beams generation and OAM multicasting at the wavelength of 1550 nm, as displayed in **Figure 20(a)** [52]. **Figure 20(b)** and **(c)** respectively depict the simulated amplitude and phase responses of y-polarized reflected light using FDTD method. Then we choose eight kinds of resonators with different geometric dimensions as shown in **Figure 21(a)**, providing a phase shift from 0 to 2π and nearly constant amplitude. The method of phase pattern replacing axicon is used to generate Bessel beams. By replacing the phase patterns with the eight chosen resonators, simulated Bessel beams are achieved as plotted in **Figure 21(b)**. The purities of generated Bessel beams with different indices are also analyzed. As shown in **Figure 21(c)**, the purities of generated Bessel beams with different indices of $l = -20, -19, \ldots, 19, 20$ are all higher than 0.93.

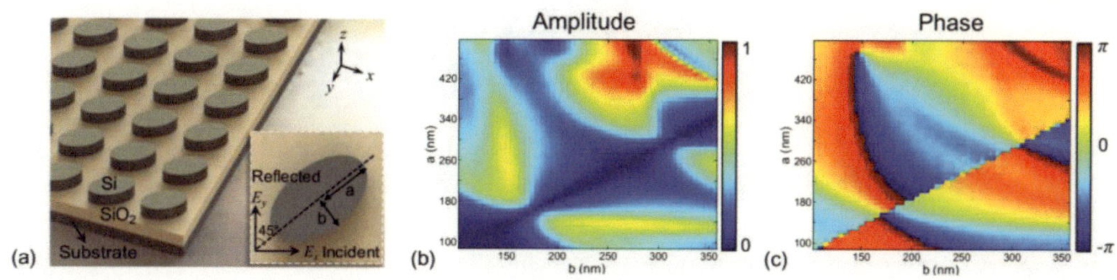

Figure 20. (a) Schematic diagram of designed dielectric ellipse metasurface with a period of 1 μm and silicon height of 500 nm. Inset is the top view of a dielectric elliptical resonator. (b) and (c) are simulated near-field amplitude and phase of y-polarized reflected light as functions of half of long axis length a and short axis length b [52].

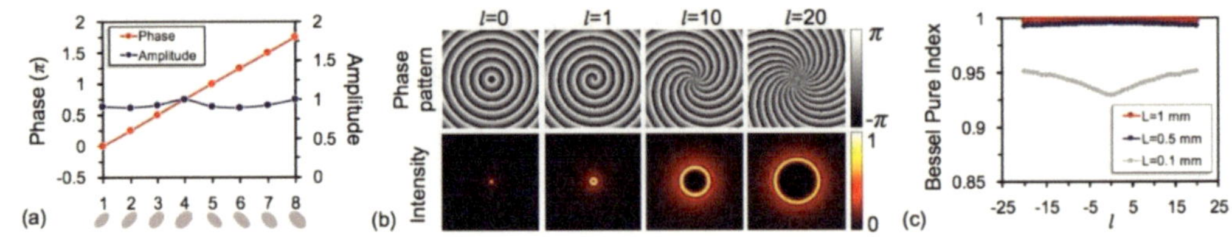

Figure 21. (a) Schematic of eight resonators chosen to generate Bessel beams, providing a phase shift from 0 to 2π and nearly constant amplitude. (b) Simulated intensity of four Bessel beams (l = 0, 1, 10, 20) by replacing the corresponding phase patterns with the eight chosen resonators. (c) Purities of generated Bessel beams with different indices versus topological number l with varied propagation length L [52].

Moreover, sixfold OAM multicasting using the designed dielectric ellipse metasurfaces is also simulated. The ideal phase pattern to generate collinearly superimposed six OAM beams (l = 1, 4, 7, 10, 13, 16) is plotted in **Figure 22(a)**. By replacing the ideal phase pattern with corresponding resonators, we design the dielectric ellipse array shown in **Figure 22(b)**. The intensity profile of the superimposed six OAM modes produced by the proposed ellipse array with a trigonal dark center is depicted in **Figure 22(d)**. **Figure 22(e)** and **(f)** shows the intensity profiles of OAM channels after demulticasting. For desired OAM channels, a bright light spot emerges at the center of beam (**Figure 22(f)**), while for needless channels, dark region will come out at the beam center (**Figure 22(e)**). The power ratios of the needed OAM channels (e.g. l = 16) to its neighboring ones (e.g. l = 15 and l = 17) are defined as cross talks, which are approximately less than −14 dB as plotted in **Figure 22(c)**.

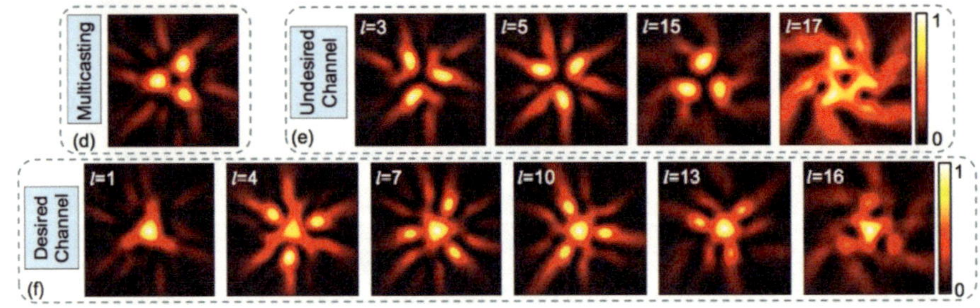

Figure 22. (a) Ideal phase pattern to generate sixfold OAM multicasting (l = 1, 4, 7, 10, 13, 16). (b) Top view of designed dielectric ellipse array to generate sixfold OAM multicasting by replacing the ideal phase pattern with corresponding resonators. (c) Power distributions of OAM channels generated by ideal phase pattern and dielectric ellipse array. (d) Intensity distribution of collinearly superimposed six OAM beams. (e) Intensity distribution of undesired OAM channels after demultiplexing. (f) Intensity distribution of multicasting OAM channels after demultiplexing [52].

http://dx.doi.org/10.5772/66319

4. Discussion

Although metasurfaces composed of plasmonic and dielectric units have only recently emerged as an active area of research, their potential as an approach to realize light manipulation devices has already been successfully implemented in flat optics such as OAM beams generator, lens, wave plates, holograms, and so on, for its superiorities of compact and flexible design. For instance, convenient optical elements are based on refraction, reflection or diffraction of light. Their ability of wavefront shaping is relay on propagation through media whose refractive indices can be designed to control the light path. As a consequence, those conventional optical elements suffer from thick materials, large volume, long diffraction distance and even low optical manipulation accuracy that they are difficult to be compatible with compact systems. Metasurface-based flat optical elements can potentially replace traditional ones. They are able to modulate wavefront of light beam by introducing a variety of nanoscaled electric dipole resonators replacing conventional propagation effect and therefore considerably reduce the whole volume of systems. Besides, the responses of metasurfaces rely on the characteristics of unit structures such as structure dimensions and materials types. It is relatively convenient to flexibly control the amplitude, phase and polarization of scattering light.

However, there are still some challenges of metasurfaces. For plasmonic metasurfaces, they suffer from ohmic loss at optical wavelength. One common solution to loss compensation is to introduce a gain medium in or around the plasmonic material. This solution can partially or fully compensate the loss in metals, but the incorporation of such active materials is very challenging [57]. Another challenge for plasmonic metasurfaces is that noble metals have large plasma frequency and cannot be adjusted, making tunable or switchable metasurface devices hard to realize [57]. What is more, noble metals are not compatible with the standard semiconductor fabrication technologies, restricting metasurface devices to the proof-of-concept stage only. To deal with those issues, some recent researches focus on new materials platforms defined new, intermediate carrier density materials as the best candidates that perform low loss, excellent tuning, and modulation capabilities and that are compatible with established semiconductor processing technologies and integration procedures [57, 58]. In addition, metasurfaces are also facing the challenge of reconfigurable. For metasurface, the unit structures must be subwavelength-scaled optical switches with a fraction of a square micrometer footprint and a volume of only approximate 10-19 m^3. The optical switches are desired to be fast and energy saving. It is challenging for unit structures to satisfy the former two conditions. To solve the problem, one can alter the dimensions of individual unit structures, or manipulate the near-field interactions between them. The nanomechanical devices platform can provide one possible ideal solution. There are several impressive works reported recently focusing on the tunable and reconfigurable metasurfaces [57, 58]: (1) by combining the Fano resonant metal nanoparticles with a single layer graphene, obviously electrically controlled damping are observed in the Fano resonances in the near-infrared regime; (2) by positioning a thin graphene sheet over asymmetric silicon nanobars, transmission spectrum of the proposed device can be efficiently adjusted at the near-infrared frequencies; (3) a metal nanomechanical metasurface is fabricated on the nanoscale-thickness dielectric membrane, and the displacement of metamolecules can be controlled by light illumination; (4) using differential thermal expansion and Lorentz force, electrothermal tuning and magnetic modulation are achieved in a reconfigurable metal metasurface; (5) using a vanadium dioxide metasurface lens, the amplitude spectrum and the focal intensity can be adjusted by changing the temperature in the terahertz regime; (6) utilizing a heating-induced vanadium dioxide planar metamaterial, the device is able to be switched between capacitive and inductive responses; (7) by combing metal nanoantenna array with indium tin oxide, phase and amplitude of reflected light can be controlled at near infrared wavelengths.

In spite of facing many challenges, metasurfaces are still one of the most potential devices applied in beam-steering, spatial light modulator, nanoscale-resolution imaging, sensing, and novel quantum optics devices. Besides, the increasing demand for large capacity of data transmission and fast signal processing promotes to eliminate the bottleneck between optical fiber communication networks and electronic data handling. Achieving this goal will require strong and fast nonlinearities for switching light with light, and much improved manipulation of the electromagnetic fields with external stimuli such as electric signals in compact space. Metasurfaces-based devices can considerably enhance these functionalities by exploiting the characteristics of metasurfaces [58]. In the future, metasurfaces-based devices would be in developing tunable, reconfigurable, nonlinear, switchable, gain and quantum metasurfaces, which might be achieved by adopting new materials (such as superconductors or graphene) and hybridizing plasmonics metal nanostructures with other functional materials such as nanocarbon, organics, nanosemiconductors, organic polymers and phase-change media [58].

5. Conclusion

In summary, we have reviewed our recent works in light manipulation on the dimension of spatial structure using metasurfaces: (1) metasurfaces consisting of multiple concentric rings in a gold film are designed to generate broadband OAM carrying vector beams in the regime from near infrared to mid infrared band; (2) a metal V-shaped antenna array is designed to realize multicasting from a single Gaussian beam to four OAM beams assisted by PSI algorithm, achieving a low-crosstalk of near −15 dB; (3) a kind of gold metasurface is designed and fabricated on the facet of a G.652 SMF to generate LP_{11} mode at the wavelength of 632.8 nm; (4) a kind of dielectric metasurface is designed and fabricated on SOI platform, realizing OAM beams generation; (5) a kind of dielectric metasurface is designed on SOI platform at the wavelength of 1550 nm, achieving Bessel beams generation with high purity and multicasting from a single Gaussian beam to six OAM beams with a relatively low crosstalk less than −14 dB. The designed and fabricated metasurfaces in these works have nanoscaled dimensions and the generated beams carrying OAM have favorable qualities, which are promising for compact and effective beam-steering system and space-division multiplexing (SDM)-assisted optical communication systems. In the future, the metasurface devices to generate optical beams carrying OAM can combine with tunable and reconfigurable functions reported in Refs. [57–58], which makes robust spatial light manipulation possible and facilitates more interesting applications.

http://dx.doi.org/10.5772/66319

Acknowledgements

This work was supported by the National Natural Science Foundation of China (NSFC) under Grants 11574001, 11274131, and 61222502, the National Basic Research Program of China (973 Program) under Grant 2014CB340004, the National Program for Support of Top-Notch Young Professionals, and the Program for New Century Excellent Talents in University (NCET-11-0182). The authors would like to thank Zhe Zhao, Xiao Hu, Nan Zhou, Long Zhu, Shuhui Li, Jun Liu, Yifan Zhao, and Hongya Wang for technical supports and helpful suggestions.

Author details

Jian Wang* and Jing Du

*Address all correspondence to: jwang@hust.edu.cn

Wuhan National Laboratory for Optoelectronics, School of Optical and Electronic Information, Huazhong University of Science and Technology, Wuhan, Hubei, China

References

[1] M. Born, W. Emil, Principles of optics: electromagnetic theory of propagation, interference and diffraction of light. Cambridge university press, Cambridge, 2000.

[2] C. A. Brackett, "Dense wavelength division multiplexing networks: Principles and applications." IEEE J. Sel. Areas Commun. 8, 948–964 (1990).

[3] S. L. Jansen, I. Morita, T. C. Schenk, H. Tanaka, "Long-haul transmission of 16× 52.5 Gbits/s polarization-division-multiplexed OFDM enabled by MIMO processing." J. Opt. Networking, 7, 173–182 (2008).

[4] S. Kawanishi, "Ultrahigh-speed optical time-division-multiplexed transmission technology based on optical signal processing: feature issue on fundamental challenges in ultrahigh-capacity optical fiber communication systems." IEEE J. Quantum Electron. 34, 2064–2079 (1998).

[5] D. J. Richardson, J. M. Fini, L. E. Nelson, "Space-division multiplexing in optical fibres." Nat. Photon. 7, 354–362 (2013).

[6] P. J. Winzer, R. J. Essiambre, "Advanced modulation formats for high-capacity optical transport networks." J. Lightw. Technol. 24, 4711–4728 (2006).

[7] I. Moreno, J. A. Davis, B. M. L. Pascoguin, M. J. Mitry, D. M. Cottrell, "Vortex sensing diffraction gratings." Opt. Lett. 34, 2927–2929 (2009).

[8] S. Bernet, A. Jesacher, S. Fürhapter, C. Maurer, M. Ritsch-Marte, "Quantitative imaging of complex samples by spiral phase contrast microscopy." Opt. Express 14, 3792–3805 (2006).

[9] M. Padgett, R. Bowman, "Tweezers with a twist." Nat. Photon. 5, 343–348 (2011).

[10] J. Leach, M. R. Dennis, J. Courtial, M. J. Padgett, "Laser beams: knotted threads of darkness." Nature 432, 165 (2004).

[11] L. Paterson, M. P. MacDonald, J. Arlt, W. Sibbett, P. E. Bryant, K. Dholakia, "Controlled rotation of optically trapped microscopic particles." Science 292, 912–914 (2001).

[12] D. R. Smith, W. J. Padilla, D. C. Vier, S. C. Nemat-Nasser, S. Schultz, "Composite medium with simultaneously negative permeability and permittivity." Phys. Rev. Lett. 84, 4184–4187 (2000).

[13] W. Cai, V. Shalaev, Optical Metamaterials: Fundamentals and Applications. Springer Science and Business Media, New York, 2009.

[14] N. Meinzer, W. L. Barnes, I. R. Hooper, "Plasmonic meta-atoms and metasurfaces." Nature Photon. 8, 889–898 (2014).

[15] J. D. Jackson, Classical Electrodynamics. Wiley, New York, 1998.

[16] V. M. Shalaev, W. S. Cai, U. K. Chettiar, H. K. Yuan, A. K. Sarychev, V. P. Drachev, A. V. Kildishev, "Negative index of refraction in optical metamaterials." Opt. Lett. 30, 3356–3358 (2005).

[17] M. W. Klein, C. Enkrich, M. Wegener, S. Linden, "Second-harmonic generation from magnetic metamaterials." Science 313, 502–504 (2006).

[18] S. Linden, C. Enkrich, M. Wegener, J. F. Zhou, T. Koschny, C. M. Soukoulis, "Magnetic response of metamaterials at 100 terahertz." Science 306, 1351–1353 (2004).

[19] N. Fang, H. Lee, C. Sun, X. Zhang, "Sub-diffraction-limited optical imaging with a silver superlens." Science 308, 534–537 (2005).

[20] E. Plum, V. A. Fedotov, A. S. Schwanecke, N. I. Zheludev, Y. Chen, "Giant optical gyrotropy due to electromagnetic coupling." Appl. Phys. Lett. 90, 223113 (2007).

[21] J. B. Pendry, D. Schurig, D. R. Smith, "Controlling electromagnetic fields." Science 312, 1780–1782 (2006).

[22] N. Yu, P. Genevet, M. A. Kats, F. Aieta, J. P. Tetienne. F. Capasso, Z. Gaburro, "Light propagation with phase discontinuities: generalized laws of reflection and refraction." Science 334, 333–337 (2011).

[23] S. Jahani, Z. Jacob, "All-dielectric metamaterials." Nat. Nanotech. 11, 23–36 (2016).

[24] C. F. Bohren, D. R. Huffman, Absorption and Scattering of Light by Small Particles. John Wiley & Sons, Hoboken, 2008.

[25] M. Decker, I. Staude, M. Falkner, J. Dominguez, D. N. Neshev, I. Brener, T. Pertsch, Y. S. Kivshar, "High-efficiency dielectric Huygens' surfaces." Adv. Opt. Mater. 3, 813–820 (2015).

[26] C. Enkrich, F. Perez-Williard, D. Gerthsen, J. Zhou, T. Koschny, C. M. Soukoulis, M. Wegener, S. Linden, "Focused-ion-beam nanofabrication of near-infrared magnetic metamaterials," Adv. Mater. 17, 2547 (2005).

[27] S. R. J. Brueck, "Optical and interferometric lithography-nanotechnology enablers," Proc. IEEE 93, 1704 (2005).

[28] S. Y. Chou, P. R. Krauss, P. J. Renstrom, "Nanoimprint lithography," J. Vac. Sci. Technol. B 14, 4129 (1996).

[29] W. Wu, E. Kim, E. Ponizovskaya, Z. Liu, Z. Yu, N. Fang, Y. R. Shen, A. M. Bratkovsky, W. Tong, C. Sun, X. Zhang, S.-Y. Wang, R. S. Williams, "Optical metamaterials at near and mid-IR range fabricated by nanoimprint lithography," Appl. Phys. A 87, 147 (2007).

[30] N. Yu, F. Aieta, P. Genevet, M. A. Kats, Z. Gaburro, F. Capasso, "A broadband, background-free quarter-wave plate based on plasmonic metasurfaces." Nano Lett. 12, 6328–6333 (2012).

[31] S. Sun, K. Y. Yang, C. M. Wang, T. K. Juan, W. T. Chen, C. Y. Liao, Q. He, S. Xiao, W. T. Kung, G. Y. Guo, L. Zhou, D. P. Tsai, "High-efficiency broadband anomalous reflection by gradient meta-surfaces." Nano Lett. 12, 6223–6229 (2012).

[32] S. Sun, Q. He, S. Xiao, Q. Xu, X. Li, L. Zhou, "Gradient-index meta-surfaces as a bridge linking propagating waves and surface waves." Nat. Mater. 11, 426–431 (2012).

[33] F. Aieta, P. Genevet, M. A. Kats, N. Yu, R. Blanchard, Z. Gaburro, F. Capasso, "Aberration-free ultrathin flat lenses and axicons at telecom wavelengths based on plasmonic metasurfaces." Nano Lett. 12, 4932–4936 (2012).

[34] D. Parker, D. C. Zimmermann, "Phased arrays—part 1: Theory and architectures." IEEE Trans. Microwave Theory Tech. 50, 678–687 (2002).

[35] A. Pors, M. G. Nielsen, R. L. Eriksen, S. I. Bozhevolnyi, "Broadband focusing flat mirrors based on plasmonic gradient metasurfaces." Nano Lett. 13, 829-834 (2013).

[36] J. J. Cowan, "The surface plasmon resonance effect in holography." Opt. Comm. 5, 69–72 (1972).

[37] L. Huang, X. Chen, H. Mühlenbernd, H. Zhang, S. Chen, B. Bai, Q. Tan, G. Jin, K. W. Cheah, C. W. Qiu, J. Li, T. Zentgraf, J. Li, "Three-dimensional optical holography using a plasmonic metasurface." Nat. Commun. 4, 2808 (2013).

[38] Z. H. Jiang, S. Yun, L. Lin, J. A. Bossard, D. H. Werner, T. S. Mayer, "Tailoring dispersion for broadband low-loss optical metamaterials using deep-subwavelength inclusions." Sci. Rep. 3, 1571 (2013).

[39] M. Kang, J. Chen, X.-L. Wang, H.-T. Wang, "Twisted vector field from an inhomogeneous and anisotropic metamaterial." J. Opt. Soc. Am. B 29, 572–576 (2012).

[40] L. Huang, X. Chen, H. Mu¨hlenbernd, G. Li, B. Bai, Q. Tan, G. Jin, T. Zentgraf, S. Zhang, "Dispersionless phase discontinuities for controlling light propagation." Nano Lett. 12, 5750–5755 (2012).

[41] M. Khorasaninejad, W. Zhu, K. B. Crozier, "Efficient polarization beam splitter pixels based on a dielectric metasurface." Optica 2, 376-382 (2015).

[42] N. Segal, S. Keren-Zur, N. Hendler, T. Ellenbogen, "Controlling light with metamaterial-based nonlinear photonic crystals." Nat. Photon. 9, 180-184 (2015).

[43] S. Kim, H. Wakatsuchi, J. J. Rushton, D. F. Sievenpiper, "Switchable nonlinear metasurfaces for absorbing high power surface waves." Appl. Phys. Lett. 108, 041903 (2016).

[44] X. Yin, Z. Ye, J. Rho, Y. Wang, X. Zhang, "Photonic spin Hall effect at metasurfaces." Science 339, 1405–1407 (2013).

[45] Y. Yang, I. I. Kravchenko, D. P. Briggs, J. Valentine, "All-dielectric metasurface analogue of electromagnetically induced transparency." Nat. Commun. 5, 6753 (2014).

[46] S. Mühlig, M. Farhat, C. Rockstuhl, F. Lederer, "Cloaking dielectric spherical objects by a shell of metallic nanoparticles." Phys. Rev. B 83, 195116 (2011).

[47] Z. Zhao, J. Wang, S. Li, A. E. Willner, "Metamaterials-based broadband generation of orbital angular momentum carrying vector beams." Opt. Lett. 38, 932–934 (2013).

[48] Z. Zhao, J. Wang, S. Li, A. E. Willner, "Selective Broadband Generation of Orbital Angular Momentum Carrying Vector Beams Using Metamaterials." Proc. CLEO2013, Optical Society of America, paper QM4A. 7, 2013.

[49] J. Du, J. Wang, "Design of on-chip N-fold orbital angular momentum multicasting using V-shaped antenna array." Sci. Rep. 5, 9662 (2015).

[50] J. Du, S. Chen, J. Liu, S. Li, L. Zhu, Y. F. Zhao, J. Wang, "Design and fabrication of metasurface on conventional optical fiber facet for linearly polarized mode (LP11) generation at visible light wavelength." Proc. CLEO2016, Optical Society of America (OSA), paper JTu5A.104, 2016.

[51] J. Du, X. Li, S. Li, L. Zhu, N. Zhou, J. Liu, S. Chen, Y. Zhao, J. Wang, "Experimental demonstration of chip-scale orbital angular momentum (OAM) beams generation and detection using nanophotonic dielectric metasurface array." Proc. OFC2016, Optical Society of America (OSA), paper W2A.13, 2016.

[52] J. Du, Z. Xu, L. Zhu, S. Li, J. Wang, "Design of On-Chip Dielectric Elliptical Meta-Reflectarray for Bessel Beams Generation and N-Fold Orbital Angular Momentum (OAM) Multicasting." Proc. CLEO2015, Optical Society of America (OSA), paper JTu5A.57, 2015.

[53] Y. F. Yu, A. Y. Zhu, R. Paniagua-Domínguez, Y. H. Fu, B. Luk'yanchuk, A. I. Kuznetsov, "High-transmission dielectric metasurface with 2π phase control at visible wavelengths. Laser Photon. Rev. 9, 412–418 (2015).

[54] Y. Yang, W. Wang, P. Moitra, I. I. Kravchenko, D. P. Briggs, J. Valentine, "Dielectric meta-reflectarray for broadband linear polarization conversion and optical vortex generation." Nano Lett. 14, 1394–1399 (2014).

[55] P. R. West, J. L. Stewart, A. V. Kildishev, V. M. Shalaev, V. V. Shkunov, F. Strohkendl, Y. A. Zakharenkov, R. K. Dodds, R. Byren, "All-dielectric subwavelength metasurface focusing lens." Opt. Express 22, 26212–26221 (2014).

[56] L. Hsu, T. Lepetit, B. Kanté, "Extremely thin dielectric metasurface for carpet cloaking." arXiv preprint arXiv1503.08486 (2015).

[57] N. I. Zheludev, "A roadmap for metamaterials." Opt. Photon. News 22, 30-35 (2011).

[58] N. I. Zheludev, Y. S. Kivshar, "From metamaterials to metadevices." Nat. Mater. 11, 917–924 (2012).

Chapter 4

Development of Metamaterial EBG Absorbers for Application of Wireless Inter/Intrachip Communication Systems

Xin Zheng, Yinchao Chen, Wensong Wang and
Shuhui Yang

Additional information is available at the end of the chapter

http://dx.doi.org/10.5772/67018

Abstract

First, the chapter presents a novel design of electromagnetic bandgap (EBG) absorber with the characteristics of broad bandwidth, low profile, and polarization-independence to a normal incident electromagnetic wave. The absorber is composed of three consecutive octagon or decagon loops, and highly-resistive frequency selective surface (FSS) layers. Second, based on the feature of the designed absorber unit, a broadband, metamaterial absorber-bounded, wireless inter/intrachip (WIIC) communication channel is constructed at the center frequency of 60 GHz. Third, in order to validate the developed methodology used in WIIC analysis, a wired channel on a conventional PCB has been measured, simulated, and analyzed. Fourth, with the extracted S-parameters of the WIIC system and wired PCB channel, the system impulse responses and transfer functions of the investigated channels have been further extracted, which are used for validation and BER analysis of the WIIC system. Finally, it has been shown that based on the derived BER results, the performance of the designed WIIC channel is close to that of an additive Gaussian white noise (AWGN) channel when the WIIC transceivers are built in with the functionalities of forward error control (FEC), channel estimation, and equalization.

Keywords: metamaterial EBG, wireless inter/intra chip communication, channel design, absorber, bit error rate

1. Introduction

As the dramatic development of high speed integrated circuits and the fast increment of operating frequency of computer systems, the dimensions of complementary metal-oxide semiconductor (CMOS) transistors have reached at the nanoscale. The very large scale integrated circuits, interconnecting communication channels and devices, and I/O pins have been becoming extremely complex, highly dense, and compact by using multilayer and complicated routing PCB technologies [1]. When the frequency of the computer system operation is greater than 20 GHz or higher, the signal integrity (SI) bottlenecks have been becoming increasingly protuberant, which are particularly resulted in signal reflection, crosstalk, trace loss and delay, parasitic resistance, inductance, and capacitance [2–4].

In order to resolve these serious SI issues, several methods have been recently investigated to improve the interconnection communication systems. For example, using a low-resistivity conductor or a low dielectric constant material, as well as multilayer stacked structures, improves the interconnection communication [5, 6]. Meanwhile, various approaches are used to eliminate traditional trace interconnection issues that have been explored for improving SI systems, including optical interconnects, electromagnetic wave solution, and radio frequency wireless communication technologies [7–11].

This chapter presents an innovative wireless inter/intrachip (WIIC) communication channel, as shown in **Figure 1**, where the major concept of the WIIC communication is illustrated to distinguish the existing high-density trace communication-based PCB and the wireless communication version developed in the chapter.

The rationale and motivation of this research are mainly resulted from the following considerations:

- The WIIC system can greatly reduce the cost of PCB boards as the number of PCB layers is reduced from 10 of layers to less than 5 layers.
- The WIIC system can fully take the advantage of the advanced and well-developed wireless and mobile communication systems.
- The wireless interconnect channel will be more convenient and flexible for layout chip distribution and for computer architecture design without concerning about wiring and routing of complicated vias and traces.

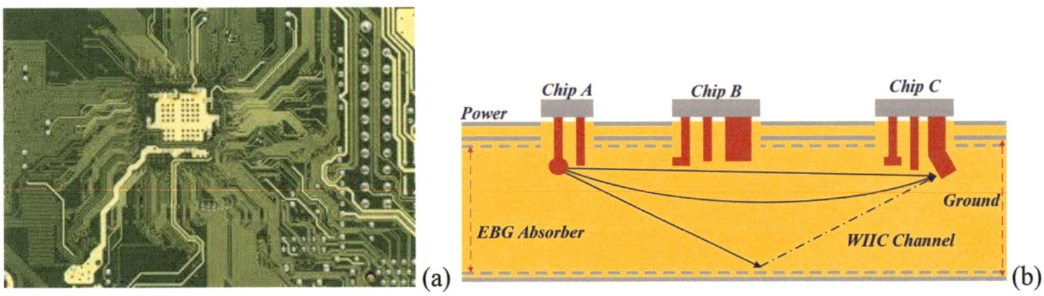

Figure 1. The concept of WIIC communication developed in this research. (a) Traditional high density, trace-based interconnect PCB, and (b) the concept of the proposed WIIC communication channel bounded by EBG absorber layers.

http://dx.doi.org/10.5772/67018

- There is no crosstalk, low distortion, dispersion, and time delay, although interferences between transceivers are unavoidable on this designed WIIC board.
- The long cycle process in designing and testing PCB trace wiring and routing will be completely eliminated.

2. Broadband EBG absorber design

2.1. Configuration of EBG absorber

The unit cell of the firstly proposed metamaterial EBG absorber, which is designed at the center frequency of 25 GHz, is displayed in **Figure 2**. The absorber unit includes a resistive layer, a dielectric substrate layer, and a ground layer. The octagon-shaped resistive layer consists of three consecutive loops of highly resistive frequency selective surface (FSS). The material of the three loops is tantalum nitride with the relative dielectric constant of $\varepsilon_r = 1$, the relative permeability of $\mu_r = 1$, and the conductivity of $\sigma = 7400$ S/m. The board substrate is foam-characterized with $\varepsilon_{rf} = 1.05$, $\mu_{rf} = 1$, and the loss tangent of tan $\delta = 0.005$. The ground is made from copper with a conductivity of $\sigma = 5.8 \times 10^7$ S/m.

Similarly, the proposed 60 GHz absorber unit is built with the substrate of foam characterized by the dielectric constant $\varepsilon_r = 1.0$. The dimensions of the absorber unit are about 4.2 × 4.2 × 1 mm [13]. The absorber unit consists of three consecutive decagon loops in order to achieve a broadband frequency band centered at 60 GHz. The thickness of the resistive metal layer is 0.02 mm, the width of the loop traces is 0.08 mm, the edge-to-edge spacing is 0.08 mm, and the side lengths of three loops are 0.3245, 0.2503, and 0.2256 mm, respectively.

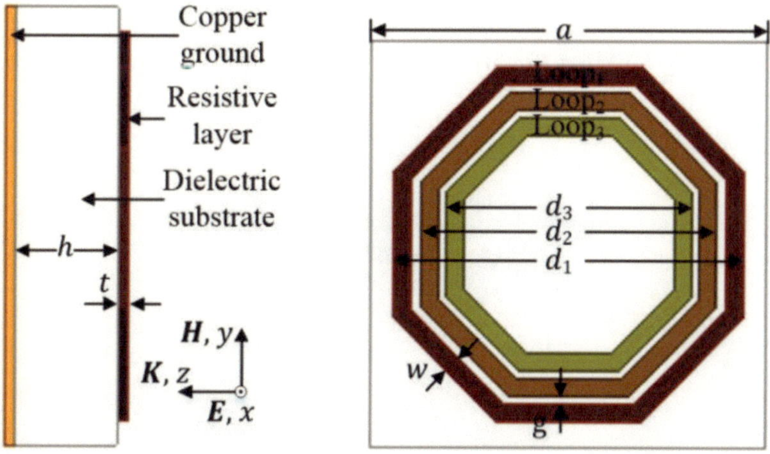

Figure 2. Side and top views of the 25 GHz EBG absorber unit, where $d_1 = 3.69$, $d_2 = 3.13$, $d_3 = 2.57$, $w = 0.18$, $g = 0.1$, $h = 1$, $t = 0.02$, $a = 4.2$, all in mm.

2.2. Broadband absorption performance

The absorber unit has been built into an HFSS (high frequency structure system) environment, as presented in **Figure 3**, where the boundary surfaces of the unit cell are curtailed with the HFSS built-in master and slave boundary conditions (BCs) in addition to two Floquet port excitations placed at the top and bottom as a wave port. In this case, the setting of the absorber unit can be considered as an equivalent, periodic, infinitely large absorbing configuration. Electromagnetic plane waves are placed to be incident onto the top surface of the absorber unit in the normal direction polarized as a transverse electric (TE) or a transverse magnetic (TM) wave.

The absorptance *A*(*f*) is a measure of the absorber's EM wave power absorption, which is precisely given as

$$A(f) = 1 - |S_{11}|^2 - |S_{21}|^2 \tag{1}$$

where S_{11} and S_{21} are the reflection and transmission coefficients estimated in the two-port network system, respectively. Because the absorber is essentially shielded by the copper ground, the EM wave cannot penetrate the structure, i.e., $S_{21} = 0$.

As shown in **Figure 4**, it is found that the proposed design of the absorptance *A*(*f*) of the 25 GHz EBG absorber unit is about 90% or greater in the frequency band of 19.9–31.2 GHz. The relative bandwidth of the designed absorber unit is approximately 44% centered at the frequency of 25.6 GHz. The predicted *A*(*f*) has been verified by an FDTD (finite difference time domain) simulation, which shows a higher agreement between those developed from the FDTD and HFSS methods [12].

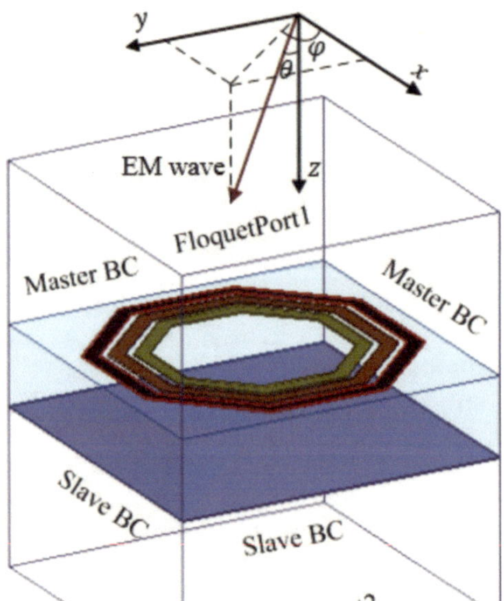

Figure 3. EBG absorber unit assessment system in an HFSS environment.

http://dx.doi.org/10.5772/67018

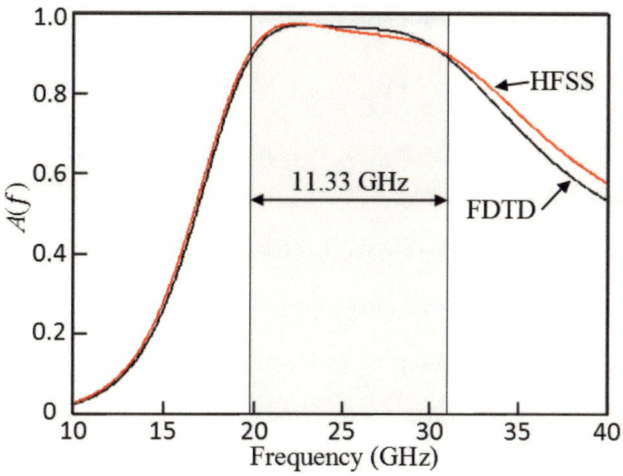

Figure 4. The simulated absorptance of the proposed 25GHz EBG absorber.

2.3. Equivalent circuit analysis

It is very often to develop an equivalent circuit model for evaluating the absorptance $A(f)$ of an EBG absorber speedily and efficiently in electrical engineering research and electronic engineering practice. An equivalent circuit for the EBG absorber unit has been presented using a transmission-line model as displayed in **Figure 5**, where C_i, L_i, and R_i (i=1, 2, 3) are the capacitance, inductance, and resistance for the ith one of the three resistive loops (i=1, 2, 3), and C_{ij}, L_{ij}, and R_{ij} (i, j =1, 2, 3, i≠j) are correspondingly the mutual capacitance and inductance as well as resistance between the ith and jth resistive loops.

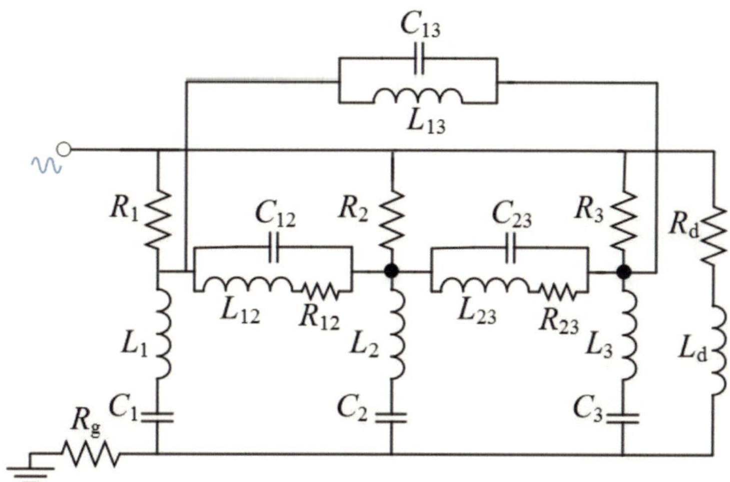

Figure 5. Equivalent circuit model for the absorber unit.

By neglecting the fringing effect, the capacitance of the parallel-plate capacitor is approximated as [14]

$$C = \frac{\varepsilon_0 \varepsilon_r wl}{h} \tag{2}$$

where l is the perimeter of the resistive loop, w is the width of the trace, and ε_r is the relative permittivity of the substrate. C_1, C_2, and C_3 are approximately evaluated by this relation.

Subsequently, the self-inductance of a microstrip etched on the substrate can be expressed as [15]

$$L = 2 \times 10^{-13} \times l \times \left(\ln\left(\frac{2l}{w+h}\right) + 0.5 + 0.2235 \times \frac{w+h}{l}\right) \tag{3}$$

where the above relation is used for gauging L_1, L_2, and L_3, respectively, for the equivalent circuit.

The resistance introduced by the transmission-line conductors can be approximated by [16]

$$\begin{cases} R_L = \dfrac{l}{\sigma w t} & (\delta \geq t) \\ R_H = \dfrac{l}{w}\sqrt{\dfrac{\pi \mu_0 \mu_r f}{\sigma}} & (\delta < t) \end{cases} \tag{4}$$

where t is the thickness of the resistive layer, σ is the conductivity of the resistive layer, and δ is the skin depth of the conductor at the specified frequency of f. The skin depth of the resistive layer in the frequency band of 10– 40 GHz is within the range of 0.0292–0.0585 mm. Because the skin depth is greater than the resistive layer thickness t, R_L in Eq. (4) should be used to calculate the resistances of the three loops. As shown in **Figure 5**, R_g, the resistance of the ground layer is also evaluated by employing Eq. (4).

The mutual capacitances and inductances among the resistive loops of the EBG absorber unit are approximated by employing the following relations [17]:

$$C_m = \frac{\pi \varepsilon_0 \varepsilon_r}{\ln(4(w+g)/w)} \times l \tag{5}$$

$$L_m = \frac{\mu_0 \mu_{r2}}{\pi} \times \frac{w}{w+g} \times l \tag{6}$$

where C_{ij} and L_{ij} $(i, j = 1, 2, 3)$ are evaluated using the above two relations. In this absorber design, due to the small gaps between any two neighboring loops, the resistances R_{12} and R_{23} are approximately equal to the characteristic impedance of air; while due to that Loop$_2$ separates Loop$_1$ and Loop$_3$, and R_{13} is very small and simply neglected here.

Given the permittivity ε_d, permeability μ_d, and conductivity σ_d of the dielectric substrate, by using the relations for the loss tangent, propagation constant and characteristic impedance [16]

$$\varepsilon_d = \varepsilon'_d - j\varepsilon''_d \tag{7a}$$

$$\tan\delta' = \frac{\omega \varepsilon''_d + \sigma_d}{\omega \varepsilon'_d} \tag{7b}$$

$$\gamma = j\omega\sqrt{\mu_d \varepsilon_d - j\sigma_d} \tag{7c}$$

$$\eta_d = \frac{j\omega \mu_d}{\gamma} \tag{7d}$$

http://dx.doi.org/10.5772/67018

The characteristic impedance in the substrate can be written as [18]

$$\eta_d = \frac{\mu_d}{\sqrt{\mu_d \varepsilon_d - j\omega(\varepsilon'_d \tan\delta' - \varepsilon''_d)}} \tag{8}$$

where the permittivity and permeability of substrate $\varepsilon_{d'} = \varepsilon_0\varepsilon_{r'}$ $\mu_d = \mu_0\mu_{r2'd}{}'' = 0$, and $\tan\delta' = 0.005$. The ground-equivalent resistance and inductance shown in **Figure 5** are initially set as $R_d = Re(\eta_d)$, and $L_d = \mathrm{Im}(\eta_d)/j\omega$, which can be easily evaluated using Eq. (8) for contribution from the substrate.

The equivalent circuit model of the EBG absorber unit has been established in the ADS with the configuration of components, as shown in **Figure 5**. The values of these circuit components are finely tuned and optimized, and the achieved results are listed in **Table 1**. The absorptance $A(f)$ simulated from the equivalent circuit model in ADS is compared to that arrived from the corresponding HFSS full wave field solver as shown in **Figure 6**. In the frequency band of 18–40 GHz, $A(f)$ results obtained from ADS equivalent circuit and HFSS field approach remain a fairly good consistent.

Similarly, for the 60 GHz EBG absorber, an equivalent circuit of the designed absorber unit has also been developed as shown in **Figure 7**.

By using the transmission line model, the self and mutual inductances, capacitances, and resistances are derived and put into a schematic in an ADS window. The values of the components are tuned in ADS as listed in **Table 2**.

The absorptance of the designed decagon absorber obtained in HFSS and the one derived from the equivalent circuit simulated in ADS are displayed in **Figure 8**, where the absorption bandwidths for both cases are approximately 20 GHz. They are fairly consistent, except a frequency shift in the front and back edges of the band. It is apparent that for this designs both the central frequency (60 GHz) and the absolute bandwidth (50–70 GHz) are the highest among the recent metamaterial EBG absorbers [19–24].

R_1 (Ω)	C_1 (pF)	L_1 (nH)	R_2 (Ω)	C_2 (pF)	L_2 (nH)
40.76	0.20	0.45	37.69	0.13	0.26
R_3 (Ω)	C_3 (pF)	L_3 (nH)	R_{12} (Ω)	C_{12} (pF)	L_{12} (nH)
26.45	0.09	0.18	371.89	1.71	23.53
R_{23} (Ω)	C_{23} (pF)	L_{23} (nH)	R_{13} (Ω)	C_{13} (pF)	L_{13} (nH)
334.71	1.66	27.35		1.44	8.82
R_d (Ω)	L_d (nH)	R_g (Ω)			
371.81	0.24	0.0025			

Table 1. Equivalent circuit components for 25 GHz EBG absorber.

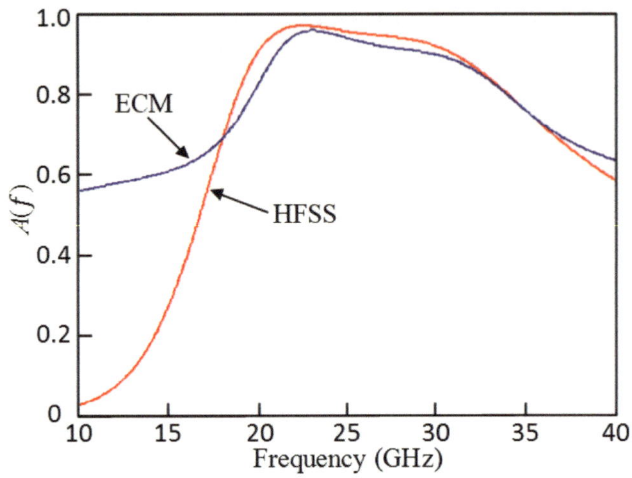

Figure 6. The equivalent circuit-simulated *A*(*f*).

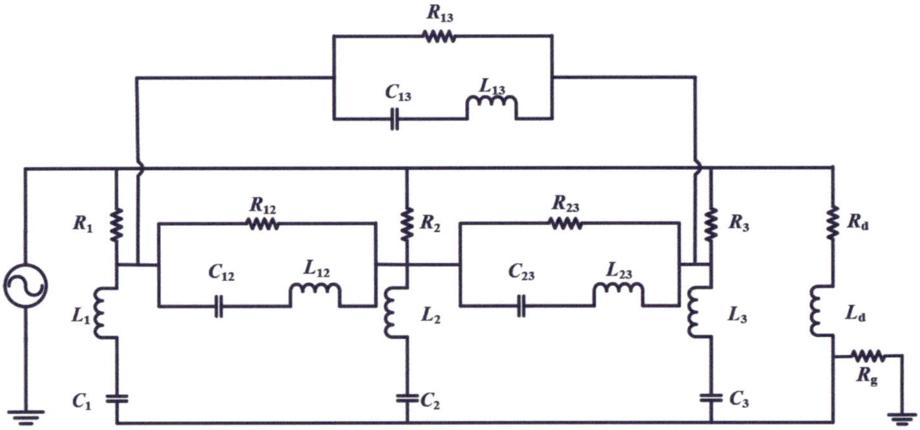

Figure 7. Equivalent circuit of the 60 GHz EBG absorber.

R_1 (Ω)	C_1 (pF)	L_1 (nH)	R_2 (Ω)	C_2 (pF)	L_2 (nH)
55.97	0.0046	0.36	1183.45	0.081	0.061
R_3 (Ω)	C_3 (pF)	L_3 (nH)	R_{12} (Ω)	C_{12} (pF)	L_{12} (nH)
997.31	0.67	0.04	278.80	0.040	
R_{23} (Ω)	C_{23} (pF)	L_{23} (nH)	R_{13} (Ω)	C_{13} (pF)	L_{13} (nH)
0	0.14	0.65	278.80	0.081	0.20
R_d (Ω)	L_d (nH)	R_g (Ω)			
0.0045	0.32	0.72			

Table 2. Equivalent circuit components for 60 GHz absorber.

http://dx.doi.org/10.5772/67018

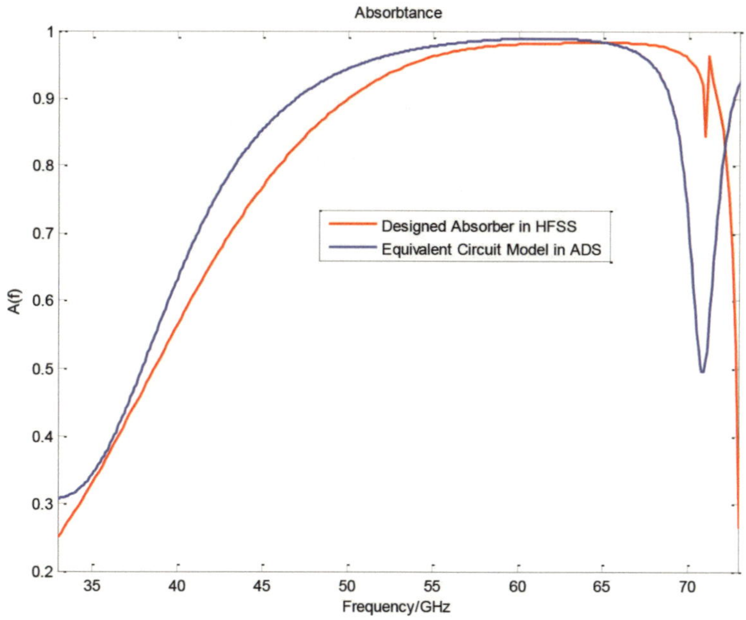

Figure 8. Absorbtance of the designed 60 GHz absorber and the equivalent circuit.

3. Wireless WIIC channel design

3.1. Construction of broadband wireless channel

Based on the design of the metamaterial absorber, a metamaterial-absorbing layer bounded WIIC channel has been further designed [13]. In the designed channel, two layers of the metamaterial EBG absorbers are placed on the top and bottom of the PCB board. Also, three antennas are placed on the top layer of the channel to perform the transceivers of chips in the WIIC system. The transceivers are used to transmit and receive signals operating at a 60 GHz frequency band. The WIIC channel is modeled in HFSS as shown in **Figure 9**.

The WIIC channel is designed with the total dimensions of 60 × 80 × 5 mm, which has been equipped with total 249 and 252 (14 × 18) absorber units on the top and bottom layers, respectively. The substrate dielectric material is filled with Foam in the designed in the WIIC channel. Three microantennas are placed on the top layer, which are fed with the coaxial lines. The antennas are about 30 mm apart from each other.

The HFSS-simulated S-parameters for the wireless PCB channel and a comparable structure, parallel plate system, are shown in **Figure 10**. As seen from the figure, the 10 dB bandwidth with acceptable insertion loss approximately ranges from 50 to 70 GHz. The insertion loss between the two transceivers ranges approximately from 22 to 35 dB in the frequency band from 50 to 70 GHz. The return loss is about -13 dB at the central frequency, with a minimum frequency of -32.06 dB at 49.62 GHz. It is noticed that the S-parameters for the corresponding parallel plates are also fluctuated to meet the bandwidth requirement, and the insertion loss is much lower in the frequency band of interest.

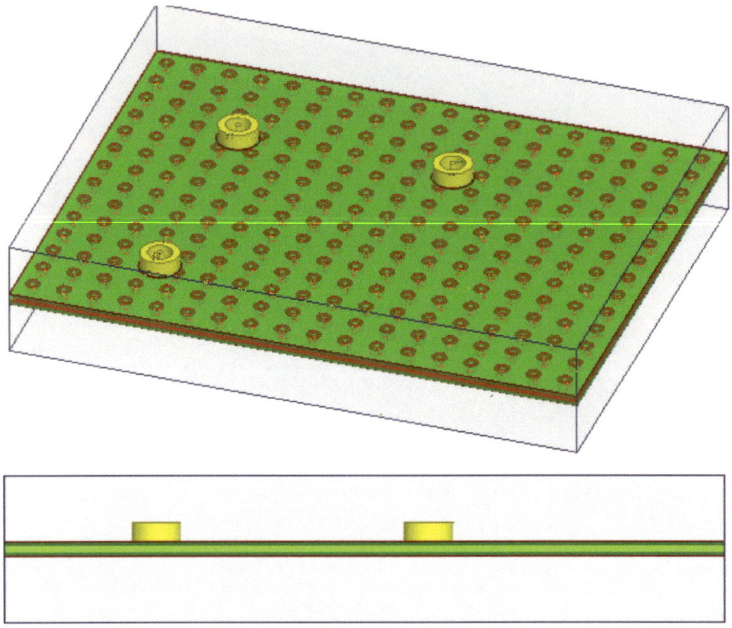

Figure 9. WIIC channel built with the metamaterial EBG absorber units: the geometry of the WIIC channel and its side view.

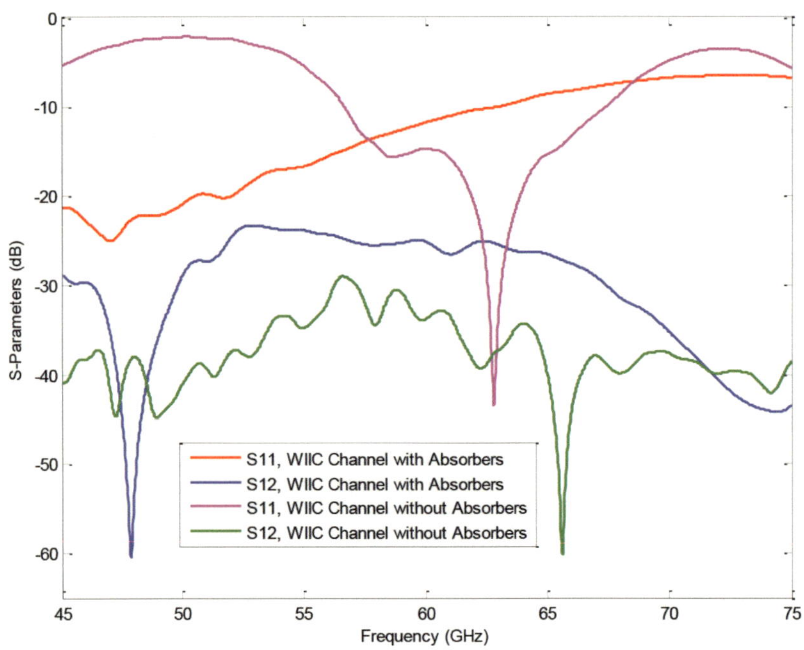

Figure 10. Extracted S-parameters for the designed 60 GHz WIIC channels with the EBG absorber layers and ground plates.

http://dx.doi.org/10.5772/67018

3.2. Numerical extraction of impulse responses and transfer functions

To characterize the WIIC channel and to simplify the channel performance analysis, the impulse response is then extracted in ADS. The schematic for impulse response extraction and the time domain waveform of the input unit impulse are, respectively, shown in **Figure 11**.

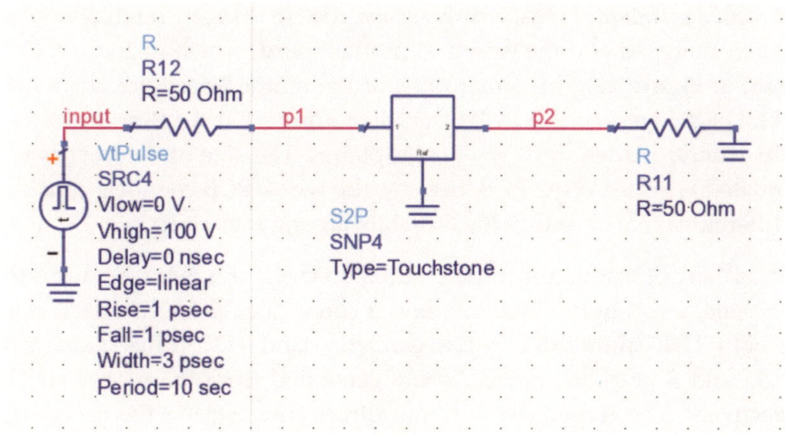

(a)

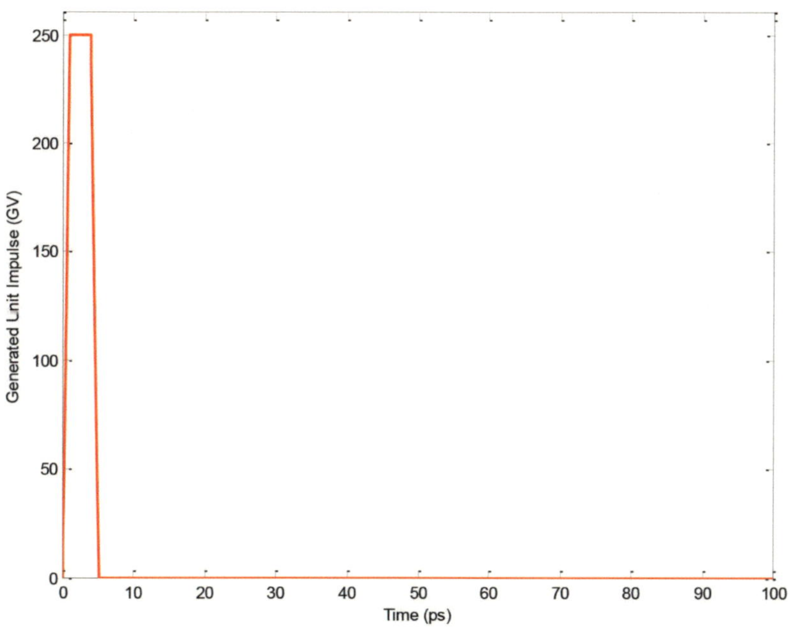

(b)

Figure 11. Impulse response extraction techniques. (a) Schematic for impulse response extraction, and (b) the input unit impulse.

The generated unit impulse is numerically determined to satisfy its definition as follows:

$$\int_{-\infty}^{\infty} \delta_W(t)dt = \frac{1}{2}\left(\Delta t_1 + \Delta t_2\right)V_H = 1\ (V \cdot s) \tag{9}$$

where V_H is the magnitude of the pulse, Δt_1 and Δt_2 are the total pulse widths at the pulse top and bottom, respectively.

To compare the wireless PCB channel analysis to the existing wired trace approach, a wired PCB channel with a number of vias, striplines, microstrip lines, through whole pins, and connectors has been analyzed and measured. A motherboard, which includes a CPU on the top layer, is shown in **Figure 12**. The dimensions of the motherboard are 508 × 218 × 2.54 mm. There are a total of 22 layers on this PCB board, including eight stripline layers, two microstrip line layers, 10 ground planes, and two power planes. The size of the boards is much larger than the simulated wireless WIIC PCB, because the wired PCB channels consist of hundreds of traces at different layers to reduce the crosstalk among these traces.

The highlighted part of the board to be studied consists of a 300-mil microstrip, two vias, a 5.3-inch stripline, a through whole pin, and a connector. The wired PCB channel is measured using both TDR (time domain reflectometry) and TDT (time domain transmission) signals. The S_{11} and S_{12} can be approximately generated from the measured TDR and TDT signals, respectively. The reason for why not directly measuring the S-Parameters using a Vector Network Analyzer (VNA) to extract the S-Parameters is primarily due to that the cost

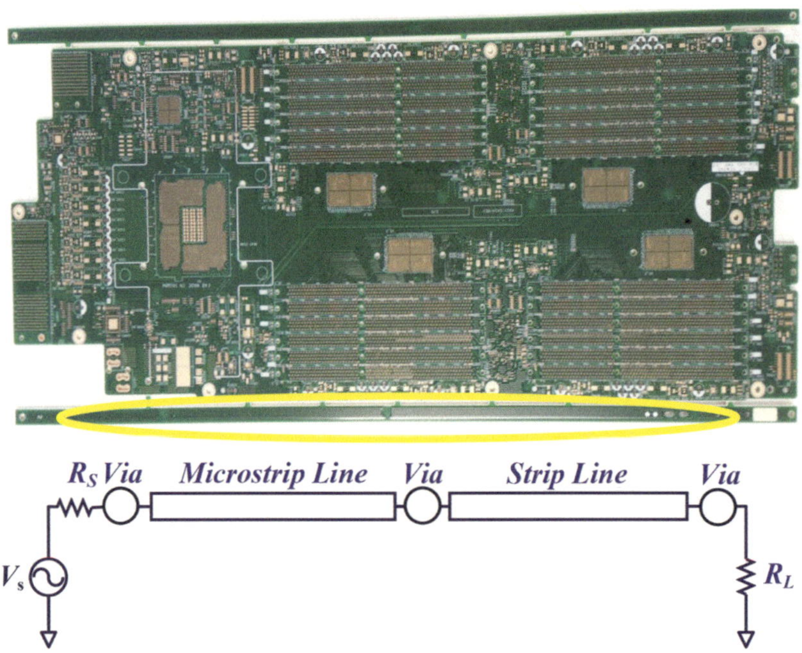

Figure 12. Wired PCB channel consisting of vias, microstrip, and striplines.

http://dx.doi.org/10.5772/67018

of the VNA capable for the desired high frequency band of 50-70 GHz is too expensive. It was unavailable when doing the measurement for the wired PCB channel. The TDR/TDT signals will be used for extraction of the impulse response and transfer function of the channel. The wired PCB channel schematic diagram is displayed in **Figure 12**. The S-parameter models of via, through whole pin, microstrip and stripline simulated in HFSS, and the connector S-parameters model provided by the vendor are cascaded in ADS to generate the overall S-parameters and to obtain the simulated TDT/TDR for the wired PCB channel.

The extracted impulse responses of the WIIC channel, measured and simulated wired PCB channels are shown in **Figure 13**. It is clearly seen that the measured and simulated impulse responses match very well.

As the frequency response of a system can be easily generated in MATLAB by performing an FFT to the impulse response. The discrete frequencies and the sampled time points have to satisfy the following relation [25]:

$$\Delta f = \frac{1}{\Delta t N} \quad (10)$$

where Δt is the sampling time interval of the impulse response, and N is the total samples of the impulse response in the time domain.

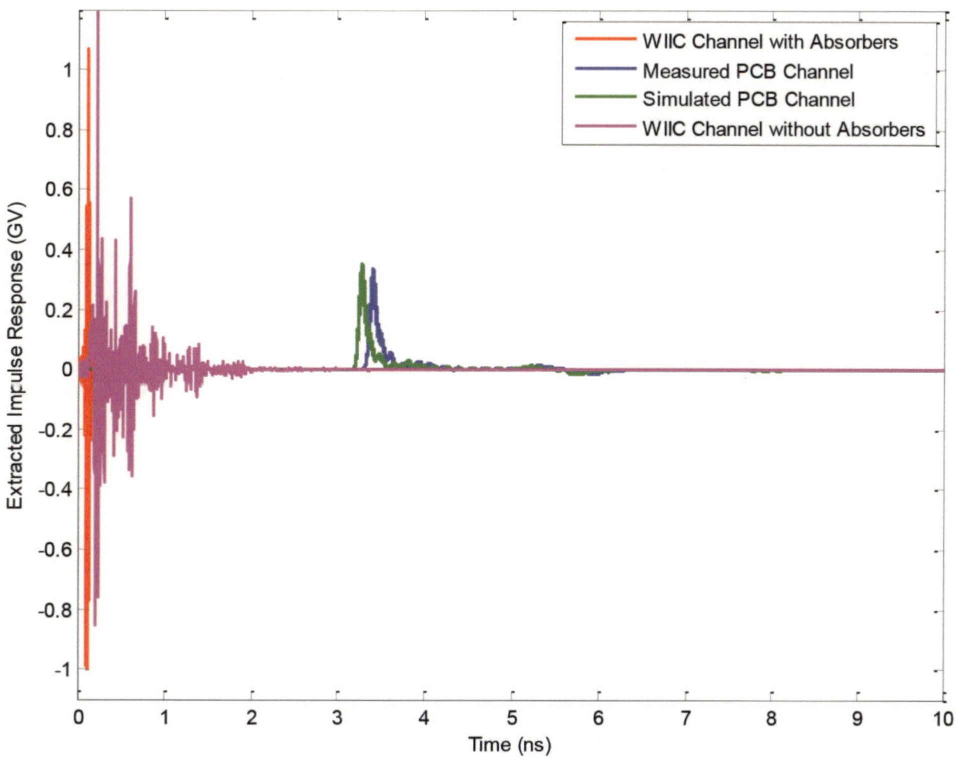

Figure 13. Extracted impulse responses.

The extracted system transfer functions for the measured and simulated wired and wireless PCB channels are shown in **Figure 14**. It is clearly seen that the transfer functions of the simulated and measured PCB channels are very consistent, and they are almost cutoff from 15 GHz. However, the proposed wireless WIIC PCB channel performs very well for the desired frequency band around the neighborhood of 60 GHz. In other words, with an identical input to the channels, the bandwidth of the wireless WIIC PCB channels will be much larger than that of the wired PCB channel. The transfer function for the wireless WIIC PCB channel with the absorbers is relatively flat in the frequency range of 45–65 GHz. It is also noticed that this bandwidth is not completely consistent with that of the insertion loss resulted from HFSS.

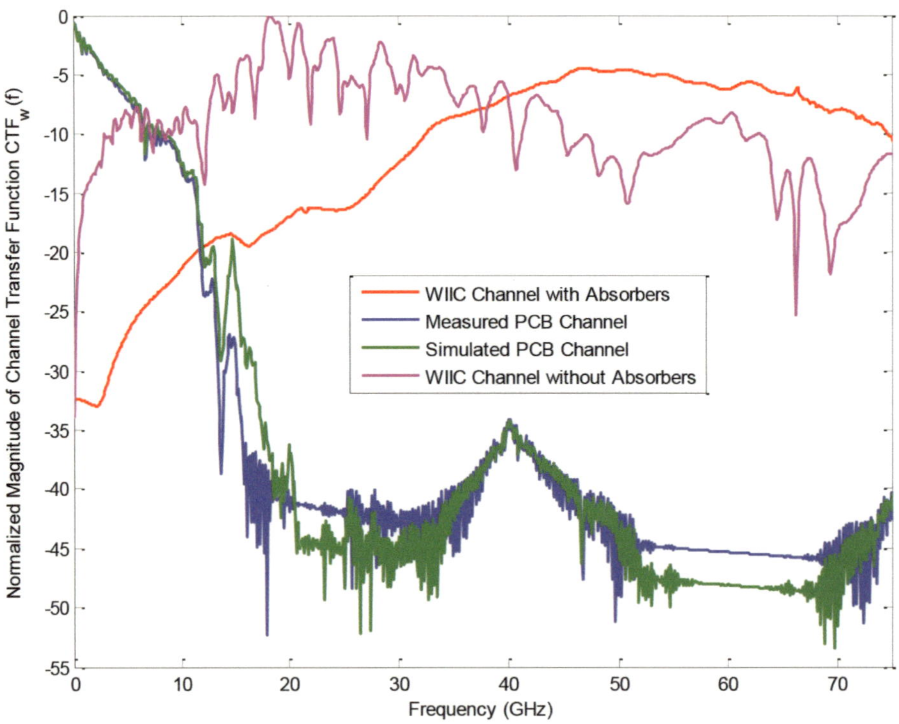

Figure 14. Transfer functions (dB) for the WIIC, measured and simulated wired PCB Channels.

4. System and channel performance analysis

4.1. WIIC transceivers

In this chapter, various advanced wireless and mobile technologies are implemented in the proposed WIIC system, including the orthogonal frequency division multiplexing (OFDM), quadrature phase shift keying (QPSK), quadrature amplitude modulation (QAM), and multiple input multiple output (MIMO). Herein, an ideal 4 × 4 antenna array with the four identical channels is assumed, which correspondingly will lead to the upper bond data rate of the

http://dx.doi.org/10.5772/67018

system. As the result of employing the technologies of OFDM, MIMO, and QPSK/QAM, the data rate and spectrum utilization of the system are dramatically improved.

The WIIC system in this work also includes forward error control (FEC), and interference mitigation to diminish the error brought by the channel and interferences from the channel and other transmitters. In this system, FEC mainly includes cyclic redundancy check (CRC) and channel coding, and interference mitigation features the techniques of scrambling and interleaving. For the FEC scheme, the system employs 16-bit CRC and 1/3 code rate tail-biting convolution coding. The WIIC system employs cyclic prefix (CP) to deal with the timing problem, and OFDM-UWB (ultra wide-band) to achieve low power consumption. The system block diagram is shown in **Figure 15**.

4.2. OFDM symbol decoding

MIMO demapping, demodulation, descrambling, deinterleaving, and channel decoding are all equipped in the OFDM symbol decoding scheme in this WIIC system [26–28]. Demodulation is able to automatically diminish the mistakes brought by the noise from the channels, and channel decoding can correct some errors in the receiving signals. Correctly MIMO demapping, demodulation, descrambling, deinterleaving, and channel decoding will provide the receiver the correct output bits, which can be used for comparison with the transmitted signals, and can accurately calculate the system BER.

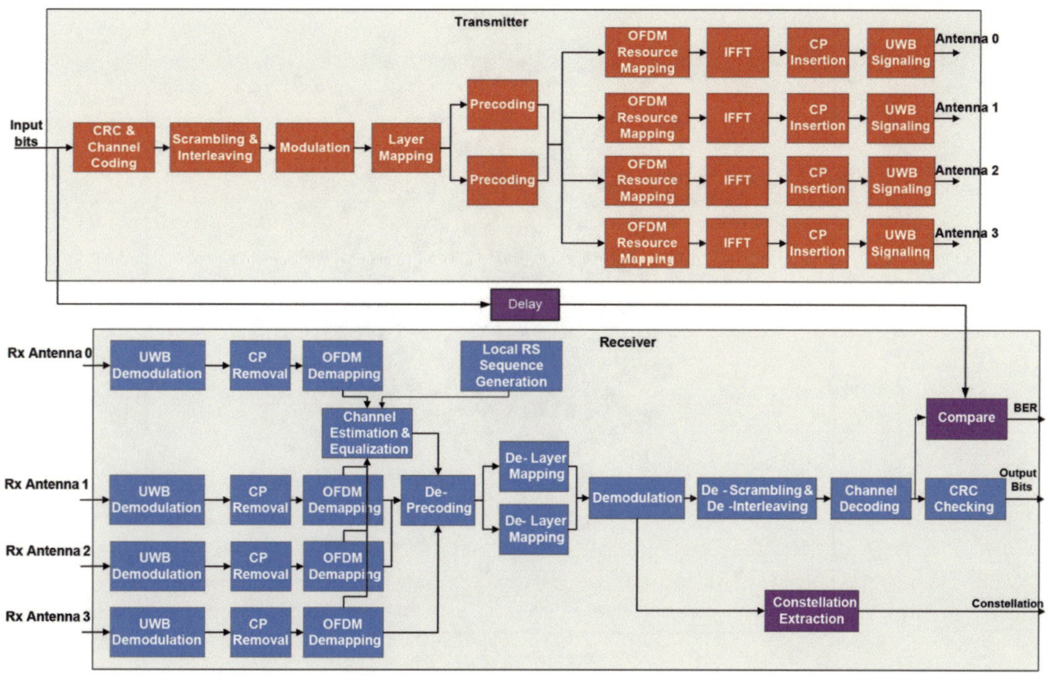

Figure 15. WIIC system block diagrams for transmitter and receiver.

As the precoding matrices employed by MIMO are orthogonal matrices, correspondingly, the inverse matrices, in other words, the deprecoding matrices are simply that are used for transposed precoding matrices. Thus, the deprecoding procedure will be the received signal matrix multiplied by the correct transposed precoding matrix. Besides, after deprecoding, the delayer mapping is the inverse procedure of the layer mapping, which is a procedure for collection and rearrangement of deprecoded data.

The constellation improvement of the SDM brought by precoding is simulated in MATLAB. The constellation diagrams collected at the receiver of a system without precoding, and that with precoding are displayed in **Figures 16** and **17**, respectively. The simulation includes 2,048,000 bits, under the AWGN channel, with SNR 5 dB.

4.3. WIIC system performance analysis

In the system and channel performance analysis, four different channels are investigated in MATLAB, which are the additive white Gaussian noise (AWGN), Rayleigh, the designed WIIC, and the measured PCB channels. The relationships of bit error rate (BER) and signal-to-noise rate (SNR) with four different channels are shown in **Figure 18**. In each SNR condition, a random binary of 441,600,000 bit stream is input to these four different channels and the BER is obtained by acquiring the number of bits in which the input and output are different.

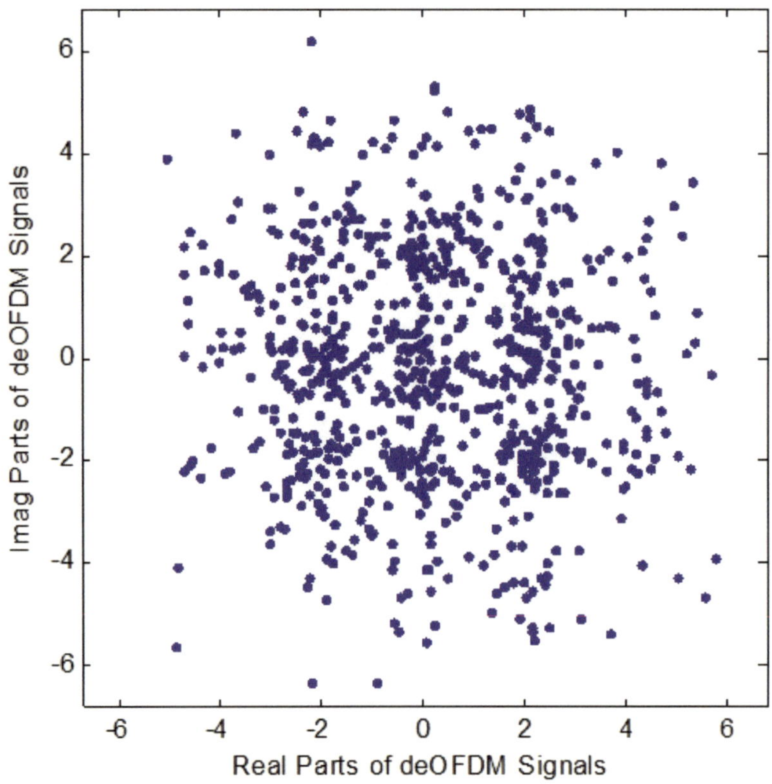

Figure 16. Constellation at receiver without MIMO.

http://dx.doi.org/10.5772/67018

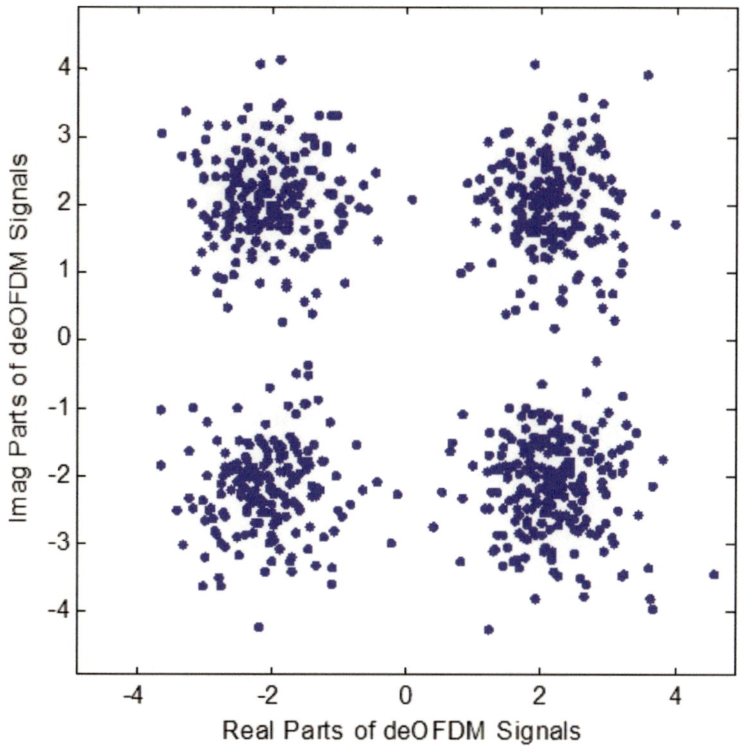

Figure 17. Constellation at receiver with MIMO.

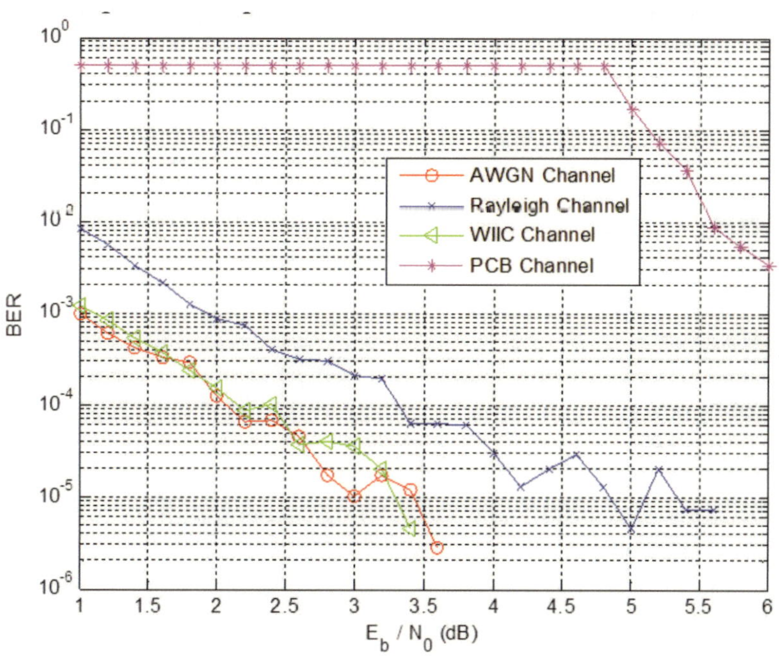

Figure 18. BER versus SNR in different channels.

Herein, BERs are extracted by directly comparing the input and output, so that the results are more reliable than obtaining BER by calculation using CRC-checking results. Besides, in the simulation 0 dB antenna gain is assumed, and without including any low-noise amplifier (LNA), as the worst-case analysis of the simulated channels. Also, AWGN and Rayleigh channels provide the best or worst case for wireless channels, which cannot validate the designed system, but can also be compared to the proposed WIIC and wired PCB channels.

As predicted in the previous analysis, the BER reaches to be less than 10^{-5} at SNR of 3, 3.4, and 5.4 dB for AWGN, WIIC, and Rayleigh channels, respectively. Also, when SNR is greater than 3.6, 3.4, and 5.6 dB, none of the single error bit is found in BER analysis, with AWGN, WIIC, and Rayleigh channels, respectively. In contrast, it is not surprising that the BER for the wired PCB channel is much higher than the other channels due to its cutoff in the frequency band of interest.

5. Conclusion

In this chapter, based on the design of an EBG absorber unit, a WIIC channel has been designed and validated in the WIIC system. Also, as a reference system, a wired PCB channel is measured, simulated, and analyzed for extracting its S-parameters and system transfer functions. It is found that the wired PCB boards are no longer qualified for transmitting signals at a 60 GHz band for the case studied, while the designed WIIC channel works properly in this band. With the extracted WIIC S-parameters, its impulse responses and the normalized transfer functions are obtained in ADS and MATLAB. The BER analysis shows that the performance of the proposed WIIC channel is close to a simple AWGN channel when the WIIC transceivers are characterized with channel coding, channel sounding, channel estimation, and channel equalization.

Author details

Xin Zheng[1,2], Yinchao Chen[1*], Wensong Wang[3] and Shuhui Yang[4]

*Address all correspondence to: chenyin@cec.sc.edu

1 Department of Electrical Engineering, University of South Carolina, Columbia, SC, USA

2 Oracle America Inc., Santa Clara, CA, USA

3 College of Electronic and Information Engineering, Nanjing University of Aeronautics and Astronautics, Nanjing, Jiangsu, PRC

4 Department of Communication Engineering, Communication University of China, Beijing, PRC

References

[1] W. Wang, Y. Chen, S. Yang, Q. Cao, H. Li, X. Zheng, and Y. Wang, "Wireless inter/intra-chip communication using an innovative PCB channel bounded by a metamaterial absorber," *IEEE Antennas and Wireless Propagation Letters*, vol. 15, Issue 1, pp. 1634–1637, January 2016.

[2] J. T. Bialasiewicz, D. Gonzalez, J. Balcells, and J. Gago, "Wavelet-based approach to evaluation of signal integrity," *IEEE Trans. Ind. Electron.*, vol. 60, no. 10, pp. 4590–4598, 2013.

[3] Y. Chen, S. Yang, L. Sun, and K. Sun, "Frequency and time domain crosstalk signal analysis for high-density circuits," *Proc. 2009 WRI Global Congress on Intel. Syst.*, pp. 334–337, 2009.

[4] Y. Take, H. Matsutani, D. Sasaki, M. Koibuchi, T. Kuroda, and H. Amano, "3D NoC with inductive-coupling links for building-block SiPs," *IEEE Trans. Computers*, vol. 63, no. 3, pp. 748–763, 2014.

[5] X. He, W. Wang, and Q. Cao, "Crosstalk modeling and analysis of through silicon-via connection in 3D integration," *Proc. Prog. Electromagn. Res. Symp.*, pp. 857–861, 2013.

[6] W. Rhett Davis, J. Wilson, S. Mick, J. Xu, H. Hua, C. Mineo, A. M. Sule, M. Steer, and P. D. Franzon, "Demystifying 3D ICs: the pros and cons of going vertical," *IEEE Des. Test Comput.*, vol. 22, no. 6, pp. 498–510, 2005.

[7] N. Miura, D. Mizoguchi, M. Inoue, T. Sakurai, and T. Kuroda, "A 195 Gb/s 1.2 W inductive inter-chip wireless superconnect with transmit power control scheme for 3D-stacked system in a package," *IEEE J. Solid-State Circuits*, vol. 41, no. 1, pp. 23–34, 2006.

[8] K. Kanda, D. D. Antono, K. Ishida, H. Kawaguchi, T. Kuroda, and T. Sakurai, "1.27 Gbps/pin, 3mW/pin wireless superconnect (WSC) interface scheme," *Proc. IEEE Int. Solid-State Circuits Conf.*, pp. 186–187, February 2003.

[9] T. J. Spencer, T. Osborn, and P. A. Kohl, "High-frequency chip connections," *Sci.*, vol. 320, pp. 756–757, May 2008.

[10] K. O Kenneth, K. Kim, B. A. Floyd, J. L. Mehta, H. Yoon, C.-M. Hung, D. Bravo, T. O. Dichson, X. Guo, R. Li, N. Trichy, J. Caserta, W. R. Bomstad, J. Branch, D.-J. Yang, J. Bohorquez, E. Seok, L. Gao, A. Sugavanam, J.-J. Lin, J. Chen, and J.E. Brewer, "On-chip antennas in silicon ICs and their application," *IEEE Trans. Electron Devices*, vol. 52, no. 7, pp. 1312–1323, 2005.

[11] Y. P. Zhang and D. Liu, "Antenna-on-chip and antenna-in-package solutions to highly integrated millimeter-wave devices for wireless communications," *IEEE Trans. Antennas Propag.*, vol. 57, no. 10, pp. 2830–2841, 2009.

[12] W. Wang, Y. Chen, S. Yang, X. Zheng, and Q. Cao, "Design of a broadband electromagnetic wave absorber using a metamaterial technology," *J. Electromagnet. Wave.*, vol. 29, no. 15, pp. 2080–2091, 2015.

[13] X. Zheng, "System Level Analysis and Design for Wireless Inter-Chip Interconnection Communication Systems by Applying Advanced Wireless Communication Technologies," PhD Dissertation, University of South Carolina, 2016.

[14] S. C. Thierauf, "High-Speed Circuit Board Signal Integrity,"Artech House, 2004, Norwood, MA, USA.

[15] D. G. Reed, *ARRL Handbook for Radio Communications*, 82nd Ed. American Radio Relay League, Newington, 2005.

[16] C. A. Balanis, *Advanced Engineering Electromagnetics*, 2nd edition, Wiley, 2012, Hoboken, NJ, USA.

[17] The Clemson University Vehicular Electronics Lab, *Expert System Algorithm Summaries*, http://www.cvel.clemson.edu/emc/expert_systems/PCB/pcb_summaries.html.

[18] D. M. Pozar, *Microwave Engineering*, 3rd edition, Wiley, 2005, Hoboken, NJ, USA.

[19] H. H. Yang, Cao XY, Cao J, Liu T, Li SJ, Zhao Y, Yuan ZD, Zhang H. "Broadband low-RCS metamaterial absorber based on electromagnetic resonance separation," *Acta Phys. Sin.*, vol. 62, Issue 21, no. 214101, pp. 1–8, October 2013.

[20] X. Han, J. S. Hong, C. M. Luo, and L. L. Zhong, "An ultrathin and broadband metamaterial absorber using multi-layer structures," *J. Appl. Phys.* vol. 114, Issue 6, no. 4109, pp. 1, August 2013.

[21] M. Yoo and S. Lim, "Polarization-independent and ultrawideband metamaterial absorber using a hexagonal artificial impedance surface and a resistor-capacitor layer," *IEEE Trans. Antennas. Propag.*, vol. 62, no. 5, pp. 2652–2658, May 2014.

[22] W. Li, T. Wu, W. Wang, P. Zhai, and J. Guan, "Broadband patterned magnetic microwave absorber," *J. Appl. Phys.*, vol. 116, Issue 4, no. 4110, July 2014.

[23] X. Xiong, C. L. Zou, X. F. Ren, and G. C. Guo, "Broadband plasmonic absorber for photonic integrated circuits," *IEEE Photonic. Tech. Lett.*, vol. 26, no. 17, pp. 1726–1729, Sept. 2014.

[24] G. Yang, X. Liu, Y. Lv, J. Fu, Q. Wy and X Gu. "Broadband polarization-insensitive absorber based on gradient structure metamaterial". *J. Appl. Phys.* vol. 115, Issue 17, pp. 17E523 (3 pp.), May 2014.

[25] Kipp Schoen, "S-parameter Measurements with a High-Speed TDR and TDT System," Picosecond Pulse Labs (PSPL), Tektronix, Boulder, CO.

[26] 3GPP. 2011. TS36.211 v10.1.0 Evolved Universal Terrestrial Radio Access (E-UTRA); Physical Channels and Modulation (Release 10). *3rd Generation Partnership Project Technical Specification*. 3GPP.

[27] X. Zheng, Y. Chen and D. Matolak, "Downlink "feedback" channel design for LTE Femtocell inter-cell interference mitigation," *Proceeding of SOUTHEASTCON 2014*, IEEE, Lexington, KY, pp.1–4, 2014

[28] X. Zheng, Y. Chen, S. Yang, and W. Wang, "Feedback channel design for LTE Femtocell inter-cell interference mitigation," *J. Beijing Inform. Sci. Technol. Univ.*, vol. 29, pp.19–25, March 2014.

Chapter 5

Microwave and THz Metamaterial-Based Devices for Potential Applications in NDE, Chem-Bio Detection and Dielectric Characterization of Complex Oxide Thin Films

Daniel Shreiber

Additional information is available at the end of the chapter

http://dx.doi.org/10.5772/65951

Abstract

The purpose of this chapter is to convey a message that the variety of applications of the classical (and the novel ones, such as Mie resonance based) metamaterials are going far beyond the originally proposed applications such as geometrical optics and antennas. In addition, it is important to mention that most of these applications are just an idea or a first proof of principle. Hence, an additional message of this chapter is that a lot of further research is required to implement these scientifically sound ideas. It is also a hope that this chapter will trigger the reader's curiosity and interest to pursue this exciting field, which will yield additional applications that have never been imagined.

Keywords: MW and THz metamaterials, NDE, chem/bio detection, dielectric properties characterization, complex oxide thin films

1. Introduction

Multiple applications of a new range of materials called metamaterials have been developed since their evolution, suggested by Veselago [1], to the development of their design, first proposed by Pendry et al. [2, 3]. Most notably, the proposed applications included antennas [4] and geometrical optics [5]. Indeed, the proposed metamaterial-based structures were designed to enhance the antenna properties such as bandwidth, or geometrical optics properties such as reduced aberrations and enhanced focusing by the metamaterial lens [6].

In the years to follow, extensive research in the field of metamaterials yielded novel metamaterial classes (metamaterials based on dielectric resonators, for example) [7] and configurations [8] in addition to the "classical" ones in the form of metallic split ring and a wire designed by Shelby et al. [5] that are shown in (**Figure 1**).

In their original definition, the metamaterials are artificial materials that respond to impinging electromagnetic waves by exhibiting negative values of the permittivity (ε) and/or the permeability (μ). Typically, the metamaterials are divided into double-negative when both values of ε and μ are negative or single-negative when one of these values is negative.

In this chapter, we discuss utilization of the "classical" and other metamaterial devices for a number of applications that go beyond the antenna and geometrical optics application. This approach will demonstrate the extent to which the field of metamaterial applications has grown in recent years, taking this exciting discovery one step further to implementation in real-life devices. Both single-negative and double-negative metamaterial devices are discussed in this chapter. Most of the ideas that are to be discussed here are on the level of proof-of-concept and some of them have a good chance to become real life application in the nearest future.

It is our desire to discuss these concepts in the context of their applications. Hence, it is prudent to provide the background for the application in order to give a complete picture of the concrete capability gaps and how the proposed metamaterial devices can address these shortcomings.

When the "classical" metamaterials are discussed, we refer to each metallized-structure unit cell as to an artificial atom that "reacts" to impinging electromagnetic wave. The dimensions of these structures and their features will determine the resonant frequency of the device.

Figure 1. Optical image of a classical metamaterial structure that operates at a resonant frequency of 10.5 GHz[8] (Reproduced with permission of American Association for the Advancement of Science via Copyright Clearance Center).

http://dx.doi.org/10.5772/65951

Typically, such a unit cell measures to about 1/10 to 1/20 of a resonant wavelength and consists of a wire and a split ring (either together for double-negative or one of the two for single-negative). An example of a two-dimensional (2D) double-negative and one-dimensional (1D) metamaterial model and transmission results obtained from the operation of the metamaterial fabricated based on this model [9] are shown in **Figure 2**.

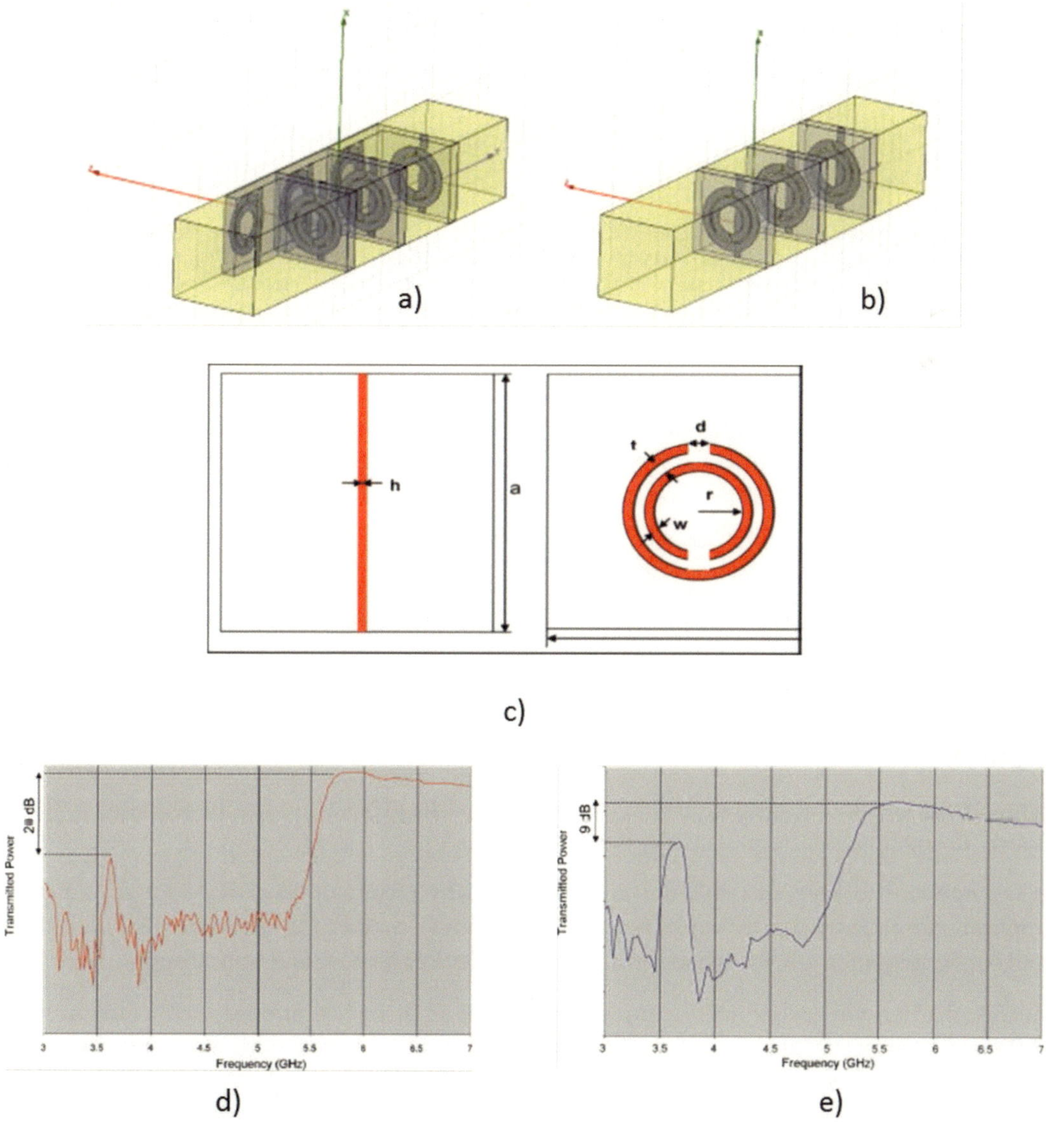

Figure 2. A few unit cells modeled in HFSS Ansoft © for (a) 2D metamaterial and (b) 1D metamaterial slab, (c) dimensions of the metallized components of the designed unit cell are as follows: FR4 ($\varepsilon = 4.4$, tan $\delta = 0.02$), metallic structures—copper (30 μm thick), $r = 1.6$ mm, $d = t = 0.2$ mm, $w = 0.9$mm, $a = 9.3$ mm, $h = 0.9$ mm. PCB sheets were 1.8 mm thick, (d) transmission scan from the structure depicted in (a) and e) transmission scan of a structure depicted in (b) [9]. (Reproduced with permission of ELSEVIER S.A. via Copyright Clearance Center).

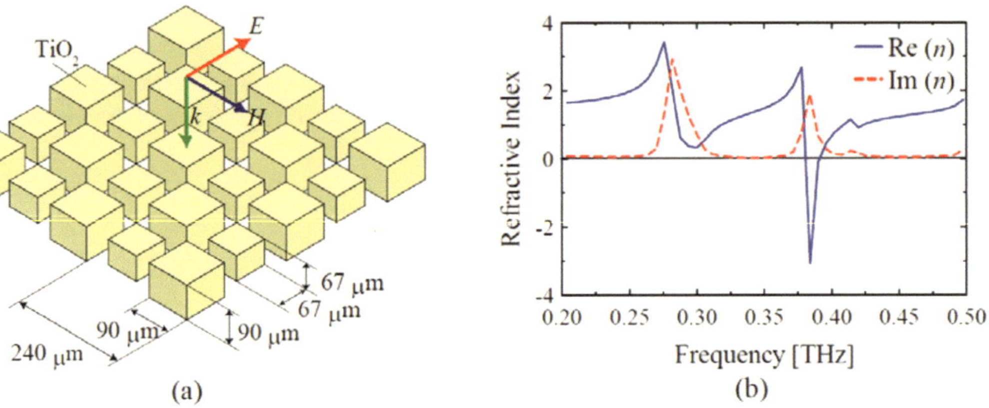

Figure 3. (a) Designed metamaterial structure in the low THz regime, (b) transmission scan showing negative index of refraction for this configuration [10] (reprinted with permission from Shibuya, K., K. Takano, N. Matsumoto, H. Miyazaki, K. Izumi, Y. Jimba, and M. Hangyo,"Terahertz metamaterials composed of TiO2 cube arrays". Proceedings of the Second International Congress on Advanced Electromagnetic Materials in Microwaves and Optics - Metamaterials, Pamplona, Spain 2008, Copyright 2008 Metamorphose IV).

From **Figure 2(d)** and **(e)**, one can see that the resonant frequency of this double-negative metamaterial is 3.65 GHz, which corresponds to a wavelength of about 8.2 cm. **Figure 2(c)** indicates that the largest dimension of the designed structure is the size of the unit cell and, hence, the length of the metallic wire which measures 9.3 mm is about 1/10 of the resonant wavelength.

The same rule of thumb can be applied to alternative metamaterial structures such as an array of TiO_2 oxide cubes used as a metamaterial in low THz range of frequencies [10] which have dimensions of about 1/10 of the lowest resonant frequency measured for this metamaterial system. The model and the transmission scan of this metamaterial structure are depicted in **Figure 3**.

Simulations of the metamaterial system can be used to predict the resonant frequency and hence, the operational frequency of a device.

It is important to keep in mind the dimensions of the metamaterial surface for a particular application/frequency. For example, it may be technically challenging (due to their dimensions and configuration) to implement some metamaterial devices for THz and optical applications [11].

In this chapter, we review the examples of metamaterial device applications in microwave (MW) and terahertz (THz) spectra of radiation. We will look into both single-negative and double-negative metamaterial devices, as well as into active and passive devices. This review is not meant to be an exhaustive list of potential metamaterial applications outside of the antenna and geometrical optics, but just a few ideas of how the very novel concept of artificial materials can be applied in a wide variety of ways.

The remainder of the chapter is divided into individual applications where the background for the application, its configuration, advantages, and limitations will be discussed.

http://dx.doi.org/10.5772/65951

2. Novel nondestructive evaluation (NDE) detector based on a metamaterial lens

The first application that we are going to discuss is a novel NDE detector based on a metamaterial lens. The proposed detector is capable of operating in a low GHz range of frequencies.

2.1. Background

There are many instances where it is important to detect a flaw or inclusion that is not visible to human eye as it is covered by a layer of nonconducting material. Detection through conductive surfaces is limited as they will reflect the waves in the radiation ranges that are typically utilized by metamaterial devices (excluding, probably, acoustical [12]). However, some work has been done to investigate carbon-fiber-reinforced structures with microwaves [13]. One of the examples where the application of such a detector could be highly beneficial is a system where a thick layer of thermal insulation covers a metallic surface (you can think of a space shuttle tile covering shuttle skin, for example). The thermally insulating layer is attached by an adhesive and we do not want to remove in order to determine whether the onset of corrosion had occurred underneath the insulating layer. In order to successfully detect this flaw, two conditions need to be satisfied: (a) the resolution of our detector has to be good enough to detect the flaw and (b) the wave that we are using has to be "strong" enough to penetrate the insulating layer to reach the flaw and to reflect back through the layer to the detector. Here we arrived at a contradiction, though longer wavelengths penetrate further, the resolution of the image will be limited by the wavelength, due to the diffraction limit. Indeed, we could defy the diffraction limit if we investigate the same flaw using the near-field detector, but in many cases, this approach is not realistic since the thickness of the insulator might be larger than the distance that will define the true near field. Additionally, it might not be realistic for real-life devices to maintain the tip of the detector so close to the surface of the sample. As an example, Tabib-Azar et al. demonstrated the resolution of 0.4 μm with 1 GHz wavelength but the tip of their detector was maintained at a very small distance away from the surface of the metal [14].

Pendry [6] analytically demonstrated that a metamaterial lens defies the diffraction limit and, in an ideal case, can transmit all the information emitted from the source. This approach can be utilized when an NDE detector is designed based on a metamaterial lens [15]. Realistically, the detector can serve as a "middle ground" between a far-field and a near-field detector where the standoff distance can be manipulated by the distance of the source from the metamaterial lens and the thickness of the lens, as can be seen from **Figure 4** where the focus spot of the resonant frequency will be significantly smaller than predicted by the diffraction limit.

The equation that shows the distance to the image in the case of an ideal metamaterial lens can be written as:

$$d_1 + d_2 = t. \tag{1}$$

Although Eq. (1) represents the case of an ideal lens which is extremely hard to achieve, the equation gives a very good approximation for the design of such a device.

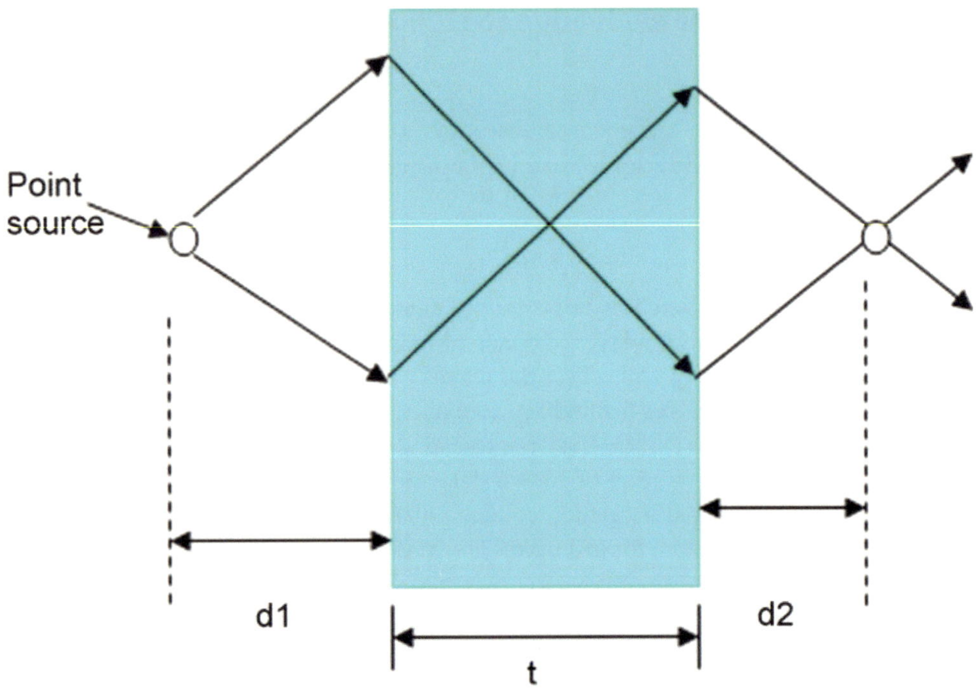

Figure 4. Ray diagram of a metamaterial lens [9] (Reproduced with permission of ELSEVIER S.A. via Copyright Clearance Center).

2.2. Experiment and results

Indeed, the original results are promising. It has been demonstrated [15] that a hole in a fiberglass block with the diameter of 3 mm that is perpendicular to the propagation direction of the electromagnetic wave (**Figures 5** and **6**) can be detected with a device operating at a resonant frequency of 3.65 GHz (8.2 cm). Therefore, the resolution of such a device is determined to be 0.037 λ, which is far below the diffraction limit predictions.

The detection of the location of the hole and of a shift in the hole location are shown in **Figure 6**.

To better understand the effect of the metamaterial lens that was utilized in this proof-of-concept experiment, one can observe that at the frequency of 6 GHz, the location of the holes is not detectable at all as the same lens does not exhibit metamaterial properties, and therefore the electromagnetic wave just passes through the structure and reflects back from the sample into the detector (although the wavelength at this frequency is shorter than at 3.65 GHz). The results of this experiment are shown in **Figure 7**.

It is also important to mention that a lot of thought should be dedicated to the design of the metamaterial lens when considering the inevitable trade-offs. As it was mentioned before, the metamaterial structures are typically very lossy. Therefore, when an NDE detector is designed based on a double-negative metamaterial lens where a high-strength transmitted signal is needed to maintain an adequate signal-to-noise ratio, one should keep in mind the limitations of the system in terms of losses. Sometimes, it is possible to trade off the resolution of the system for better transmission. For example, a 1D structure of parallel (in the direction of the

http://dx.doi.org/10.5772/65951

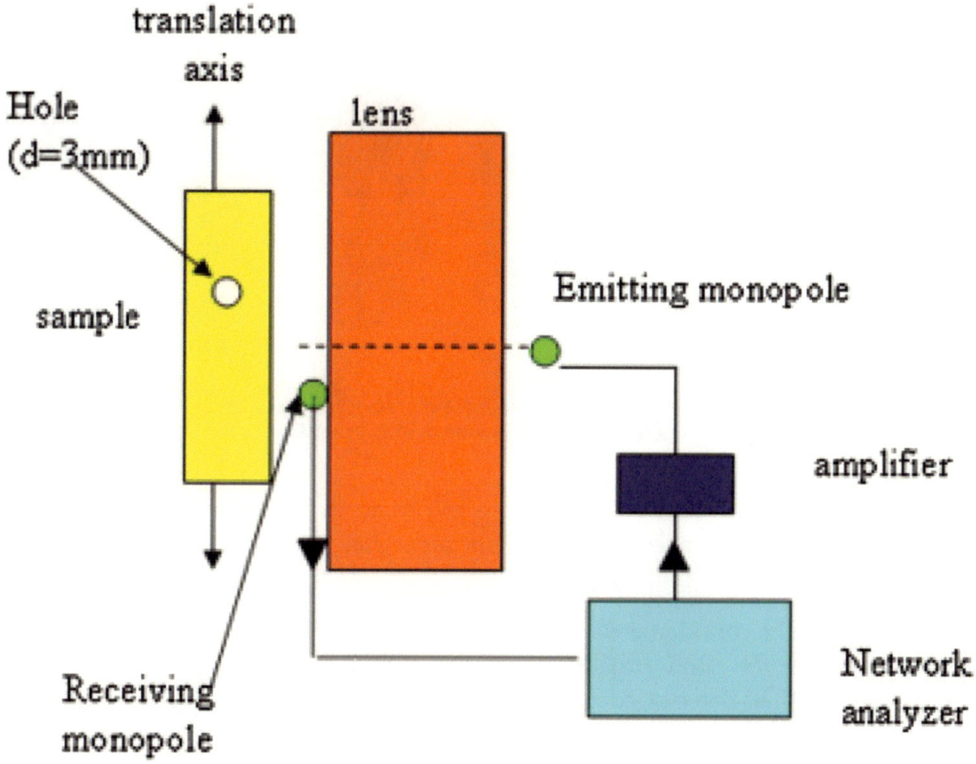

Figure 5. Sketch of the NDE sensor system used to detect the hole defect [15] (Reproduced with permission of ELSEVIER S.A. via Copyright Clearance Center).

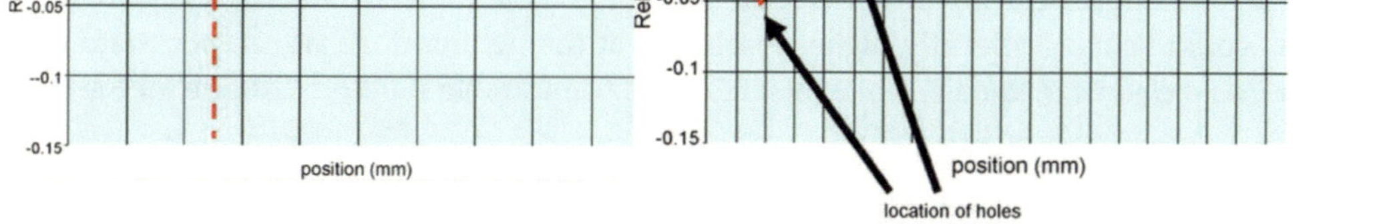

Figure 6. Relative power scan of a sample with one hole and when two holes were sealed with a filler alternately ($f = 3.65$ GHz). The second hole was drilled 3 cm away from the first one [15] (Reproduced with permission of ELSEVIER S.A. via Copyright Clearance Center).

electromagnetic wave propagation) plates with metallized metamaterial unit cells (**Figure 2(b)**) could be used instead of a 2D wine-crate structure for the metamaterial lens. The comparative losses obtained through a power scan are depicted in **Figure 2(d)** and **(e)**. The transmission

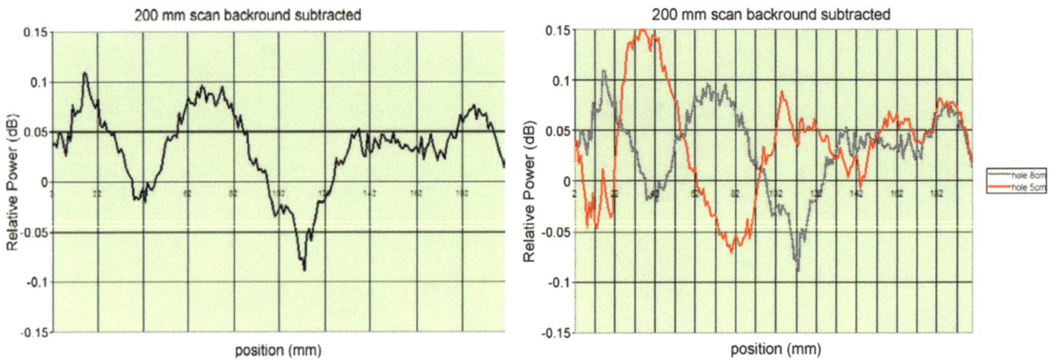

Figure 7. Relative power scan of a sample with a hole and when two holes were filled alternately ($f = 6$ GHz). The second hole was drilled 3 cm away from the first one [15] (Reproduced with permission of ELSEVIER S.A. via Copyright Clearance Center).

peak for a 2D lens is at –20 dB compared to the transmission through the same lens at frequencies where the lens does not exhibit metamaterial properties (at about 6 GHz). In contrast the same peak for a 1D lens is at –9 dB.

Another trade-off to be considered is how much penetration we really need into the sample to successfully perform the nondestructive evaluation versus the required resolution. In other words, maybe, an NDE detector could be designed to work at a higher frequency, hence, it might not penetrate as far into the dielectric, but the resolution of the metamaterial lens operating at the appropriate frequency will be much better. In that case, we need to consider the fabrication limitations of the metamaterial lens since it is possible that the designed operating frequency of our metamaterial device forces the use of more complicated fabrication methods, such as those that are utilized for devices operating in the THz spectrum.

Also, as we previously mentioned, one can consider alternative metamaterial structures such as oxide metamaterials for these applications where the losses are smaller.

In conclusion, utilization of a metamaterial lens as a tool to significantly enhance capabilities of the nondestructive evaluation detectors has been demonstrated to be successful. This initial proof-of-concept can serve as a very solid foundation from which intensive research and development can build upon to bring these detectors to wide acceptance. The main advantage of these detectors would be a significantly higher resolution at the frequencies that can penetrate relatively far into dielectric materials, while maintaining a comfortable standoff distance for the user.

3. Detection of chem/bio hazards with a metamaterial-based device in THz spectrum

One of the important benefits of the recently discovered THz spectrum of radiation is that most large chemical molecules are resonant in this spectrum as shown, for example, in [16]. There is a significant body of work dedicated to the characterization of DNA molecules [17] in the THz spectrum. In fact, it was suggested that these molecules have multiple unique resonant frequencies in the THz spectrum [18]. Hence, it is possible to uniquely identify these chemical hazards by the combination of these unique resonant frequencies in the THz spectrum. It is also desired to detect these substances in very low concentrations. It was suggested [18] that one approach that exhibits the highest sensitivity is a detector constructed with a single-negative metamaterial surface. This single-negative metamaterial surface, where the metallic split-ring resonators (SRRs) are resonating at the same frequency as the molecule under investigation, could have the effect of overlapping resonances when this molecule "lands" in the gap of the SRR. The overlapping resonance has a nonlinear effect and, hence, could enhance the sensitivity of the method by 10^6–10^7. The original metamaterial structure for this approach is depicted in **Figure 8**.

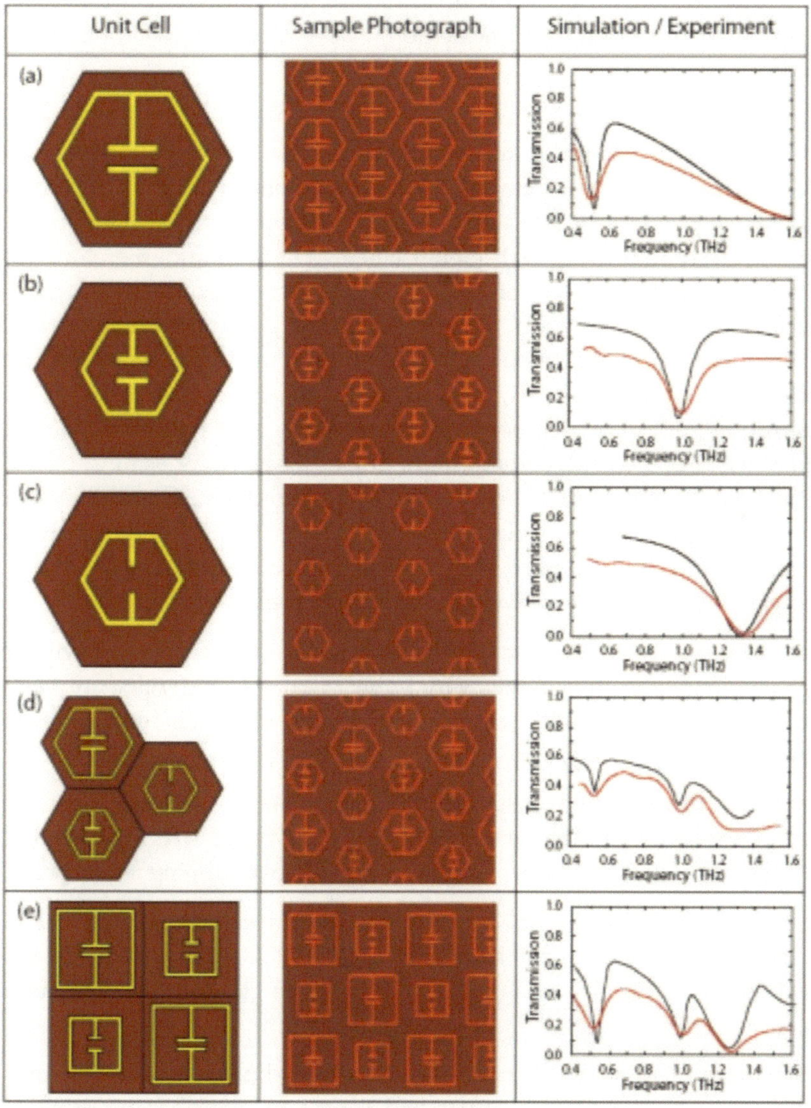

Figure 8. A set of structures designed to act as a metamaterial detector in the THz spectrum for bio substances [18] (Reproduced with permission of OSA).

The simulated and experimental transmission spectra clearly demonstrate a shift in the resonant frequency of the metamaterial device when the design of the metamaterial unit cell is changed. Introduction of a metamaterial surface that contains all three designs will yield multiple resonant peaks.

The issue that the proposed approach faces when a metamaterial structure is designed for real applications is that, as was mentioned before, the classical metamaterial SRRs are resonant in a very narrow band of frequencies. Hence, in order to uniquely detect even one substance (based on the measured values of resonant frequencies) we need to fabricate a device containing a few unique SRR designs.

However, the proposed structure could be optimized if the dielectric properties of the substrate surface (where the SRRs are fabricated on) can be tuned by the application of a simple electrical bias.

It could be suggested that tunable complex oxide thin films can be used to address this challenge. Indeed, the tunability of these films that are typically about 200 nm thick is over 50% [19] when a bias field of about 500 kV/cm is applied. For practical purposes we are restricted to utilize the film of 200 nm or similar thickness as the fields required for the tunability effect are prohibitively high (about 600 kV/cm [20]) and, therefore, thin films are necessary so that practical voltages of about 5–10 V can be used. Fabrication of the tunable complex oxide thin films is not trivial and their dielectric properties are much more challenging to stabilize when compared to bulk materials due to the dependence of these properties (such as dielectric constant, tunability, losses, and leakage current) on factors as coefficient of thermal extension (CTE) mismatch between the substrate and the film yielding a stress that affect the dielectric properties [21]. In addition, proper bias electrodes must be used in such a structure. For example, transparent conductive oxide electrodes such as indium tin oxide (ITO) can be used, however, certain measures have to be taken during the thin film growth to extend the transparency of this material into the THz spectrum from the optical. A typical response (frequency scan) from a single-negative metamaterial is presented in **Figure 9**.

If the resonant frequency of the metamaterial surface could be matched with the resonant frequency of a molecule resulting in the overlapping resonance, we anticipate an even stronger response on the frequency scan at the resonant frequency, which will indicate the presence of a substance under investigation.

Alternative metamaterial structures such as ceramic metamaterials [10] could be considered for the same purpose. Again, the benefit of using these novel structures could be reduced losses and, therefore, a possible "sharper" reaction to the overlapping resonances.

Alternative methods for metamaterial device utilization in the THz spectrum for detection and analysis of chem/bio molecules have been proposed in [22]. Ding et al. suggested the use of asymmetrical split-ring resonators as a unit cell for a single-negative metamaterial surface (**Figure 10**).

This approach produced a metamaterial device that, in a response to an impinging THz wave, yielded simultaneously high Q quadrupole and Fano resonances. Both are ultra-sharp resonances and could be used in highly sensitive detection of chem/bio substances (**Figure 11**).

http://dx.doi.org/10.5772/65951

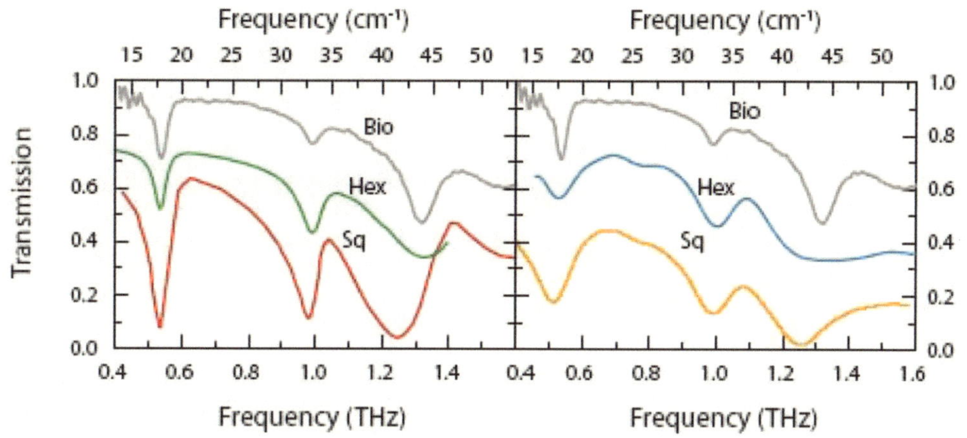

Figure 9. Computational (left) and experimental (right) measurements of the $n = 3$ hexagonal metamaterial, and the $n = 2$ square checkerboard metamaterial, compared to experimental measurements of the molecule biotin. Simulated and experimental transmission spectra of the hexagonal metamaterial has been shifted up by 20% for clarity and $T(\omega)$ of biotin is in arbitrary units [18] (Reproduced with permission of OSA).

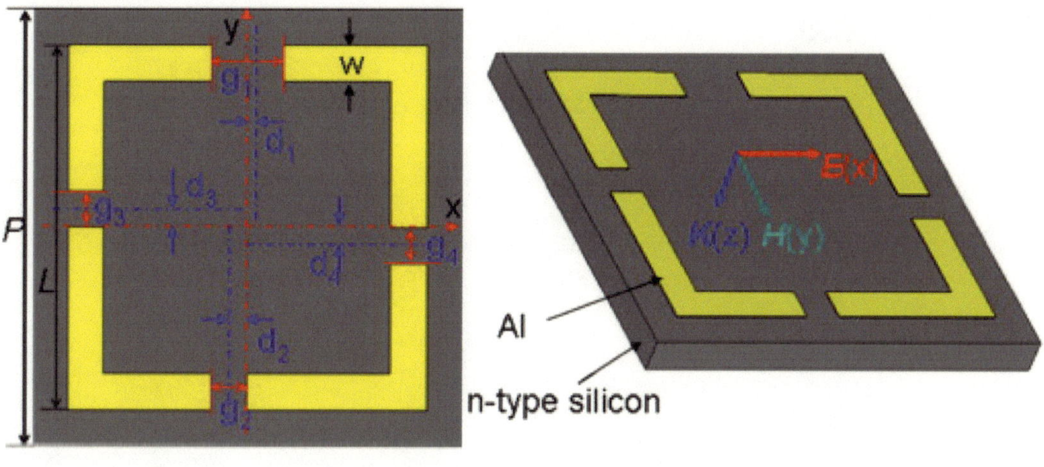

Figure 10. The front and oblique views of the unit cell of the metamaterial with asymmetrical gaps $g_1 = g_2 = g_3 \neq g_4$ [22] (Reprinted with permission from Elsevier via Copyright Clearance Center).

Again, it can be suggested that applying a layer of easily tunable complex oxide thin film can significantly enhance the agility of the proposed device since such a device would be capable of handling multiple resonant frequencies on the same surface due to the dependence of the resonance response of the device on dielectric properties of the substrate.

Metamaterial surfaces have a high potential to detect chem/bio substances in the THz spectrum. As it was mentioned before, many of these substances exhibit at least three characteristic resonant frequencies in the THz spectrum. This would make it possible to uniquely identify many materials. Unfortunately, there is no library yet compiled that can identify these unique resonant frequencies for the materials of interest.

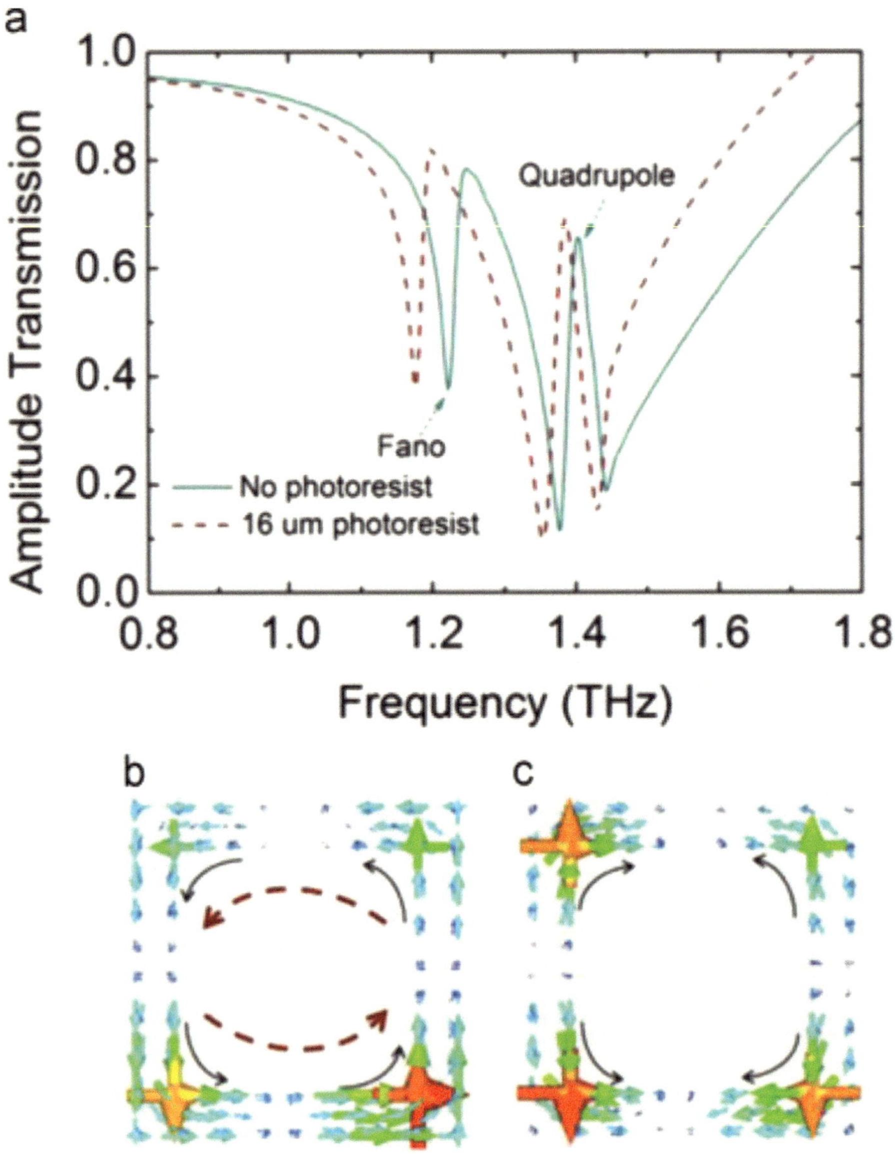

Figure 11. (a) The simulated transmission spectra of the metamaterial with (the wine line) and without (the dark cyan line) a 16 μm thick photoresist overlay. (b) and (c) The surface current distributions of the metamaterial at 1.236 and 1.408 THz, respectively [22] (Reprinted with permission from Elsevier via Copyright Clearance Center).

4. Determination of dielectric properties of complex oxide thin films in THz

Recent developments in the field of communications require higher data transmission speeds [23]. This cannot be achieved at the frequencies in which current 4G wireless systems are operating. Instead, frequencies in the range of 100 GHz are required to meet this requirement. Applications where high data transmission speeds are crucial and include holographic teleconferences, improved medical sensor implants, and fast internet of things. Hence, transition to a 5G wireless system is anticipated. While optical frequencies could be a choice in terms of energy transmission in open space, this is generally not acceptable due to the eye safety concerns, power limitations, atmospheric effects, challenges in maintaining alignment between the transmitter and receiver, and, in the case of a free-space optical (FSO) communications system, the very large size of some proposed components. In a recent design, the optical front end for such a system had dimensions of 12 cm ×12 cm ×20 cm, and a weight of almost 1 kg. Therefore, low THz would be a natural choice for these applications. Indeed, Facebook © has already deployed the first experimental 5G system in a neighborhood in San Jose [24] that operates at 60 GHz. However, as an example of the difficulties associated with these frequencies, it is worthwhile to mention that the receivers for the home units are located outside the house as the low-power signal penetrate the house walls. Also, the transducers for this system were needed on each light pole in the neighborhood due to maintenance of the strength of the signal.

There are two main technical challenges that need to be overcome to make 5G widely acceptable. First, there are no sources that are capable to emit a powerful enough signal. Second, the attenuation in the air for the electromagnetic waves in the frequencies of interest is very significant [25]. Clearly, in order to overcome these issues, a lot of thought is invested in developing more powerful sources for the electromagnetic waves at low THz frequencies.

One of the obvious ways to enhance the power of a low THz source is to design and implement suitable phased array antennas. A phased array antenna consists of multiple single antennas that operate at the same frequency, but at different phases. These antennas are used in broadcasting, radars, space probes, weather research, and other applications. A suitable design would need to be compact enough to be deployed in multiple locations at relatively short distances, but large enough to contain multiple individual antennas that operate at different phases to generate a tight "focused" beam (**Figure 12**) [26].

This approach allows the beam to be condensed (as shown in **Figure 12**), maintain certain radiation pattern, and the ability to steer the beam if necessary. It was suggested [20] that thin tunable complex oxide films such as $BaSrTiO_3$ (BST) can be used in the design and fabrication of a new generation of the phased array antenna components, such as phase shifters and filters. Indeed, such films exhibit substantial tunability of their dielectric constant upon application of a simple electrical bias (**Figure 13**).

One can see from **Figure 13** that the required fields in these materials are very high, hence, thin films with the thickness of about 200 nm must be implemented to operate with the low voltages found in realistic applications.

In order to utilize these thin film components in low THz applications such as 5G, one needs to know with a very high fidelity the dielectric properties of the films in the low THz regime. Unfortunately, it is uncertain if the methods that are used to determine the dielectric properties of the tunable complex oxide thin films at lower frequencies, such as MW probes, are going to be relevant in THz range of the electromagnetic spectrum. For example, when these thin films have been investigated at low GHz, it was necessary to replace the metal-insulator-metal (MIM) structure which was used to determine the dielectric and other properties of the complex oxide thin films at lower frequencies, with a different configuration [27]. In

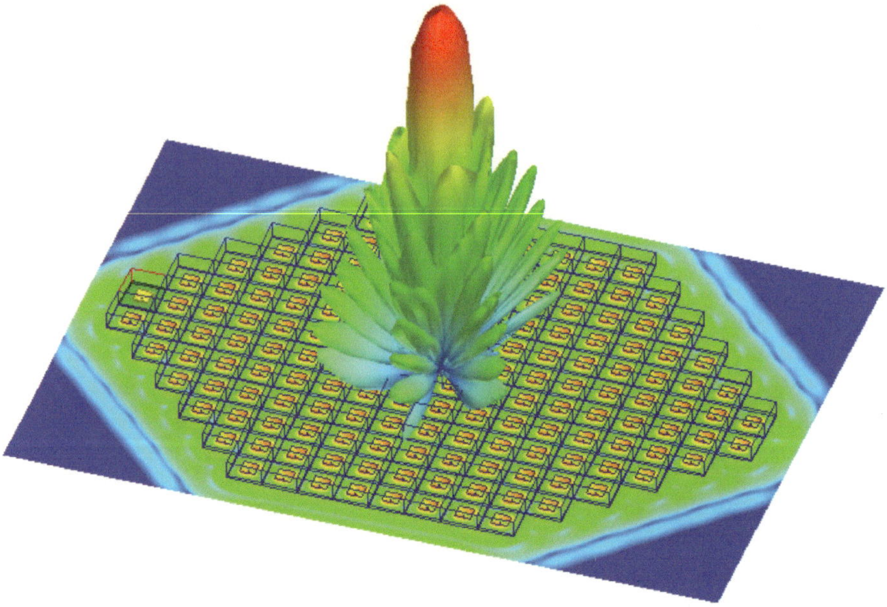

Figure 12. A model of a phased array antenna in Ansoft HFSS © commercially available software [26] (copyrighted material is reprinted with the permission of ANSYS INC©).

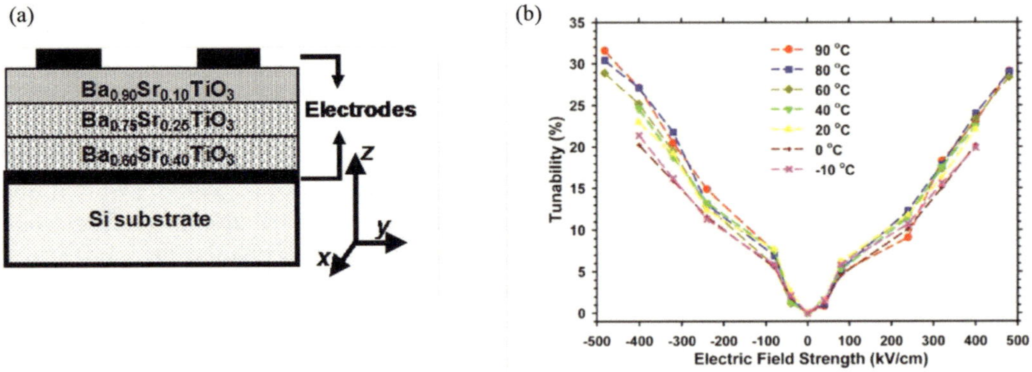

Figure 13. (a) Schematic cross section image of a BST thin film deposited on a sapphire substrate and (b) tunability plot of the BST thin film biased up to 6 V [20] (Reproduced from "Cole, M., et al., Dielectric properties of MgO-doped compositionally graded multilayer barium strontium titanate films. Appl. Phys. Lett., 2008. 92(7): p. 72906–72906", with the permission of AIP Publishing).

addition, the equipment required to perform all necessary measurements is currently very expensive. In addition, free space measurements in the THz range can help to determine the dielectric properties of thin films that are 400 nm or thicker. Therefore, a new method to measure the dielectric constant and tunability of complex oxide thin films needs to be developed. It was demonstrated in [28] that the dielectric constant of an 800 nm $BaSrTiO_3$

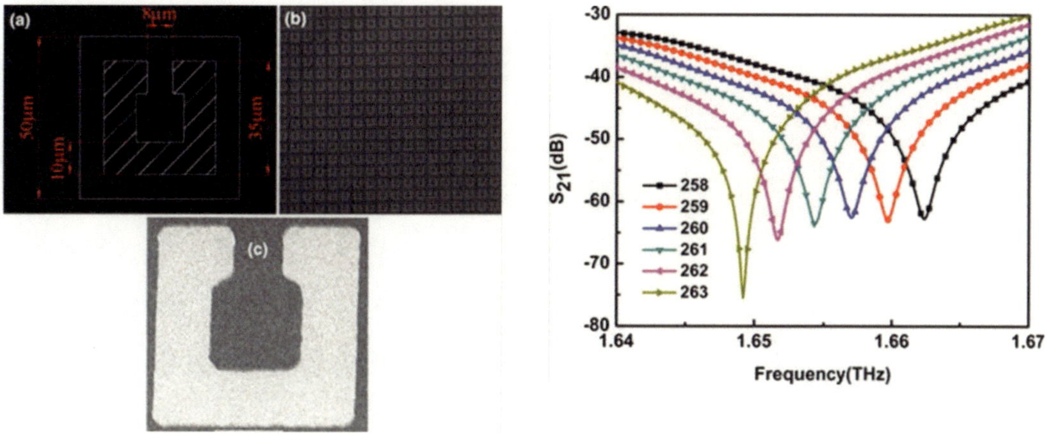

Figure 14. Schematic diagram and SEM image of a SRR unit cell with structural parameters and simulated transmission spectra of the SRRs with different dielectric constants [28] (Reproduced with permission of AMERICAN CERAMIC SOCIETY via Copyright Clearance Center).

(BST) film at low THz can be determined when a metamaterial structure is implemented on the surface of the complex oxide thin film. To do this, the resonant response of the metamaterial surface on top of the thin film is matched with a numerical model that predicts the location of this resonant peak on the frequency axis as a function of the dielectric constant of the complex oxide thin film. By matching the resonance with the numerical model, the dielectric properties (specifically, the dielectric constant) of this complex oxide thin film (**Figure 14**) can be predicted.

However, the characterization parameters that are important for the applications discussed above also include the tunability of the thin film. For the purposes of phase shifting, we need to know how much we are able to tune the dielectric properties of the complex-oxide thin films, namely, the percentage of change in the dielectric constant as a function of the applied electrical bias. A potential path to obtain this information has been demonstrated through the design and fabrication of an active metamaterial [29]. The schematics of the device are presented in **Figure 15**. One can notice that the $BaSrTiO_3$ thin film is sandwiched between two bias electrodes and, hence, "inserted" in the capacitor gap of the standing split-ring resonator (SRR) by the bottom bias electrode (**Figure 15**). The dielectric constant of the inserted BST film can be tuned by the applied bias and the resonant frequency of this single-negative metamaterial device is shifted as a function of the dielectric properties of the thin film.

Again, the model has been developed to determine the dependence of the resonant frequency of the proposed device on the dielectric properties of the $BaSrTiO_3$ thin film situated in the SRR gap. The developed model has been adjusted to match the experimental data from the fabricated sample (**Figures 16** and **17**).

From **Figure 17**, one can conclude that the frequency shift demonstrated by the fabricated metamaterial device when the $BaSrTiO_3$ thin film is biased by up to 6 V corresponds to the change of the dielectric constant of the $BaSrTiO_3$ thin film from 180 to 170. The result is based

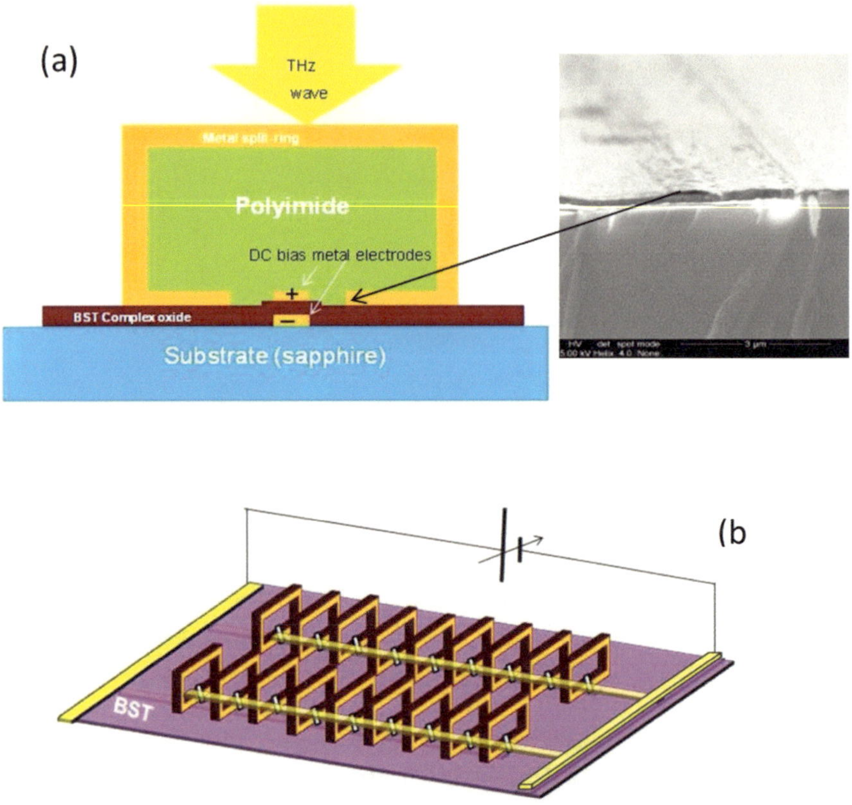

Figure 15. (a) Schematics of a tunable three-dimensional (3D) THz metamaterial unit cell with a complex oxide film as a tunable capacitor (inset: SEM image of the $BaSrTiO_3$ thin film deposited over the bias electrode). (b) 3D illustration of a complete tunable THz metamaterial device [29].

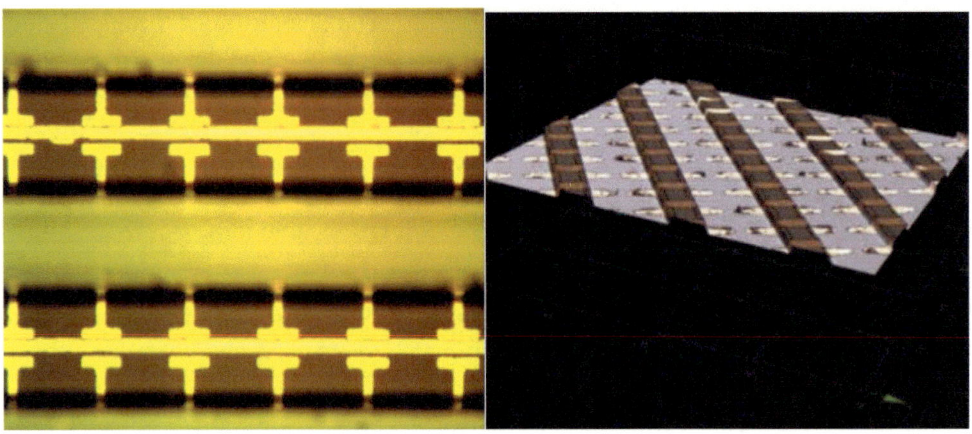

Figure 16. Left: microscopy image of the split-ring arrays focused on the bottom tunable capacitor with metal contact line; right: 3D optical microscope picture of the split-ring metamaterial device [29].

http://dx.doi.org/10.5772/65951

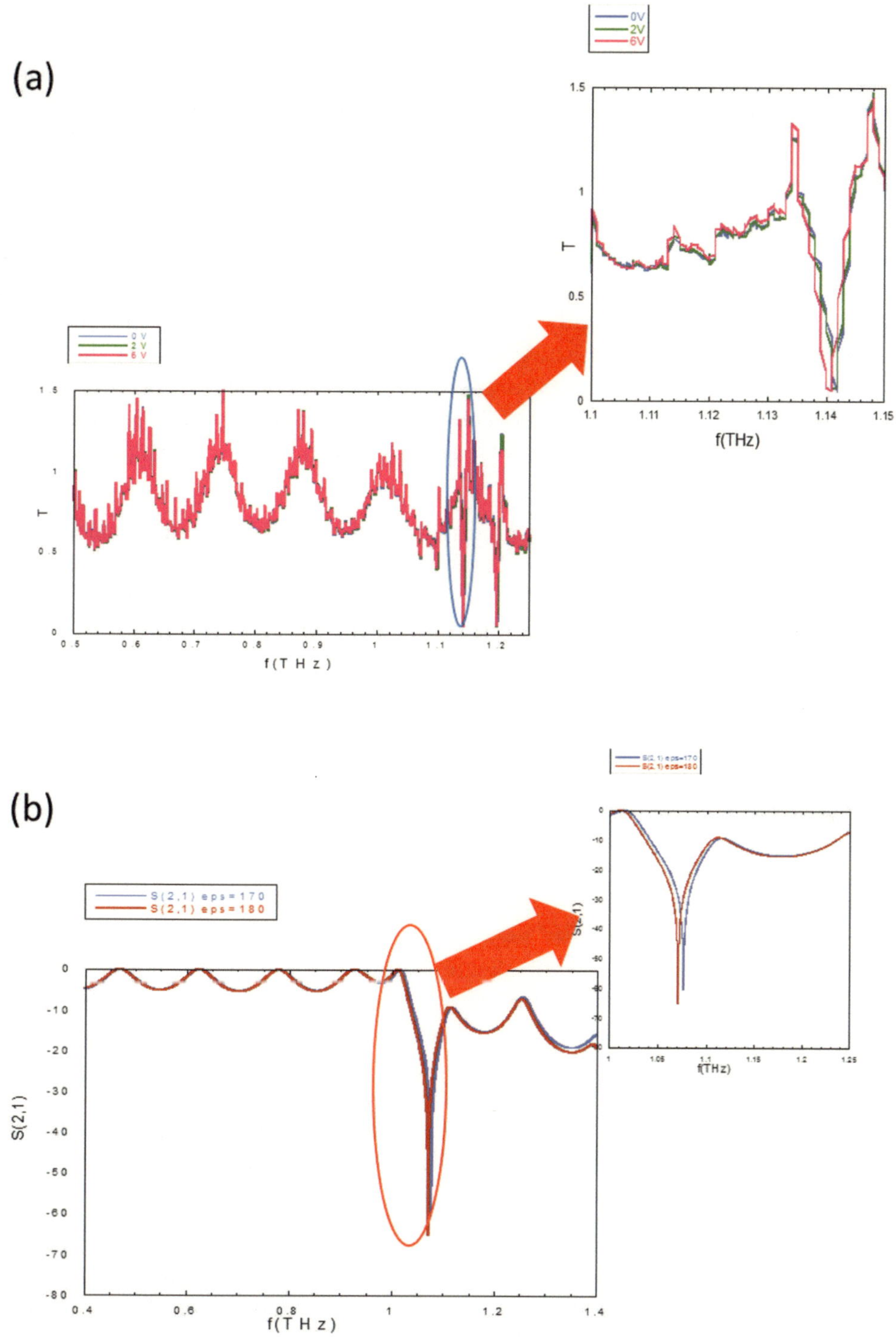

Figure 17. (a) Experimental and (b) modeled results for the metamaterial device shown in Fig. 16. A 2 GHz shift in the resonant frequency of the device is demonstrated upon application of the 6 V electrical bias [29].

on a comparison of the experimental data with the model of the single-negative metamaterial structure in HFSS© Ansoft software. This result is even more remarkable when we take into consideration the fact that in the fabricated structure (**Figure 16**), the bottom bias Pt electrode has been fabricated to the height of only 50 nm, as opposed to anticipated 200 nm to fill the gap in the metamaterial SRR completely. Hence, the fabricated metamaterial device did not operate to its utmost capacity.

The proposed approach presents an attempt to resolve an important problem of how to determine the dielectric properties of thin, tunable, complex oxide films in the THz region of radiation. Clearly, other ways to obtain this information are the source of on-going research by the scientific community. However, this approach provides a means to design devices in the frequencies of interest and serve as a standard when evaluating the fidelity of other methods, such as on-probe measurements. It was mentioned above, that alternative characterization methods exhibit certain fidelity challenges that need to be resolved in order to utilize these methods for the design and development of components operating in THz frequencies.

In addition, the presented structure constitutes an active single-negative metamaterial device. The design of the device enables reconfiguration of the structure into a double-negative metamaterial device with relative ease as opposed to other active metamaterial structures. Such a device could be used for possible THz communications or imaging applications.

5. Conclusions

This chapter attempts to expose the reader to a broader spectrum of potential metamaterial device applications, beyond antennas and geometrical optics. We have demonstrated just three potential applications that include nondestructive evaluation, chem/bio detection, and 5G communications. The intent was not to present a complete list of potential applications of metamaterial devices, as many more exist, but to give an idea of how vast the potential application space is. Metamaterials are slowly maturing from simply interesting science to becoming a very useful tool to help address multiple technical challenges in real life. Again, the presented application ideas are still at their initial point of development—either a proof-of-concept or just an idea that has been substantiated by some data. Hence, a lot of additional effort has to be invested in these areas described above to fully realize them in real-life applications. Clearly, the presented effort requires a multidisciplinary approach where fields such as electrical engineering, physics, materials science, biology, chemistry, and others are intertwined to produce a viable product.

This chapter does not attempt to present metamaterial devices as an area in which all problems have been resolved. These materials have many unresolved issues, requiring significant research investment, such as reducing transmission losses of dual-negative metamaterials. However, it is our hope that the variety of potential applications presented in this chapter will convey to the reader the extent of the impact that metamaterials can make in the future on multiple areas of technology research and development.

http://dx.doi.org/10.5772/65951

Acknowledgements

The author would like to thank Michael Golt of the US Army Research Lab for useful suggestions regarding the presented work.

This research was sponsored by the Army Research Laboratory and was accomplished under Cooperative Agreement Number W911NF-12-2-0019. The views and conclusions contained in this document are those of the authors and should not be interpreted as representing the official policies, either expressed or implied, of the Army Research Laboratory or the U.S. Government. The U.S. Government is authorized to reproduce and distribute reprints for Government purposes notwithstanding any copyright notation herein.

Author details

Daniel Shreiber

Address all correspondence to: daniel.shreiber.civ@mail.mil

U.S. Army Research Laboratory, Aberdeen, MD, USA

References

[1] Veselago, V., *The electrodynamics of substances with simultaneously negative values of ε and μ*. Soviet Physics Uspekhi. 1968. **10**(4): p. 509–514.

[2] Pendry, J. B., A. J. Holden, W. J. Stewart, and I. Youngs, *Extremely low frequency plasmons in metallic meso structures*. Phys. Rev. Lett, 1996. **76**: p. 4773–4776.

[3] Pendry, J. B., A. J. Holden, D. J. Robbins, and W. J. Stewart, *Magnetism from conductors and enhanced nonlinear phenomena*. IEEE Trans. Microw. Theory Tech., 1999. **47**(11): p. 2075–2084.

[4] Dong, Y and T. Itoh, *Metamaterial-based antennas*. Proc. IEEE, 2012. **100**(7): p. 2271–2285.

[5] Shelby, R. A., D. R. Smith, and S. Schultz, *Experimental verification of a negative index of refraction*. Science, 2001. **292**(5514): p. 77–79.

[6] Pendry, J. B., *Negative refraction makes a perfect lens*. Phys. Rev. Lett., 2000. **85**(18): p. 3966–3969.

[7] Ueda, T and T. Itoh, *Mu-negative, double-negative, and composite right/left handed metamaterials based on dielectric resonators*. IEICE Electron. Express, 2012. **9**(2): p. 65–80.

[8] Lai, A., T. Itoh, and C. Caloz, *Composite right/left-handed transmission line metamaterials*. IEEE Microw. Mag., 2004. **5**(3): p. 34–50.

[9] Shreiber, D., M. Gupta, and R. Cravey, *Comparative study of 1-D and 2-D metamaterial lens for microwave nondestructive evaluation of dielectric materials*. Sens. Actuators A: Phys., 2011. **165**(2): p. 256–260.

[10] Shibuya, K., K. Takano, N. Matsumoto, H. Miyazaki, K. Izumi, Y. Jimba, and M. Hangyo, "Terahertz metamaterials composed of TiO_2 cube arrays". Proceedings of the Second International Congress on Advanced Electromagnetic Materials in Microwaves and Optics - Metamaterials, Pamplona, Spain 2008.

[11] Soukoulis, C. M., S. Linden, and M. Wegener, *Negative refractive index at optical wavelengths*. Science, 2007. **315**(5808): p. 47–49.

[12] Cummer, S.A., J. Christensen, and A. Alù, *Controlling sound with acoustic metamaterials*. Nat. Rev. Mater., 2016. **1**: p. 16001.

[13] Kharkovsky, S., et al., *Dual-polarized near-field microwave reflectometer for noninvasive inspection of carbon fiber reinforced polymer-strengthened structures*. IEEE Trans. Instrum. Meas., 2008. **57**(1): p. 168–175.

[14] Tabib-Azar, M., et al., *Nondestructive superresolution imaging of defects and nonuniformities in metals, semiconductors, dielectrics, composites, and plants using evanescent microwaves*. Rev. Sci. Instrum., 1999. **70**(6): pp. 2783–2792.

[15] Shreiber, D., M. Gupta, and R. Cravey, *Microwave nondestructive evaluation of dielectric materials with a metamaterial lens*. Sens. Actuators A: Phys., 2008. **144**(1): pp. 48–55.

[16] Walther, M., et al., *Far-infrared vibrational spectra of all-trans, 9-cis and 13-cis retinal measured by THz time-domain spectroscopy*. Chem. Phys. Lett., 2000. **332**(3–4): pp. 389–395.

[17] Fischer, B., M. Walther, and P.U. Jepsen, *Far-infrared vibrational modes of DNA components studied by terahertz time-domain spectroscopy*. Phys. Med. Biol., 2002. **47**(21): p. 3807.

[18] Bingham, C. M., et al., *Planar wallpaper group metamaterials for novel terahertz applications*. Opt. Express, 2008. **16**(23): p. 18565–18575.

[19] Cole, M. W., et al., *The fabrication and material properties of compositionally multilayered $Ba_{1-x}Sr_xTiO_3$ thin films for realization of temperature insensitive tunable phase shifter devices*. J. Appl. Phys., 2007. **102**(3): p. 034104.

[20] Cole, M., et al., *Dielectric properties of MgO-doped compositionally graded multilayer barium strontium titanate films*. Appl. Phys. Lett., 2008. **92**(7): p. 72906–72906. http://doi.org/10.1063/1.2870079

[21] Shreiber, D., et al., *Some unusual behavior of dielectric properties of $SrTiO_3$ metal organic chemical vapor deposition grown thin films*. J. Appl. Phys., 2014. **116**(9): p. 094101.

[22] Ding, C. F., et al., *Dual-band ultrasensitive THz sensing utilizing high quality Fano and quadrupole resonances in metamaterials*. Opt. Commun., 2015. **350**: p. 103–107.

[23] Akyildiz, I.F., J.M. Jornet, and C. Han, *Terahertz band: next frontier for wireless communications*. Phys. Commun., 2014. **12**: p. 16–32.

http://dx.doi.org/10.5772/65951

[24] https://code.facebook.com/posts/1072680049445290/introducing-facebook-s-new-terrestrial-connectivity-systems-terragraph-and-project-aries/

[25] D'Amico, C., et al., *Conical forward THz emission from femtosecond-laser-beam filamentation in air*. Phys. Rev. Lett., 2007. **98**(23): 235002 p.

[26] *ANSYS web site*.

[27] Sheng, S., et al., *Characterization of microwave dielectric properties of ferroelectric parallel plate varactors*. J. Phys. D: Appl. Phys., 2009. **42**(1): p. 015501.

[28] Qi, P., et al., *Dielectric Properties of* $Ba_{0.7}Sr_{0.3}TiO_3$ *Film at terahertz measured by metamaterials*. J. Am. Ceram. Soc., 2012. **95**(4): p. 1167–1169.

[29] Weimin Zhou, G.D., Monica Taysing-Lara, Grace Metcalfe, Nathaniel Woodward, Amir Zaghloul, Daniel Shreiber, Melanie Cole, Eric Ngo, Matt Ivill, *Metamaterial and Metastructural Architectures for Novel C4ISR Devices and Sensors*. US Army Research Lab Tech Report, 2015.

Chapter 6

Metamaterial Based Ultra-Wideband Antennas for Portable Wireless Applications

Mohammad Alibakhshikenari,
Mohammad Naser-Moghadasi,
Ramazan Ali Sadeghzadeh, Bal Singh Virdee and
Ernesto Limiti

Additional information is available at the end of the chapter

http://dx.doi.org/10.5772/66674

Abstract

Antennas are essential for wireless communication systems. The size of a conventional antenna is dictated mainly by its operating frequency. With the advent of ultra-wideband systems (UWB), the size of antennas has become a critical issue in the design of portable wireless devices. Consequently, research and development of suitably small and highly compact antennas are challenging and have become an area of great interest among researchers and radio frequency (RF) design engineers. Various approaches have been reported to reduce the physical size of RF antennas including using high permittivity substrates, shorting pins, reactive components, and more recently, metamaterials (MTM) based on composite right-/left-handed transmission-lines (CRLH-TLs). MTM exhibit unique electromagnetic response that cannot be found in the nature. In this chapter, the properties of CRLH-TL are used to synthesize novel and highly compact planar UWB antennas with radiation properties suitable for wireless mobile devices and systems.

Keywords: antennas, ultra-wideband, metamaterials, composite right-/left-hand transmission-lines, microstrip

1. Introduction

Electronic circuitry in portable wireless devices is incorporated inside a small and highly integrated transceiver unit [1–4]. Miniaturization of such a system is precluded by the size of the antenna as its dimensions are related to the operating frequency. The most challenging aspect in the development of such systems is the design of miniature antennas [5–8]. In this chapter, this issue is addressed by using metamaterial technology. Several different antenna configurations are presented based on CRLH-TL/MTM unit cells [9–22].

In part 2, antenna designs are described based on simplified CRLH-TLs that are implemented using F-shaped and T-shaped dielectric slits embedded on the antenna's ground-plane and radiating arms, respectively. This antenna is shown to operate across 0.65–9.2 GHz with a maximum gain and radiation efficiency of 3.5 dBi, and 70% at 4.5 GHz, respectively. This antenna has dimensions of 25 × 15 × 1.6 mm^3. In another example, how size reduction and bandwidth extension can be achieved using CRLH-TL unit cells that are composed of a U-shaped slit and grounded spiral microstrip stub are shown.

In part 3, the antenna is implemented by cascading together in series several identical MTM unit cells. The unit cell is composed of a transmission line that has engraved on its patch a mirror image of E-shaped slits and high impedance spiral stub that is grounded through a metal via-hole. Two antennas implemented using this technique have dimensions of $0.017\lambda_0 \times 0.006\lambda_0 \times 0.001\lambda_0$ and $0.028\lambda_0 \times 0.008\lambda_0 \times 0.001\lambda_0$, where λ_0 is free space wavelength at 500 and 650 MHz, respectively. The respective antennas have bandwidths of 850 MHz (0.5–1.35 GHz) and 1.2 GHz (0.65–1.85 GHz), which correspond to fractional bandwidths of 91.9% and 96.0%, respectively. Besides the small dimensions and wide bandwidth characteristics, the measured gain and efficiency of one antenna at 1 GHz are 5.3 dBi and 85%, respectively; and the second antenna has a gain and efficiency of 5.7 dBi and 90%, respectively, at 1.4 GHz.

In part 4, antenna size reduction and bandwidth extension is realized with MTM unit cells that are composed of U-shaped dielectric slit and spiral conductor that is grounded using via-holes. The design of the antenna presented here functions over the frequency range of 5.8–7.3 GHz, i.e. it has a fractional bandwidth of 23%. The antenna's performance was measured to verify it has a wide bandwidth, high gain, and high radiation efficiency properties. At 6.6 GHz, the antenna is shown to exhibit a radiation gain of 4.8 dBi, fractional bandwidth of 23%, and efficiency of 78%. Furthermore, the proposed antenna is very compact and has dimensions of $0.39\lambda_0 \times 0.13\lambda_0 \times 0.015\lambda_0$.

In part 5, the MTM unit-cell is composed of T-shaped dielectric slit that is etched on the radiating patch and includes a grounded conductive spiral stub. The T-shaped slit antenna is shown to operate over 1.1–6.85 GHz (fractional bandwidth ~ 145%) with a maximum gain and efficiency of 7.1 dBi and 91%, respectively, at 3.7 GHz. This antenna has an electrical size of $0.05\lambda_0 \times 0.02\lambda_0 \times 0.002\lambda_0$.

The aforementioned MTM antennas exhibit superior performance compared to conventional antennas in terms of fractional bandwidth, gain, and efficiency. These antennas are suitable for UWB wireless communication systems, portable microwave handsets, and transceivers.

2. UWB antenna based on simplified CRLH-TL

The simplified composite right-/left-handed transmission-lines (SCRLH-TL) used here in the design are a novel planar antenna. The SCRLH-TL is implemented by loading the radiation patch and ground-plane with dielectric T-shaped and F-shaped slits, respectively, as shown in **Figure 1**. With the assistance of 3D full-wave electromagnetic simulator (HFSS™), the dimensions of the dielectric slits can be optimized. The design of Antenna#1 consists of two identical structures that can be considered to be a mirror image, where one of the structures has been flipped sideways, as shown in **Figure 1(a)**. The F-shaped dielectric slits embedded in the ground-plane essentially behave like left-handed series capacitance; and the T-shaped dielectric slits embedded in the radiation patch are used to enhance the radiation characteristics of the antenna. To improve the impedance matching of the feed-line to the antenna, it is necessary to load the feed-line with an H-shaped microstrip stub, as shown in **Figure 1**. The two SMA connectors on the opposite sides of antenna are used to excite the antenna through the conductor-backed coplanar waveguide (CPW) transmission-lines.

The gain and radiation efficiency of any antenna can be improved by simply increasing the aperture of the antenna. To increase the magnitude of these two characterizing parameters, the number of patches in the proposed design was increased from two to four, as shown in **Figure 1(b)**. The advantage of the technique presented here in comparison to conventional methods is that it does not affect the physical size of the antenna as the additional patches are contained within the antenna structure. To enhance the antenna's impedance bandwidth, the number of F-shaped slits in the ground-plane was halved from four to two, thus effectively reducing the left-handed series capacitance.

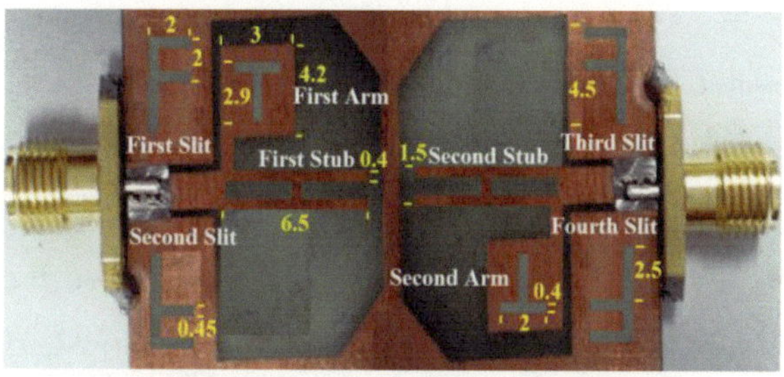

(a)

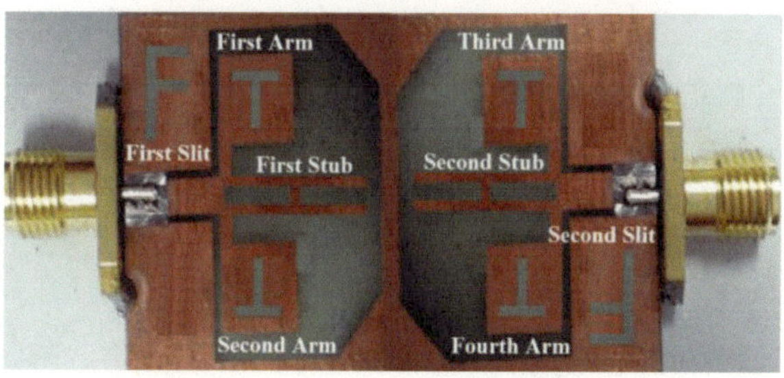

(b)

Figure 1. Antenna configurations, (a) Antenna#1 and (b) Antenna#2. Design parameters of both antennas are identical. Dimensions (in mm) are annotated.

The antennas were fabricated on RD/duroid® RO4003 substrate with dielectric constant of 3.38, thickness of 1.6 mm, and tanδ of 22×10^{-4}. The dimensions of the two antennas are: $25 \times 15 \times 1.6\ mm^3$ which is equivalent to an electrical size of $0.054\lambda_0 \times 0.032\lambda_0 \times 0.003\lambda_0$, where λ_0 corresponds to

0.65 GHz. The antenna's reflection-coefficient response (simulated and measured) is shown in **Figure 2**. The antenna's frequency of operation and bandwidth is given in **Table 1**. Antenna#1 has a measured bandwidth of 7.4 GHz and Antenna#2 has a bandwidth of 8.55 GHz. The fractional bandwidth of Antenna#2 is 16% greater than Antenna#1. **Figure 3** shows the measured antenna gain and efficiency response of both antennas. The antenna gain and efficiency values at spot frequencies are also tabulated in **Table 2**. Results show that Antenna#2 has a maximum gain and radiation efficiency of 3.5 dBi and 70% at 4.5 GHz, which is higher than Antenna#1 by 25% and 27%, respectively.

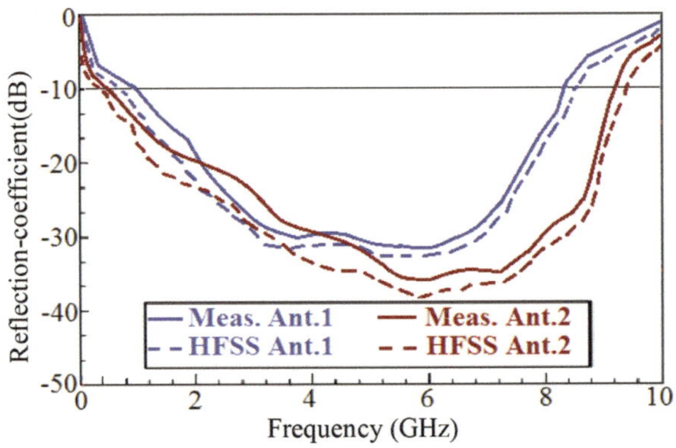

Figure 2. Simulated and measured reflection-coefficient response of the two antennas.

	Operating frequency range (fractional bandwidth)	
	Simulated	Measured
Antenna#1	0.82–8.6 GHz (165.2%)	1–8.4 GHz (157.4%)
Antenna#2	0.5–9.45 GHz (179.9%)	0.65–9.2 GHz (173.6%)

Table 1. Operating frequency range and impedance bandwidth of the proposed antennas.

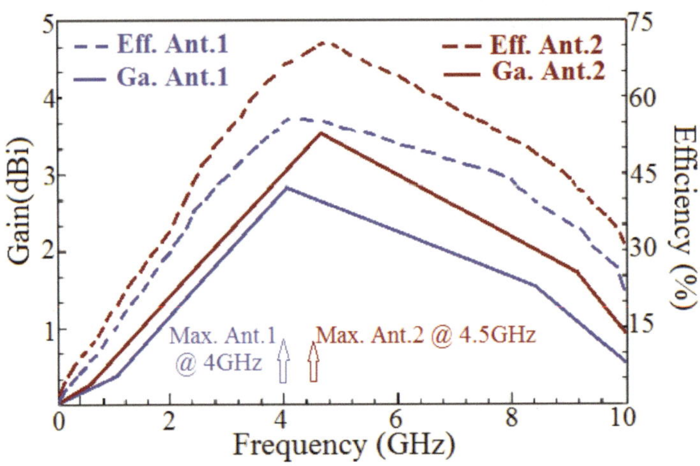

Figure 3. Measured gain and efficiency response of both antennas.

http://dx.doi.org/10.5772/66674

Antennas	Gain (dBi)	Efficiency (%)
Ant.#1 @ 1, 4, and 8.4 GHz	0.4, 2.8, and 1.5	15, 55, and 40
Ant.#2 @ 0.65, 4.5, and 9.2 GHz	0.2, 3.5, and 1.7	12, 70, and 42

Table 2. Gain and radiation characteristics of the two antennas.

Radiation characteristics (copolarization and crosspolarization) of the two antennas in the E-plane and H-plane at 2.5 GHz, 4 GHz, and 4.5 GHz are shown in **Figure 4**. The results show both antennas radiate omnidirectionally in the E-plane and bidirectionally in the H-plane. The low profile UWB antennas facilitate easy integration in wireless systems and can be flush mounted on various structures.

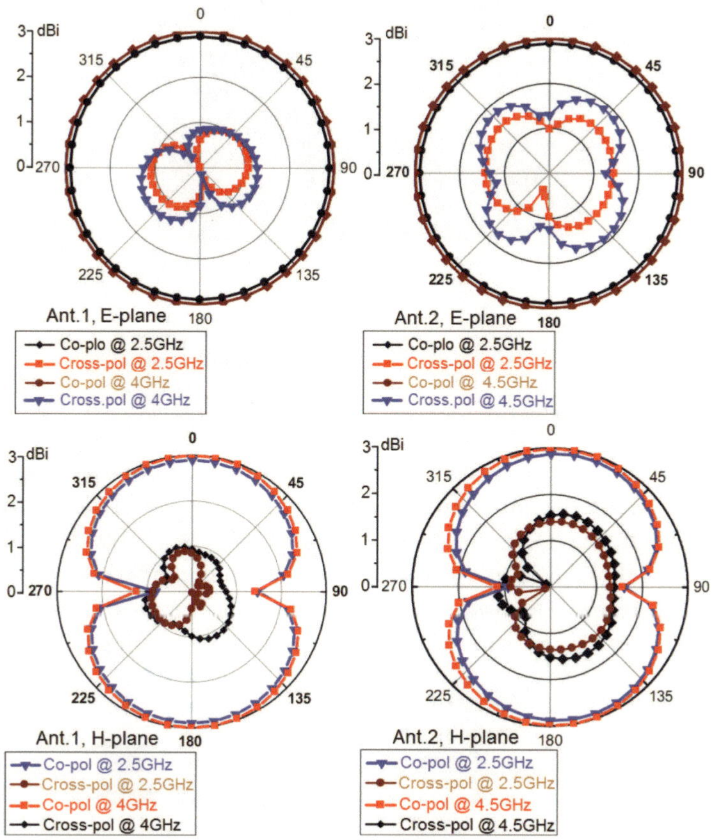

Figure 4. Measured radiation patterns at 2.5, 4, and 4.5 GHz.

3. Wideband antenna using E-shaped slit MTM unit cells

The goal in this section is to design antennas that can fit within an area of 15 × 5 mm^2 for flush mounting on various structures including vehicles and portable wireless devices, and possess characteristics of wide operational bandwidth with good radiation properties. The unique antenna design is based on MTM unit cell, shown in **Figure 5**, and comprises a microstrip patch on which is etched mirror image of E-shaped dielectric slits and the unit cell is grounded using a conductive spiral stub. The E-shaped slits and spiral stub act as left-handed (LH) capacitance (C_L) and shunt inductance (L_L), respectively, corresponding to the negative permeability ($\mu < 0$) and the negative permittivity ($\varepsilon < 0$), respectively, and are represented by [23–26]:

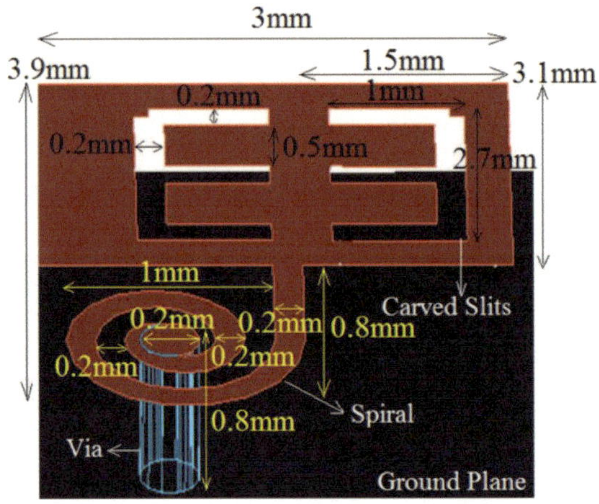

Figure 5. MTM unit cell geometry.

$$`\mu = \frac{Z}{j\omega} = \frac{1}{\omega^2 C_L}`+ L_R \ (Z\text{: } \textit{series impedence}) \quad (1)$$

$$\varepsilon = \frac{Y}{j\omega} = \frac{1}{\omega^2 L_L} + C_R \ (Y\text{: } \textit{shunt admittance}) \quad 2)$$

Surface current flow on the patch constituting the unit cell introduces right-handed (RH) series inductance (L_R), and the gap between patch and the ground plane introduces shunt RH capacitance (C_R), which are related to positive permeability ($\mu > 0$) and positive permittivity ($\varepsilon > 0$), respectively, defined by Eqs. (3) and (4). The loss in the unit cell structure can be modeled by series RH resistance (R_R), shunt LH resistance (R_L), shunt RH conductance (G_R), and series LH conductance (G_L). The equivalent circuit model of the unit cell is shown in **Figure 6**.

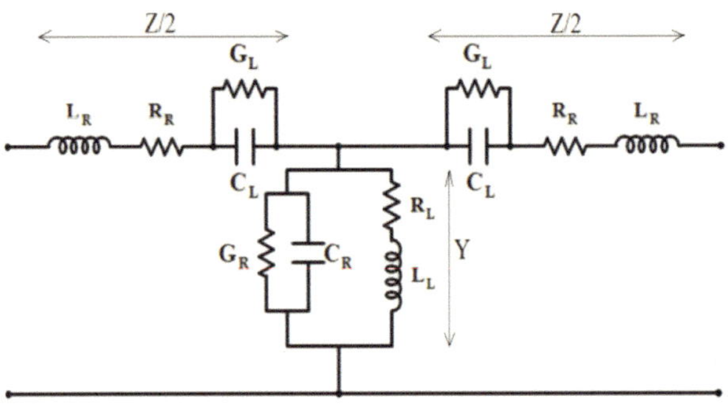

Figure 6. Equivalent circuit model of the MTM unit cell.

The values for the electrical parameters C_L, L_L, C_R, L_R, R_R, R_L, G_R, and G_L are 4.2 pF, 5.4 nH, 2.1 pF, 2.9 nH, 1.2 Ω, 1.85 Ω, 0.85 Ω, and 1.15 Ω, respectively.

http://dx.doi.org/10.5772/66674

$$\mu = \frac{Z}{j\omega} = L_R - \frac{1}{\omega^2 C_L} \tag{3}$$

$$\varepsilon = \frac{Y}{j\omega} = C_R - \frac{1}{\omega^2 L_L} \tag{4}$$

Restriction in the antenna size required its characteristics to be optimized by full wave simulators, i.e., Keysight Technologies Advanced Design System (ADS) and Ansys High Frequency Structure Simulator (HFSS™). The antenna's characteristics were tuned by modifying the dimensions of the E-shaped slit, the gap between slits, the dimensions of the spiral, and the number of spiral turns. The optimized MTM unit cell is shown in **Figure 5**. The two antenna designs use two and three MTM unit cells, respectively, that are cascaded together in series. The antennas are constructed using standard manufacturing techniques on FR4 substrate with dielectric constant of 4.6, thickness of 0.8 mm, and tanδ of 0.01. The prototype antenna designs are shown in **Figure 7**. Port 1 is used to excite both antennas, and port 2 is matched to 50 Ω load impedance (SMD1206).

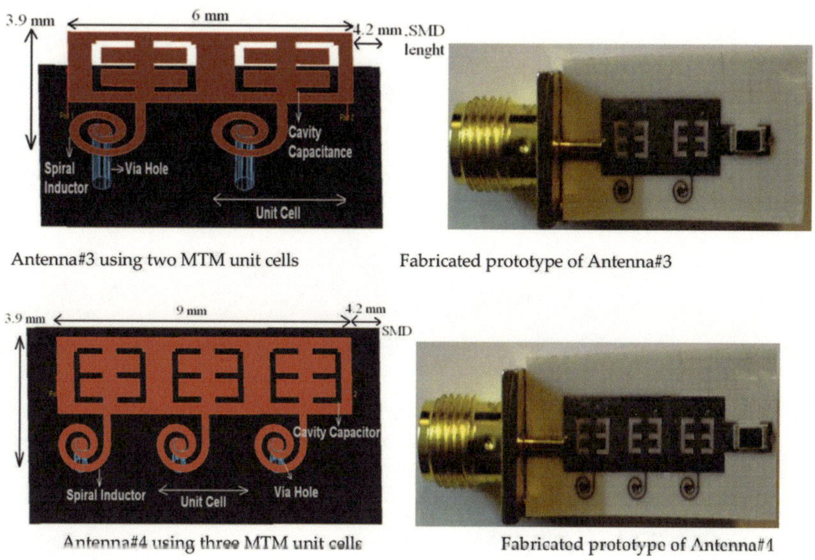

Figure 7. Fabricated MTM antennas.

The physical size of the Antenna#3 (two unit cell) is 10.2 × 3.9 × 0.8 mm³ ($0.017\lambda_0 \times 0.006\lambda_0 \times 0.001\lambda_0$, where λ_0 is the free space wavelength at 500 MHz); and the size of the Antenna#4 (three unit cell) is 13.2 ×3.9 × 0.8 mm³ ($0.028\lambda_0 \times 0.008\lambda_0 \times 0.001\lambda_0$, where λ_0 is the free space wavelength at 650 MHz. The simulated and measured bandwidth of both antennas is given in **Table 3**, and its reflection coefficient response is shown in **Figures 8** and **9**.

The simulated and measured gain and efficiency response of the Antenna#3 and Antenna#4 are shown in **Figures 10** and **11**, as well as tabulated in **Tables 4** and **5**. The results show the three unit cell antenna offers better gain and efficiency than the two unit cell; however, this is at the cost of slightly larger antenna size. The radiation pattern of both antennas in the E- and H-planes at various spot frequencies is shown in **Figure 12**. Both antennas radiate power omnidirectionally; however, Antenna#4 radiates bidirectionally at 1.85 GHz. The radiation properties were measured using a network analyzer in a standard microwave anechoic chamber. Properties of the two antennas are compared with other planar MTM antennas in **Table 6**.

Table 6 shows that Antennas#3 and #4 are smaller in size and have a high gain and efficiency than other reported antennas.

	Antenna#3	Antenna#4
ADS	1.4 GHz (200 MHz–1.60 GHz)	1.6 GHz (400 MHz–2 GHz)
HFSS™	1.2 GHz (350 MHz–1.55 GHz)	1.55 GHz (350 MHz–1.9 GHz)
Measured	0.85 GHz (500 MHz–1.35 GHz)	1.2 GHz (650 MHz–1.85 GHz)

Table 3. Simulated and measured impedance bandwidth for Antenna#3 and #4.

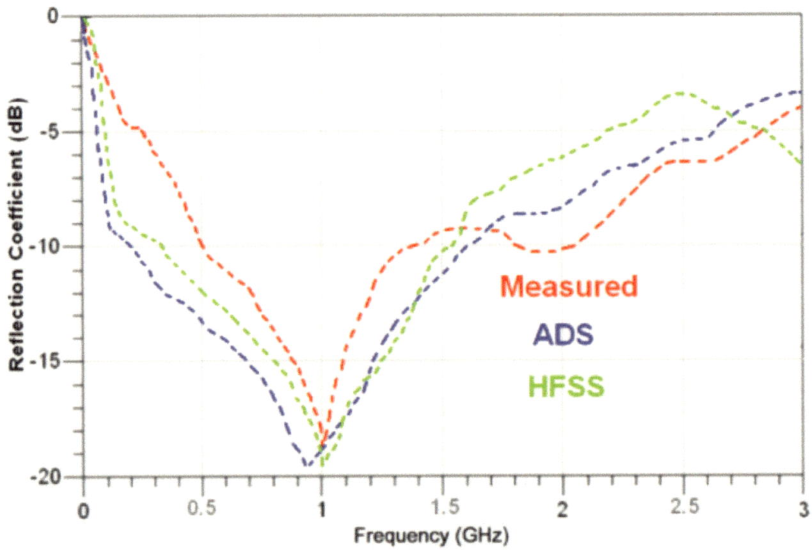

Figure 8. Simulated and measured S_{11} of Antenna#3.

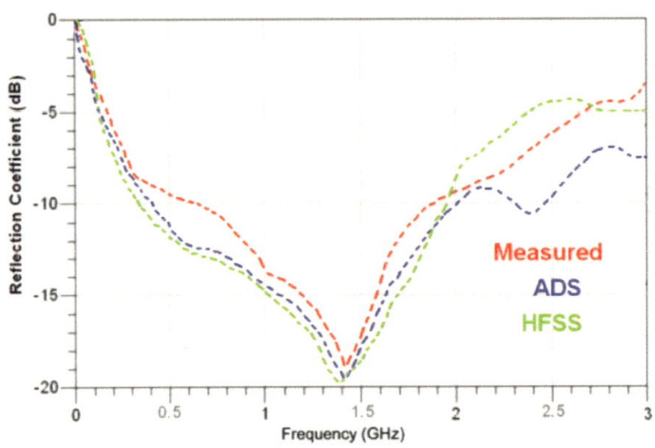

Figure 9. Simulated and measured S_{11} of Antenna#4.

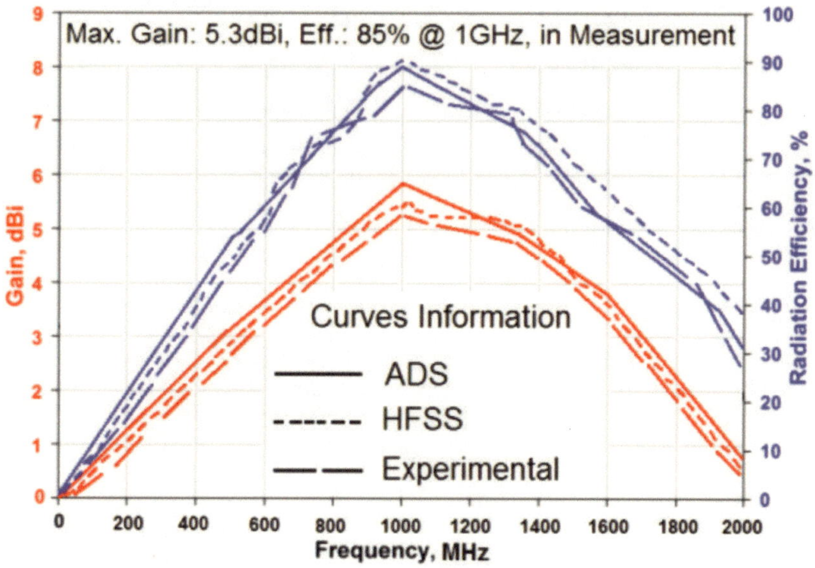

Figure 10. Gain and efficiency of Antenna#3.

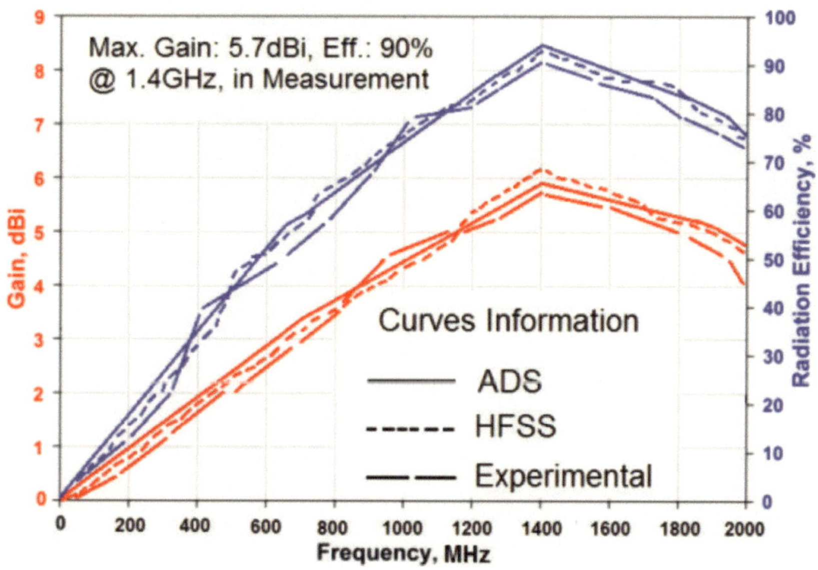

Figure 11. Gain and efficiency of Antenna#4.

Freq. (GHz)	0.5	1	1.35
ADS	3.1 dBi/53%	5.8 dBi/88%	4.9 dBi/75%
HFSS™	2.8 dBi/51%	5.5 dBi/90%	5.1 dBi/79%
Measured	2.6 dBi/48%	5.3 dBi/85%	4.7 dBi/73%

Table 4. Radiation characteristics (gain/efficiency) of Antenna#3.

Freq. (GHz)	0.65	1.4	1.85
ADS	3.2 dBi/57%	5.9 dBi/94%	5.3 dBi/81%
HFSS™	3.0 dBi/55%	6.1 dBi/93%	5.2 dBi/80%
Measured	2.8 dBi/51%	5.7 dBi/90%	4.9 dBi/78%

Table 5. Radiation characteristics (gain/efficiency) of Antenna#3.

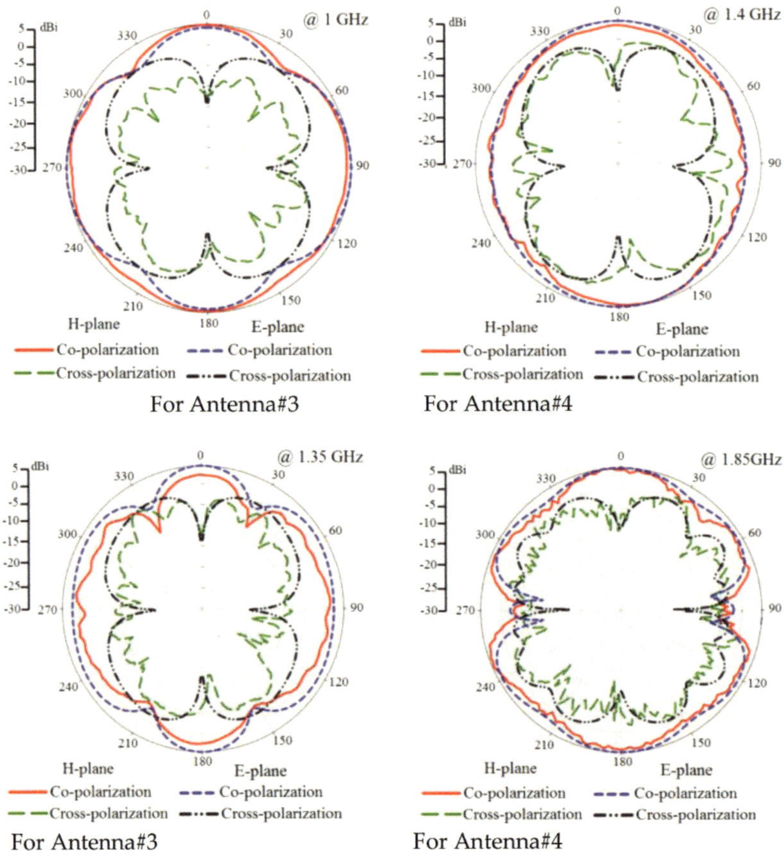

Figure 12. Measured E- and H-plane co- and cross-polarization radiation patterns.

Reference	Dimensions	Bandwidth	Gain (max)	Eff. (max)
[27] b-shaped antenna with 4 × UC	*ES:* $0.047\lambda_0 \times 0.021\lambda_0 \times 0.002\lambda_0$ at 1GHz *PHS:* 14.2 × 6.32 × 0.8 mm^3	104.76% (1–3.2 GHz)	2.3 dBi	62%
[27] b-shaped antenna with 6 × UC	*ES:* $0.051\lambda_0 \times 0.016\lambda_0 \times 0.002\lambda_0$ at 800 MHz *PHS:* 19.2 × 6.32 × 0.8 mm^3	123.8% (0.8–3.4 GHz)	2.8 dBi	70%
[28] J-shaped antenna with 8 × UC	*ES:* $0.564\lambda_0 \times 0.175\lambda_0 \times 0.02\lambda_0$ at 7.5GHz *PHS:* 22.6 × 7 × 0.8 mm^3	84.23% (7.25–17.8 GHz)	2.4 dBi	48.2%

http://dx.doi.org/10.5772/66674

Reference	Dimensions	Bandwidth	Gain (max)	Eff. (max)
[28] I-shaped antenna with 7 × UC	*ES:* $0.556\lambda_0 \times 0.179\lambda_0 \times 0.041\lambda_0$ at 7.7GHz *PHS:* $21.7 \times 7 \times 1.6$ mm^3	87.16% (7.8–19.85 GHz)	3.4 dBi	68.1%
E-shaped slit antenna with 2 × UC	*ES:* $0.017\lambda_0 \times 0.006\lambda_0 \times 0.001\lambda_0$ at 500 MHz *PHS:* $10.2 \times 3.9 \times 0.8$ mm^3	91.89% (0.5–1.35 GHz)	5.3 dBi	85%
E-shaped slit antenna with 3 × UC	*ES:* $0.028\lambda_0 \times 0.008\lambda_0 \times 0.001\lambda_0$ at 650 MHz *PHS:* $13.2 \times 3.9 \times 0.8$ mm^3	96% (0.65–1.85 GHz)	5.7 dBi	90%

UC, unit cells; ES, electrical size; PHS, physical size.

Table 6. Antenna characteristics.

4. Wideband antenna using U-shaped slit MTM unit cells

A novel and compact planar antenna is described that promotes size reduction and bandwidth extension. The antenna is essentially a rectangular patch which is loaded with six MTM unit cells, where each unit cell consists of a U-shaped dielectric slit and a spiral conductor that is grounded using a metallic via-hole, as shown in **Figure 13**. The antenna is terminated on the right-hand side to a matched load of 20 Ω using SMD1206. The load is terminated to the ground-plane through a via-hole.

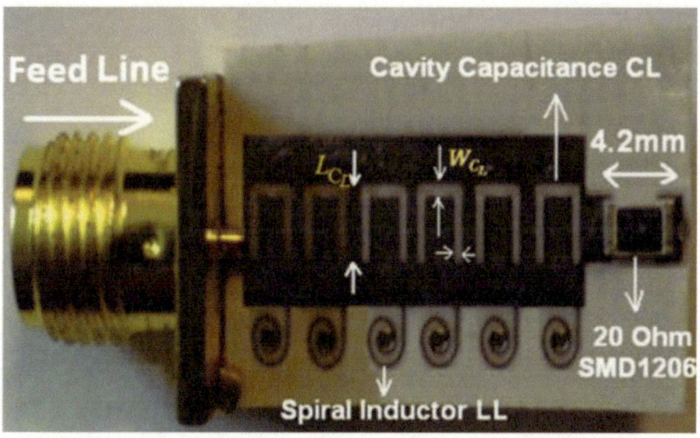

Figure 13. Fabricated prototype MTM antenna.

The U-shaped slit and spiral act like left-handed series capacitance (C_L) and shunt inductance (L_L), respectively. The current flow on the antenna structure and the voltage gradient created between the antenna and the ground-plane induces parasitic right-handed (RH) series inductance (L_R) and shunt capacitance (C_R), respectively. The equivalent circuit model of the MTM unit-cell is shown in **Figure 14**. In addition to the four reactive components (C_L, L_L, L_R, and C_R), right-handed lossy components R_R and G_R and left-handed lossy components G_L and R_L are included, which account for the dielectric loss associated with C_L and the ohmic loss associated with L_L. The metamaterial antenna was fabricated on a RT/duroid™ RO4003 substrate with dielectric constant of 3.38, thickness of 0.8 mm, and tanδ = 0.0022. The magnitudes of the unit-cell parameters were determined from simulation and these are: C_L = 3.2 pF, L_L = 4.5 nH, C_R = 1.5 pF, L_R = 3.44 nH, G_L = 5.6 S, G_R = 3.2 S, R_L = 6 Ω, and R_R = 4.2 Ω.

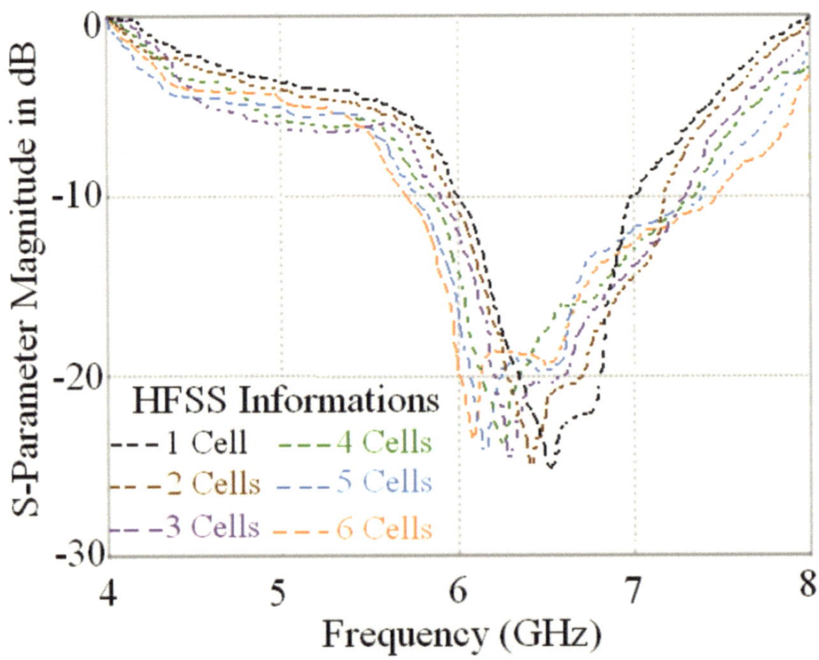

Figure 14. Reflection-coefficient response of the MTM antenna for various number of unit cells.

Trade-off between the antenna size, bandwidth, and radiation properties was used to determine the number of MTM unit cells. In the example presented here, the aim was to design and construct an antenna that had a length of 20 mm and operated over a wide frequency range with good unidirectional radiation characteristics. It was necessary to use HFSS™ to optimize the antenna design. The antenna's reflection-coefficient response as a function of number of unit cells is shown in **Figure 15**. Six unit cells provide the widest impedance bandwidth of 1.75 GHz for a reflection-coefficient of 10 dB. Therefore, six unit cells were used here in the antenna design.

The antenna's reflection-coefficient (simulated and measured) is shown in **Figure 16**. The impedance bandwidth of the antenna is 29%, 26.8%, and 26.6% using ADS, HFSS™, and CST MWS, respectively. The measured impedance bandwidth is 1.5 GHz (5.8–7.3 GHz) for $|S11|$ <–10 dB, which corresponds to 23.7%. There is 13.7% differential between the averaged simulation and measurement results.

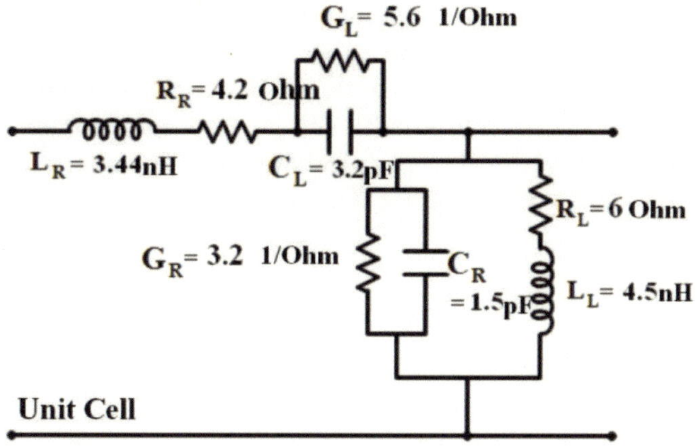

Figure 15. Equivalent circuit model of the MTM unit cell.

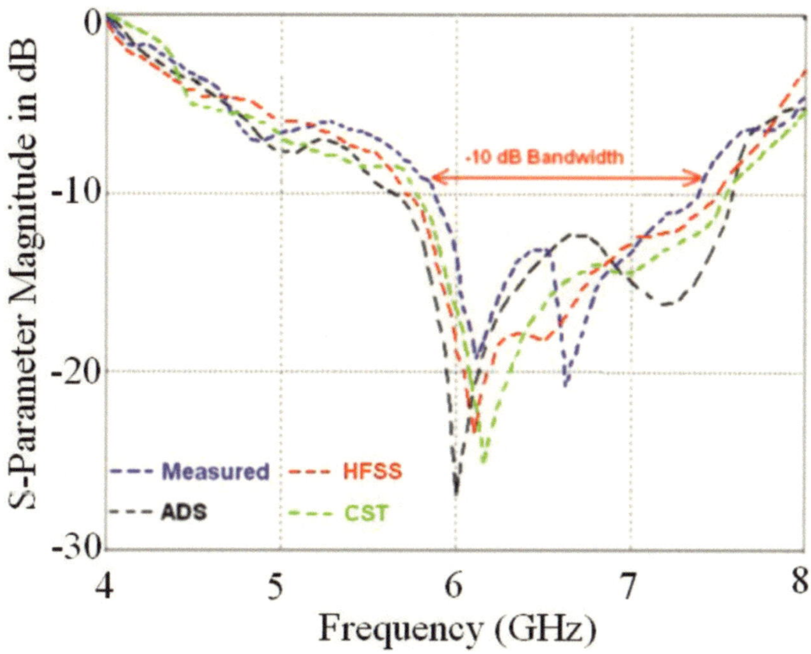

Figure 16. Measured and simulated of $|S_{11}|$ of the MTM antenna.

Future wireless systems require antennas that possess good radiation characteristics such as gain and efficiency in addition to compact size and wide bandwidth. The effective aperture of the antenna determines its gain and radiation efficiency. The magnitude of these two characterizing parameters can be improved simply by increasing the antenna's effective cross-sectional area. However, this traditional technique increases the dimensions of the antenna which is undesirable. **Figure 17** shows the effective aperture of the antenna can be increased by loading it with several MTM unit cells without increasing the antenna size. Results show that with four unit cells the gain and efficiency of the antenna at 7 GHz are 4.94 dBi and 74%, respectively. In fact, by increasing in the number of MTM unit cells from four to six improves the gain to 6.1 dBi and the efficiency to 85%. **Table 7** provides the parameters of the optimized antenna and its equivalent electrical circuit. The measured E-plane and H-plane radiation pattern at 5.8 GHz, 6.6 GHz, and 7.3 GHz is shown in **Figure 18**. The antenna radiates unidirectionally with 3 dB angular beamwidth of 90 degrees.

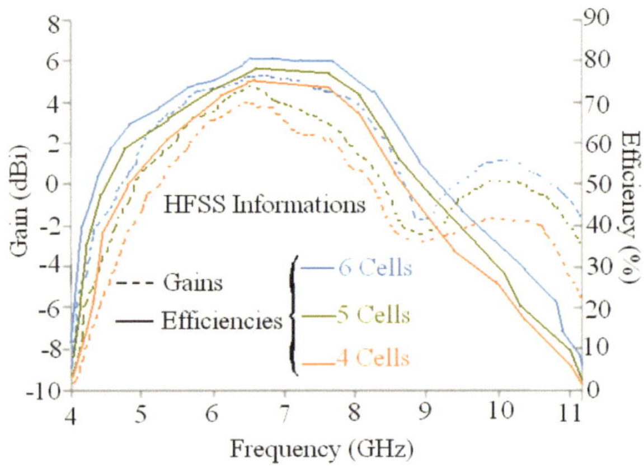

Figure 17. Gain and efficiency performance as a function of number of MTM unit cells.

Number of unit cells	**6**
Length (L_{C_L})	2.50 mm
Width (W_{C_L})	0.50 mm
Gap between slits	0.60 mm
Width of spiral	0.25 mm
Spacing of spirals	0.25 mm
Spiral turns	2
Height of via-hole	0.80 mm
Length of SMD1206	4.20 mm
SMD1206	20 Ω
C_L	3.2 pF
L_L	4.5 nH
C_R	1.5 pF
L_R	3.4 nH
G_L	5.6 S

http://dx.doi.org/10.5772/66674

Number of unit cells	6
G_R	3.2 S
R_L	6.0 Ω
R_R	4.2 Ω

Table 7. Dimensions of the MTM antenna and parameter values.

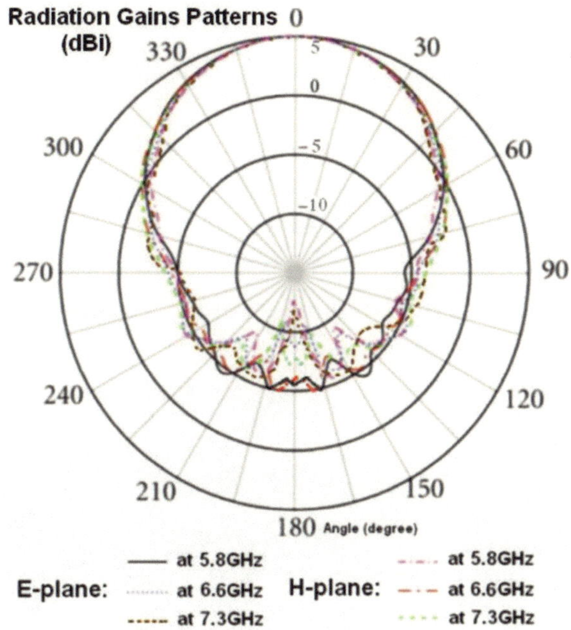

Figure 18. E- and H-plane radiation patterns at 5.8, 6.6, and 7.3 GHz.

5. UWB antenna using T-shaped slit MTM unit cells

In this section, a miniature and compact antenna design is described for integration in UWB wireless systems. The antenna exhibits good overall performance in terms of radiation patterns, gain, and efficiency. The proposed antennas are based on metamaterial unit cells that are implemented using T-shaped dielectric slits that were etched directly on the radiating patch, and include a spiral stub which is grounded using a via-hole. The T-shaped slit behaves as a left-handed series capacitance, and the grounded spiral acts as a left-handed shunt inductance. Results show just two unit cells were sufficient to realize the desired antenna performance.

The equivalent circuit model of the T-shaped slit antenna is based on the composite right-/left-handed transmission-line structure shown in **Figure 19**. Standard printed circuit board manufacturing techniques were employed in the fabrication of the left-handed series capacitors (C_L) and the left-handed shunt inductors (L_L). The T-shaped dielectric slits and the spiral-shaped conductors were implemented on the radiation patch. The MTM unit-cell consists of two T-shaped dielectric slits etched on a rectangular patch with a conductive spiral located between the two slits. The spiral is grounded using a metallic via-hole. This technique substantially reduces the size of the UWB antenna because, unlike conventional antennas, its size is not dependent on the operating wavelength. The parasitic right-handed series inductance (L_R) and shunt capacitance (C_R) result from the current flowing over the antenna and the voltage gradient created between the microstrip and the ground-plane.

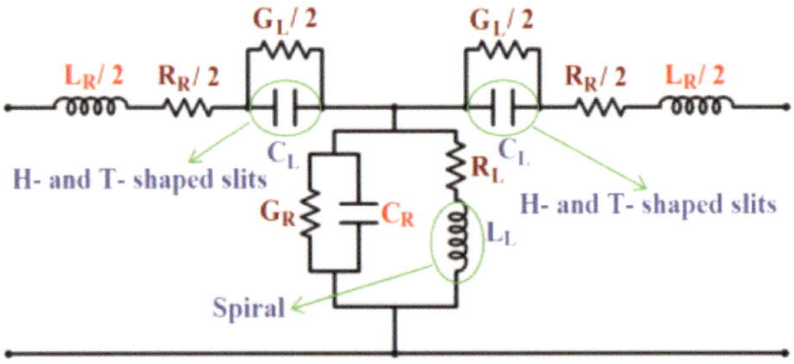

(a) Equivalent circuit model of the T-shaped slit MTM unit cell.

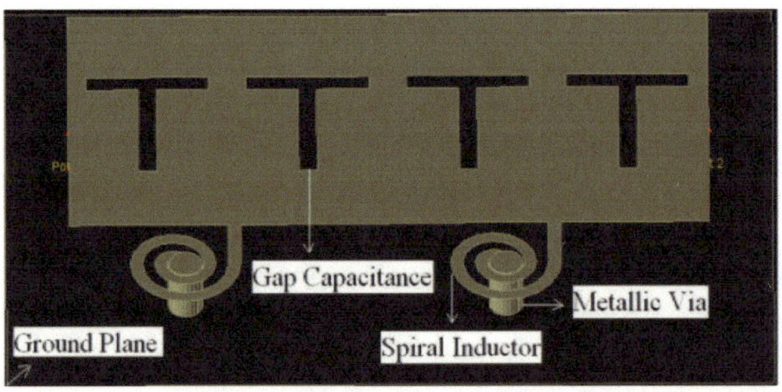

(b) Simulation model of the T-shaped slit antenna

(c) Fabricated antenna prototype

Figure 19. (a) Equivalent circuit model of the T-shaped slit MTM unit cell (b) Simulation model of the T-shaped slit antenna (c) Fabricated antenna prototype.

The proposed antenna was designed and fabricated on RT/duroid® RO4003 substrate with dielectric constant of 3.38 and 0.8 mm thickness. **Figure 19(c)** shows the configuration of the antenna, which is excited from left-hand side through a 50 Ω microstrip feed-line. The right-hand side of the patch is terminated with a matched load of 50 Ω (SMD1206) that is connected to the ground-plane through a metallic via-hole. The total electrical length, width, and height of antenna are $0.05\lambda_0$, $0.02\lambda_0$, and $0.002\lambda_0$, which correspond to 15.5 mm, 6.9 mm, and 0.8 mm, respectively. The optimized parameters of the antenna are given in **Table 8**, whose equivalent circuit parameters are C_L, L_L, C_R, and L_R are 5 pF, 6.4 nH, 1 pF, and 2.8 nH, respectively.

Length of T slits	3.0 mm
Width of T slits	0.4 mm
Distance between slits	0.4 mm
Width of spirals	0.2 mm
Spacing of spirals	0.2 mm
Turns of spirals	2
Height of via-hole	0.8 mm
Length of 50 Ω load (SMD1206)	4.2 mm

Table 8. T-shaped slit MTM antenna parameters.

The antenna resonates at three frequencies, i.e., 2, 3.7, and 5.8 GHz, as shown in **Figure 20**. The measured gain and efficiency of antenna have a maximum value of 7.1 dBi and 91%, respectively, at 3.7 GHz.

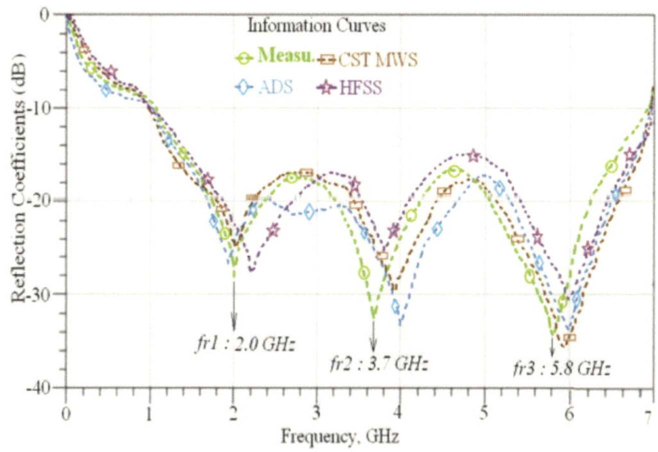

Figure 20. Simulated and measured reflection-coefficient response of the T-shaped slit antenna.

The measured radiation patterns at three spot frequencies in **Figure 21** show the antenna radiates directionally. The measured gain and efficiency response of the antenna in **Figure 22** show the antenna operates from 0 to 7 GHz with a maximum gain and efficiency of 7.1 dBi and 91%, respectively, at 3.7 GHz.

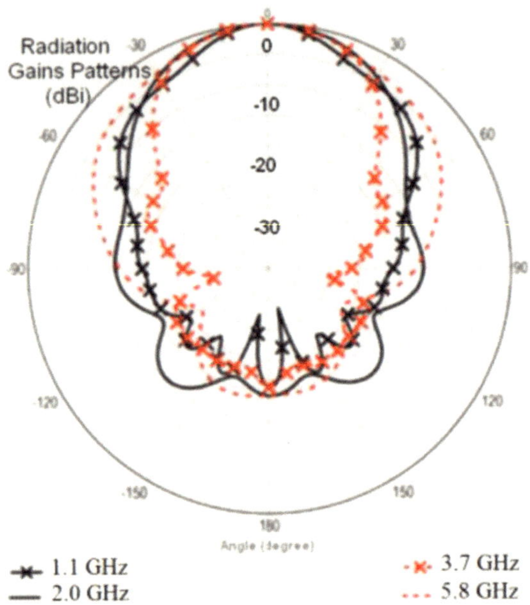

Figure 21. Measured radiation patterns of the T-shaped slit antenna.

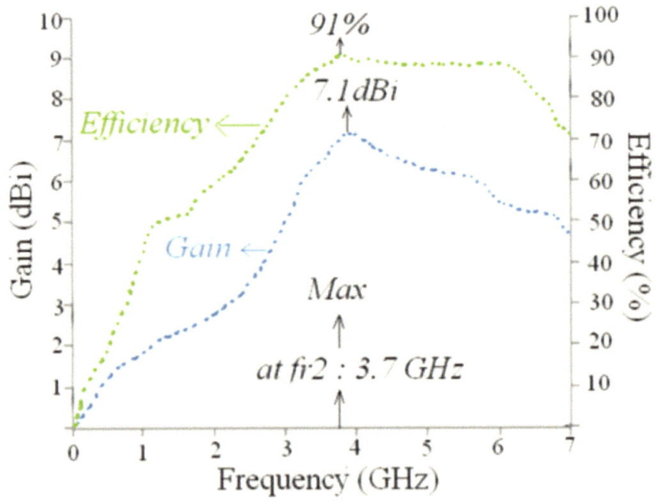

Figure 22. Measured gain and efficiency response of the T- shaped slit MTM antenna.

Comparison of the proposed antennas with other conventional antennas is given in **Table 9**. It is evident the T-shaped antenna offers superior performance.

The effect of the slit dimensions on the antenna characteristics was investigated. It is evident from **Figure 23** that increasing the length and width of slits improves the antenna's impedance bandwidth and matching performance. In fact, the bandwidth improves by 25% from 119 to 149% for increase in slit length from 1 to 3 mm, and width from 0.2 to 0.4 mm.

http://dx.doi.org/10.5772/66674

The number of slits and spiral were also investigated. By increasing the number of slits in each unit-cell, the number of spirals had a positive impact on the antenna's bandwidth and matching properties, as shown in **Figure 24**. It is also observed that by increasing the number of slits causes the number of resonance frequencies to increase as well as its bandwidth. The gain and radiation efficiency of the antenna as a function of slit dimensions, the number of slits in the unit-cell, and number of spirals were also investigated. From **Figures 25** and **26**, it is evident that by increasing the length and width of the slits, and increasing the number of slits, the gain and radiation efficiency increase considerably. This is attributed to increase in the antenna aperture.

Ref.	Dimensions	Fractional bandwidth	Max. gain	Max. eff.
[27]-*a*	$0.04\lambda_0 \times 0.021\lambda_0 \times 0.002\lambda_0$	105%	2.3 dBi	62%
[27]-*b*	$0.05\lambda_0 \times 0.01\lambda_0 \times 0.002\lambda_0$	124%	2.8 dBi	70%
[28]-*a*	$0.45\lambda_0 \times 0.17\lambda_0 \times 0.02\lambda_0$	74%	2.1 dBi	44%
[28]-*b*	$0.42\lambda_0 \times 0.17\lambda_0 \times 0.041\lambda_0$	83%	3.1 dBi	59%
[19]	$0.44\lambda_0 \times 0.22\lambda_0 \times 0.008\lambda_0$	18%	2.2 dBi	17%
[29]	$0.24\lambda_0 \times 0.3\lambda_0 \times 0.009\lambda_0$	8%	1.5 dBi	58%
T-slit ant.	$0.05\lambda_0 \times 0.02\lambda_0 \times 0.002\lambda_0$	144%	7.1 dBi	91%

Table 9. Comparison of the proposed antennas.

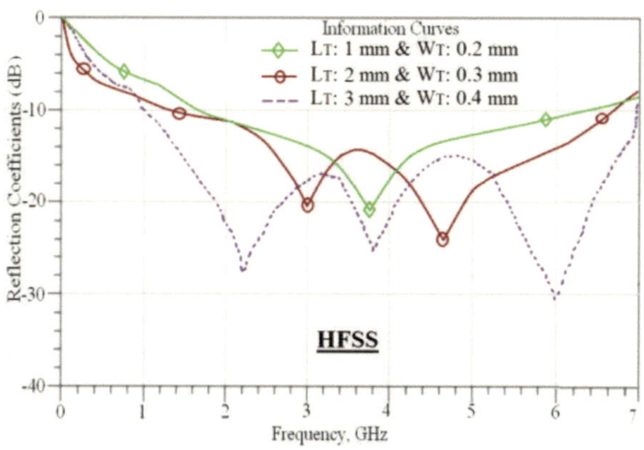

Figure 23. Effect of slit length and width on the antenna bandwidth.

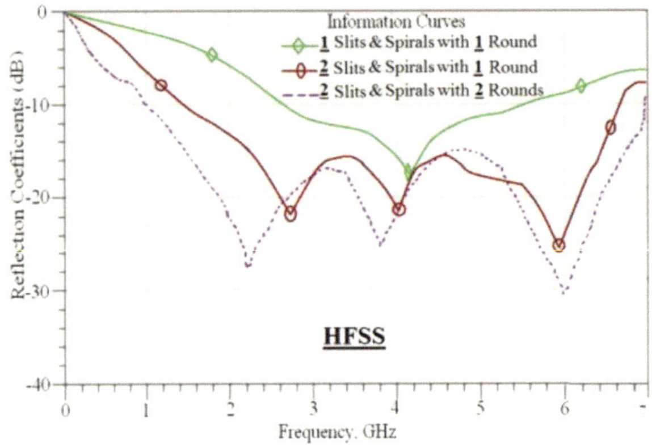

Figure 24. Impedance bandwidth as a function of number of slits in each of the unit cells and number of spirals.

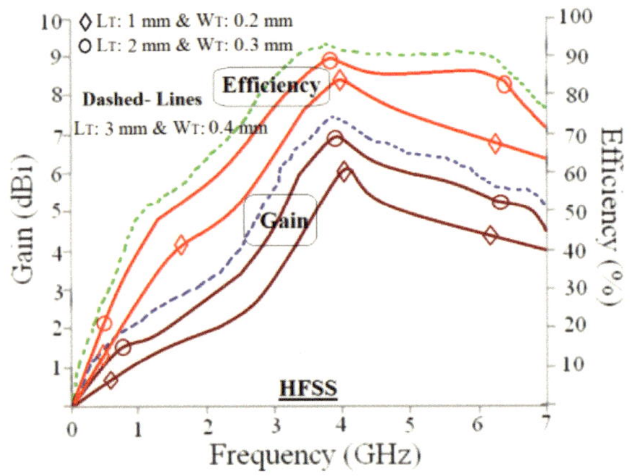

Figure 25. Gain and radiation efficiency as a function of slit dimensions.

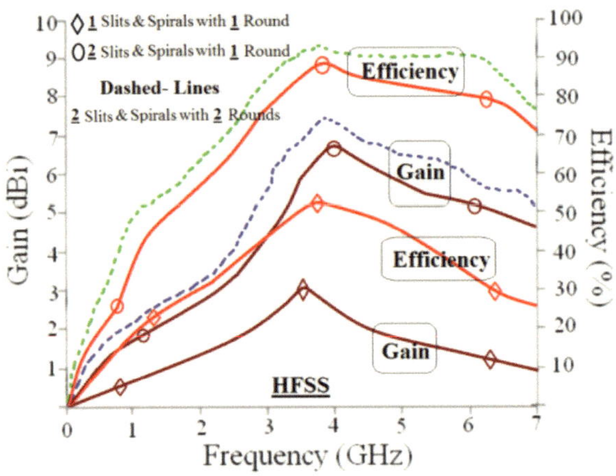

Figure 26. Gain and radiation efficiency as a function of number of slits and number of spirals.

http://dx.doi.org/10.5772/66674

Acknowledgements

The authors would like to give their special thanks to faculty of microelectronics for the financial supports.

Author details

Mohammad Alibakhshikenari[1*], Mohammad Naser-Moghadasi[2], Ramazan Ali Sadeghzadeh[3], Bal Singh Virdee[4] and Ernesto Limiti[1]

*Address all correspondence to: Alibakhshikenari@ing.uniroma2.it

1 Department of Electronic Engineering, University of Rome Tor Vergata, Rome, Italy

2 Faculty of Engineering, Science and Research Branch, Islamic Azad University, Tehran, Iran

3 Faculty of Electrical Engineering, K. N. Toosi University of Technology, Tehran, Iran

4 London Metropolitan University, Center for Communications Technology, London, UK

References

[1] P. Gao, S. He, X. Wei, Z. Xu, N. Wang, and Y. Zheng, "Compact Printed UWB Diversity Slot Antenna with 5.5 GHz Band-Notched Characteristics", *IEEE Antennas Wireless Propag. Lett.,*" 2014; 13: 376–379. DOI: 10.1109/LAWP.2014.2305772

[2] A.A. Kalteh, G.R. Dadash-Zadeh, M. Naser-Moghadasi, and B.S. Virdee, "Ultra-Wideband Circular Slot Antenna With Reconfigurable Notch Band Function," *IET Microwave Antennas Propag.*, 2012; 6: 108–112.

[3] L. Li, Z.L. Zhou, J.S. Hong, and B.Z. Wang, "Compact Dual-Band Notched UWB Planar Monopole Antenna with Modified SRR," *Electron Lett.*, 2011; 47: 950–951.

[4] M. Alibakhshi-Kenari, M. Naser-Moghadasi, and R.A. Sadeghzadah, "The Resonating MTM Based Miniaturized Antennas for Wide-Band RF-Microwave Systems," *Microwave Opt. Technol. Lett.*, 2015; 57: 2339–2344.

[5] J. Long, and D.F. Sievenpiper, "A Compact Broadband Dual-Polarized Patch Antenna for Satellite Communication/Navigation Application," *IEEE Antennas Wireless Propag. Lett.*, 2015; 14: 273–276.

[6] Nasimuddin, Y.S. Anjani, and A. Alphones, "A Wide-Beam Circularly Polarized Asymmetric-Microstrip Antenna," *IEEE Trans. Antennas Propag.*, 2015; 62(12): 6506–6509.

[7] Nasimuddin, X.M. Qin, and Z.N. Chen, "A Compact Circularly Polarized Slotted Patch Antenna for GNSS Applications," *IEEE Trans. Antennas Propag.*, 2015; 63(8): 3764–3768.

[8] K. Agarwal, Nasimuddin, and A. Alphones, "Triple-Band Compact Circularly Polarised Stacked Microstrip Antenna Over Reactive Impedance Meta-Surface for GPS Applications," *IET Microwaves Antennas Propag.*, 2014; 8(13): 1057–1065.

[9] C. Zhou, G. Wang, Y. Wang, B. Zong, and J. Ma, "CPW-Fed Dual-band Linearly and Circularly Polarized Antenna Employing Novel Composite Right/Left-Handed Transmission-Line," *IEEE Antennas Wireless Propag. Lett.* 2013; 12: 1073–1076.

[10] M. Alibakhshi-Kenari, M. Naser-Moghadasi, B.S. Virdee, A. Andujar, and J. Anguera, "Compact Antenna Based on a Composite Right/Left Handed Transmission Line," *Microwave Opt. Technol. Lett.*, 2015; 57:1785–1788.

[11] M. Alibakhshi-Kenari and M. Naser-Moghadasi, "Novel UWB Miniaturized Integrated Antenna Based on CRLH Metamaterial Transmission Lines," *AEU Int. J. Electron Commun.*, 2015; 69: 1143–1149.

[12] M. Alibakhshi-Kenari, M. Naser-Moghadasi, and R.A. Sadeghzadah, "On the Bandwidth and Radiation Specifications Enhancement of Monopole Antennas Loaded with Split Ring Resonators," *IET Microwaves Antennas Propag.*, 2015; 9(14): 1487–1496. DOI: 10.1049/iet-map.2015.0172

[13] R.W. Ziolkowski and A.D. Kipple, "Application of Double Negative Materials to Increase the Power Radiated by Electrically Small Antennas," *IEEE Trans. Antennas Propag.*, 2003; 51(10): 2626–2640.

[14] G.V. Eleftheriades, A. Grbic, and M. Antoniades, "Negative-Refractive-Index Transmission-Line Metamaterials and Enabling Electromagnetic Applications," *Proc. IEEE Antennas Propag. Int. Symp.*, 2004: 1399–1402.

[15] R.W. Ziolkowski, P. Jin, and C.-C. Lin, "Metamaterial-Inspired Engineering of Antennas," *Proc. IEEE*, 2011; 99(10): 1720–1731.

[16] M. Alibakhshi-Kenari, "Design and Modeling of New UWB Metamaterial Planar Cavity Antennas with Shrinking of the Physical Size for Modern Transceivers," *Int. J. Antennas Propag.*, 2013; Article ID 562538: 12p. DOI: 10.1155/2013/562538

[17] Mohammad Alibakhshi-Kenari, Masoud Movahhedi and Hadi Naderian, "A New Miniature Ultra Wide Band Planar Microstrip Antenna Based on the Metamaterial Transmission Line" 2012 *IEEE Asia- Pacific Conference on Applied Electromagnetics* (APACE 2012), December 11-13, 2012, Melaka, Malaysia.

[18] M. Alibakhshi-kenari and M. Naser-Moghadasi, "Novel CRLH-CP Antenna with the Capability to be Integrated Inside RF Components for RF Electronic Devices and Embedded Systems," *J. Appl. Comput. Electromagn. Soc.*, 2015; 30(6): 653–659.

[19] C.J. Lee, M. Achour, and A. Gummalla, "Compact Metamaterial High Isolation MIMO Antenna Subsystem," *Microwave Conference*, 2008. APMC 2008. Asia-Pacific, 6-20 Dec. 2008, pp, 1-4.

[20] C.J. Lee, K.M.K.H. Leong, and T. Itoh, "Broadband Small Antenna for Portable Wireless Application," *International Workshop on Antenna Technology: Small Antennas and Novel Metamaterials*, 2008: 10–13.

[21] A. Lai, C. Caloz, and T. Itoh, "Composite Right/Left-Handed Transmission Line Metamaterials," *IEEE Microwave Mag.*, 2004; 5(4): 34–50.

[22] C.-J. Lee, K.M.K.H. Leong, and T. Itoh, "Composite Right/Left-Hand-Handed Transmission Line Based Compact Resonant Antennas for RF Module Integration," *IEEE Trans. Antennas Propagat.*, 2006; 54(8): 2283–2291.

[23] C. Caloz and T. Itoh, "Application of the Transmission Line Theory of Left-Handed (LH) Materials to the Realization of a Microstrip "LH Line"," *Proc. IEEE Antennas Propag. Soc. Int. Symp.*, 2002; 2: 412–415.

[24] Christophe Caloz and Tatsuo Itoh, "Electromagnetic Metamaterials: Transmission Line Theory and Microwave Applications", ISBN: 978-0-471-66985-2, 376 pages, December 2005, Wiley-IEEE Press.

[25] A. Sanada, C. Caloz, and T. Itoh, "Planar Distributed Structure with Negative Refractive Index," *IEEE Trans. Microwave Theory Tech.*, 2004; 52(4): 1252–1263.

[26] C. Caloz, T. Itoh, and A. Rennings, "CRLH Traveling-Wave and Resonant Metamaterial Antennas," *Antennas Propagat. Mag.*, 2008; 50(5): 25–39.

[27] Mohammad Alibakhshi-Kenari, "Printed planar patch antennas based on metamaterial", *International Journal of Electronics Letters*, Volume 2, Issue 1, Jan 2014, pp 37-42.

[28] M. Alibakhshi-Kenari, "Introducing the New Wide Band Small Plate Antennas with Engraved Voids to Form New Geometries Based on CRLH MTM-TLs for Wireless Applications," *Int. J. Microwave Wireless Tech.*, 2014; 6(06): 629–637. DOI: http://dx.doi.org/10.1017/S1

[29] C.-C. Yu, M.-H. Huang, L.-K. Lin, Y.-T. Chang, "A Compact Antenna Based on Metamaterial for WiMAX," *Asia-Pacific Microwave Conference*, 16-20 Dec. 2008, pp. 1–4.

Chapter 7

Homogenization of Thin and Thick Metamaterials and Applications

Abdelwaheb Ourir, Yao Gao, Agnès Maurel and Jean-Jacques Marigo

Additional information is available at the end of the chapter

http://dx.doi.org/10.5772/66035

Abstract

The wave propagation in structures involving metamaterials can be described owing to homogenization approaches which allow to replace the material structured at the subwavelength scale by an equivalent and simpler, effective medium. In its simplest form, homogenization predicts that the equivalent medium is homogeneous and anisotropic and it is associated to the usual relations of continuity for the electric and magnetic fields at the boundaries of the metamaterial structure. However, such prediction has a range of validity which remains limited to relatively thick devices and it is not adapted to more involved geometries (notably three-dimensional). The following two aspects are considered: (i) we study how the homogenization at the leading order can be improved when the thickness of the device becomes small and (ii) we propose a heuristic extension of the solution given by the leading order homogenization in order to deal with a complex geometry; in the latter case, an application to a demultiplexer device is proposed.

Keywords: metamaterial, homogenization, surface waves, spoof plasmons, guided waves, multiplexing, subwavelength devices

1. Introduction

Metamaterials are artificial materials composed by the periodic arrangement of a unit cell; among the different materials in the unit cell, stratified or layered media involving metallic layers have been extensively studied and throughout this chapter, metallic material is thought in the far infrared, thus in a frequency range where metal behaves as an opaque medium, at the boundaries of which Neumann boundary condition applies. Such metallic arrays enter in the design of metallo-dielectric structures, as the artificial magnetic conductors used in the design of antenna and there is currently renewed interest in such array since they are the key piece of so-called metasurfaces (with a typical resonant behavior as in the mushroom structure of Sievenpiper [1]). In addition to be able to produce unexpected scattering of electromagnetic waves, these structures can support guided waves and it is for this property that they have been proposed by Pendry et al. [2]. In this context, these guided waves have been called "spoof plasmons" since they mimic, in the far infrared regime, the behavior of plasmons observed in the visible range (plasmons are the wave guided at the flat interface between air and metal and this requires a negative permittivity).

Because of their periodic subwavelength structuration, homogenization approaches are ideal tools to predict within a rigorous mathematical framework the scattering properties of these devices. In this chapter, we will present results coming from homogenization techniques. Our starting point is the simplest homogenization; owing to the small parameter $\eta = kh$, with k the wavenumber and h the typical spacing of the periodic structuration, we call simplest homogenization the homogenization performed at the leading order in η, which is the 0 order. Such homogenization predicts that a stratified medium can be replaced by an equivalent homogeneous and anisotropic medium associated to the usual conditions of continuity of the electromagnetic fields at the boundaries between two media; it has been used for the practical realization of several metamaterial devices, notably in the context of cloaking [3, 4]. Starting from this classical homogenization, which regards the effect of wave propagation in the bulk of the stratified medium, we will focus on two refinements that may be needed to get accurate results: (i) the case of a metallic array of small thickness for which boundary layer effects can become significant or even dominant when compared to the effect of propagation in the bulk; this case requires to conduce the homogenization up to order 1, which does not affect the equation of propagation in the bulk but makes new conditions at the boundaries to appear (these new conditions are called jump conditions), (ii) the case where the metamaterial device is composed of a succession of metallic elements arranged in a three-dimensional geometry. Although homogenization techniques could be used considering the whole device, the resolution may become tricky. Rather, a heuristic extension of the result coming from the classical homogenization is proposed, based on the analysis of the dispersion relations in each part of the whole device.

These two aspects are presented theoretically and the main theoretical results (in terms of the scattering properties or in terms of the dispersion relations of guided waves) are supported by experimental results. As a practical application, the realization of a demultiplexer is proposed, allowing for the frequency selection of an incoming large band signal into different "colored" channels.

2. The starting point: homogenization of metallic arrays at the leading order (order 0)

The present derivation considers first a stratified medium, with the layers made of a transparent media, afterwards the limiting case of metallic arrays in the far infrared is considered (thus, with metallic layers being associated to Neumann boundary conditions at its boundaries). Also, we restrict ourselves to polarized TM waves, for which the transverse magnetic field $H(x)$ ($x = (x_1,x_2)$) is polarized along $\mathbf{e}_3$ and the whole system is invariant by translation along

http://dx.doi.org/10.5772/66035

x_3 (thus, $\partial_{x_3} = 0$). In the harmonic regime, the time dependence of H is $e^{-i\omega t}$ and it is omitted in the following.

For a succession of layers made of transparent media, the wave equation reads

$$\text{div}\left[\frac{1}{\epsilon}\nabla H\right] + k^2\mu H = 0, \tag{1}$$

with $k = \sqrt{\epsilon_0\mu_0}\omega$ the wavenumber in the air ((ϵ_0,μ_0) are the permeability and permittivity of the air and ω the frequency). In Eq. (1), (ϵ,μ) denote the relative permeability and permittivity and they are spatially dependent. At each boundary between two layers, the continuities of H and of $1/\epsilon\ \partial_n H$ apply (with ∂_n the normal derivative).

Next, Eq. (1) can be written in an equivalent form, introducing the field $\boldsymbol{C}$

$$\text{div}\,\boldsymbol{C} + k^2\mu H = 0, \quad \boldsymbol{C} \equiv \frac{1}{\epsilon}\nabla H, \tag{2}$$

with boundary conditions being the continuity of H and of $\mathbf{C.n}$ at the boundaries between two layers (note that the field $\boldsymbol{C}$ is linked to the electric field, see [5]).

2.1. The asymptotic analysis

The idea is to define three regions where different asymptotic expansions will be used, Eq. (3), with respect to the small parameter $\eta = kh$ (with h the periodic spacing of the layers). The inner region contains the boundary between the stratified medium and the air. The two outer regions for $x_1 > 0$ and $x_1 < 0$ are the regions far enough the interface, where the evanescent field can be neglected, while the inner region contains the evanescent field. Next, the inner region and the outer regions are connected using so-called matching conditions, which will constitute the boundary conditions for the outer solutions (see **Figure 1**).

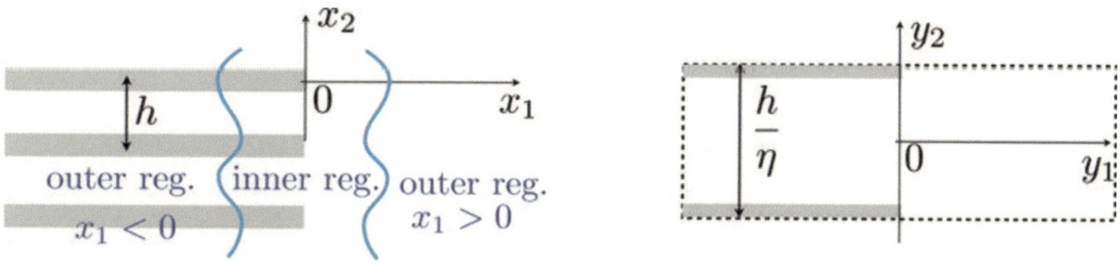

Figure 1. On the left, configuration in the real space with $\mathbf{x} = (x_1,x_2)$ coordinate; the periodicity along x_2 is h; the inner region corresponds to the neighborhood of the boundary between the stratified medium (which occupies the half-space $x_1 < 0$) and the substrate, being the air ($x_1 > 0$). On the right, the unit cell (inner region) in $\mathbf{y} = (y_1,y_2)$ coordinate, with $\mathbf{y} = \mathbf{x}/\eta$. The size of the unit cell along y_2 is h/η with $\eta \equiv kh$.

2.1.1. *The outer and inner expansions*

The asymptotic expansions are thought with spatial dependences on a *macroscopic* coordinate $\mathbf{x}$ associated with low variations of the fields (with the typical scale $1/k$ of the wave) and a *microscopic* coordinate **y** associated with rapid variations (with the typical scale h of the layers) and in each region, we keep the coordinates that are relevant to describe the variations of the field. To do so and with $\eta \equiv kh \ll 1$, we define $\mathbf{y} \equiv \mathbf{x}/\eta$ and we assume that (H, **C**) can be expanded by using the following asymptotic expansions

$$\begin{cases} \text{outer region } x_1 > 0, & H = H^0(\mathbf{x}) + \eta H^1(\mathbf{x}) + \ldots, \\ & \mathbf{C} = \mathbf{C}^0(x) + \eta \mathbf{C}^1(\mathbf{x}) + \ldots, \\ \text{outer region } x_1 < 0, & H = H^0(\mathbf{x},y_2) + \eta H^1(\mathbf{x},y_2) + \ldots, \\ & \mathbf{C} = \mathbf{C}^0(\mathbf{x},y_2) + \eta \mathbf{C}^1(\mathbf{x},y_2) + \ldots, \\ \text{inner region,} & H = h^0(x_2,\mathbf{y}) + \eta h^1(x_2,\mathbf{y}) + \ldots, \\ & \mathbf{C} = \mathbf{c}^0(x_2,\mathbf{y}) + \eta \mathbf{c}^1(x_2,\mathbf{y}) + \ldots, \end{cases} \tag{3}$$

with the outer terms (H^n, $\mathbf{C}^n$) for $x_1 < 0$ and the inner terms (h^n, $\mathbf{c}^n$) being periodic *w.r.t.* y_2. The differential operator reads, in the different regions, as

$$\begin{cases} x_1 > 0, & \nabla \to \nabla_{\mathbf{x}}, \\ x_1 < 0, & \nabla \to \nabla_{\mathbf{x}} + \dfrac{1}{\eta}\dfrac{\partial}{\partial y_2}\mathbf{e}_2, \\ \text{inner region,} & \nabla \to \dfrac{\partial}{\partial x_2}\mathbf{e}_2 + \dfrac{1}{\eta}\nabla_{\mathbf{y}}. \end{cases} \tag{4}$$

2.1.2. The matching conditions

The inner and outer problems have to be associated with boundary conditions or radiation conditions which ensure that the problems are well-posed. Since the outer expansions hold true only far away from the interface, the outer terms do not have to satisfy the continuity conditions at $x_1 = 0$. Reversely, the conditions at infinity satisfied by the inner terms are unknown *a priori*. These missing conditions are provided by the matching conditions, which read at leading order

$$H^0(0^+,x_2) = \lim_{y_1 \to +\infty} h^0(x_2,\mathbf{y}), \tag{5a}$$

$$H^0(0^-,x_2,y_2) = \lim_{y_1 \to -\infty} h^0(x_2,\mathbf{y}), \tag{5b}$$

$$\mathbf{C}^0(0^+,x_2) = \lim_{y_1 \to +\infty} \mathbf{c}^0(x_2,\mathbf{y}), \tag{5c}$$

$$\mathbf{C}^0(0^-,x_2,y_2) = \lim_{y_1 \to -\infty} \mathbf{c}^0(x_2,\mathbf{y}). \tag{5d}$$

2.2. The homogenized wave equation at the leading order

We want to establish the wave equation satisfied by the mean fields $\langle H^0\rangle(x)$ and $\langle \mathbf{C}^0\rangle(x)$ for $x_1 < 0$, where we have defined the average over $y_2 \in Y = (-1/2, 1/2)$. The homogenized wave equation is sought for $x_1 < 0$, only. For $x_1 > 0$, the wave equation is obviously

$$\mathrm{div}_{\mathbf{x}}\boldsymbol{C}^0 + H^0 = 0, \quad \boldsymbol{C}^0 = \nabla_{\mathbf{x}} H^0, \quad \text{for } x_1 > 0, \tag{6}$$

being the same at each order and the fields equal their averages. Eq. (2), at leading order ($1/\eta$), read $\partial_{y_2} \boldsymbol{C}_2^0 = 0 = \partial_{y_2} H^0$, whence

$$\langle H^0 \rangle(\mathbf{x}) = H^0(\mathbf{x}), \quad \langle \boldsymbol{C}_2^0 \rangle(\mathbf{x}) = C_2^0(\mathbf{x}). \tag{7}$$

Now, we establish the relation between $\langle \mathbf{C}^0 \rangle$ and H^0 (this latter being equal to its average). Eq. (2) at order η^0 in the outer problems $x_1 < 0$ give

$$\boldsymbol{C}^0(\mathbf{x}, y_2) = \frac{1}{\epsilon(y_2)} \left[\nabla_{\mathbf{x}} H^0(x) + \frac{\partial H^1}{\partial y_2}(\mathbf{x}, y_2) \mathbf{e}_2 \right], \tag{8}$$

and

$$\mathrm{div}_{\mathbf{x}} \boldsymbol{C}^0(\mathbf{x}, y_2) + \frac{\partial C_2^1}{\partial y_2}(\mathbf{x}, y_2) + \mu(y_2) H^0(\mathbf{x}) = 0. \tag{9}$$

Averaging both equations, with $C^0(\mathbf{x}, y_2) = C_1^0(\mathbf{x}, y_2)\mathbf{e}_1 + C_2^0(\mathbf{x})\mathbf{e}_2$ and owing to the periodicity of H^1 and of C_2^1 w.r.t. y_2 (thus, $\langle \partial_{y_2} H^1 \rangle = 0 = \langle \partial_{y_2} C_2^1 \rangle$), we easily get the homogenized wave equation at the first order

$$\begin{cases} \langle C^0 \rangle(\mathbf{x}) = \left\langle \dfrac{1}{\epsilon} \right\rangle \dfrac{\partial H^0}{\partial x_1}(\mathbf{x})\mathbf{e}_1 + \langle \epsilon \rangle^{-1} \dfrac{\partial H^0}{\partial x_2}(\mathbf{x})\mathbf{e}_2, \\ \mathrm{div}_{\mathbf{x}} \langle \mathbf{C}^0 \rangle + \langle \mu \rangle H^0 = 0. \end{cases} \tag{10}$$

2.3. The continuity relations at the leading order

To the homogenized wave equation (10), we have to associate continuity (or discontinuity) conditions at the interface $x_1 = 0$. To that aim, we have to consider the inner solution and its matching with the two outer solutions. We are looking for the quantities $[\![H^0]\!]$ and $[\![\langle C_1^0 \rangle]\!]$. Eq. (2) for the inner problem tell us, at the leading order (in $1/\eta$), that $\nabla_{\mathbf{y}} h^0 = 0$ from which h^0 does not depend on $\mathbf{y}$. From the previous section, we already know that $H^0(\mathbf{x})$ does not depend on y_2, from which the matching conditions, Eqs. (5a) and (5b), give

$$H^0(0^+, x_2) = H^0(0^-, x_2) = h^0(x_2), \quad \text{and } [\![H^0]\!] = 0. \tag{14}$$

Next, Eq. (2) in the inner region gives also, at the leading order, $\mathrm{div}_{\mathbf{y}} c^0 = 0$; by integrating this equation on $Y \times]-\infty, +\infty[$, we get

$$\int_Y dy_2 [c_1^0(x_2, +\infty, y_2) - c_1^0(x_2, -\infty, y_2)] = 0, \tag{11}$$

(we have used the periodicity of $\boldsymbol{c}^0$ w.r.t. y_2). From the matching conditions Eqs. (5c) and (5d) integrated over Y, we get

$$C_1^0(0^+,x_2) = \langle C_1^0 \rangle (0^-,x_2), \quad \text{and } [\![\langle C_1^0 \rangle]\!] = 0. \tag{12}$$

At the first order, the usual continuities of the electromagnetic fields are found.

2.4. The homogenized problem for metallic layers in the far infra red

For rigid layers in the air and denoting φ the filling fraction of air within the stratified medium, we have: in the air $\epsilon = 1 = \mu$; the metal in the far infrared regime can be considered as an opaque medium at the boundaries of which Neumann boundary condition applies; this is correctly accounted for by considering the limiting values $1/\epsilon = 0 = \mu$ (see e.g., [6]). Thus, the homogenized problem reads

Homogenization at order 0

$$\begin{cases} \text{div}\boldsymbol{C} + k^2\varphi H = 0, \quad \boldsymbol{C} = \begin{pmatrix} \varphi & 0 \\ 0 & 0 \end{pmatrix} \nabla H, \\ H \text{ and } \boldsymbol{C}.\mathbf{n} \text{ continuous at the boundary of the metallic array.} \end{cases} \tag{13}$$

It is worth noting that the above continuity relation means notably that (i) at the interface with the air, $\varphi \nabla H.\mathbf{n}_{|\text{array}} = \nabla H.\mathbf{n}_{|\text{air}}$ while (ii) at the boundary with a ground plane, the usual Neumann boundary condition applies $\nabla H.\mathbf{n}_{|\text{array}} = 0$.

3. Weakness of the first-order homogenization for small thickness devices

In this section, we inspect the validity of the homogenized problem, Eq. (14). This is done in the configuration of **Figure 2**, where a metallic array is placed in the air and illuminated by a plane wave at oblique incidence θ. We measured experimentally the transmission coefficient T^{exp} for the array of thickness e (and the array structuration is characterized by φ and h) and compared T^{exp} to the transmission coefficient T obtained in the homogenized problem. To anticipate, we

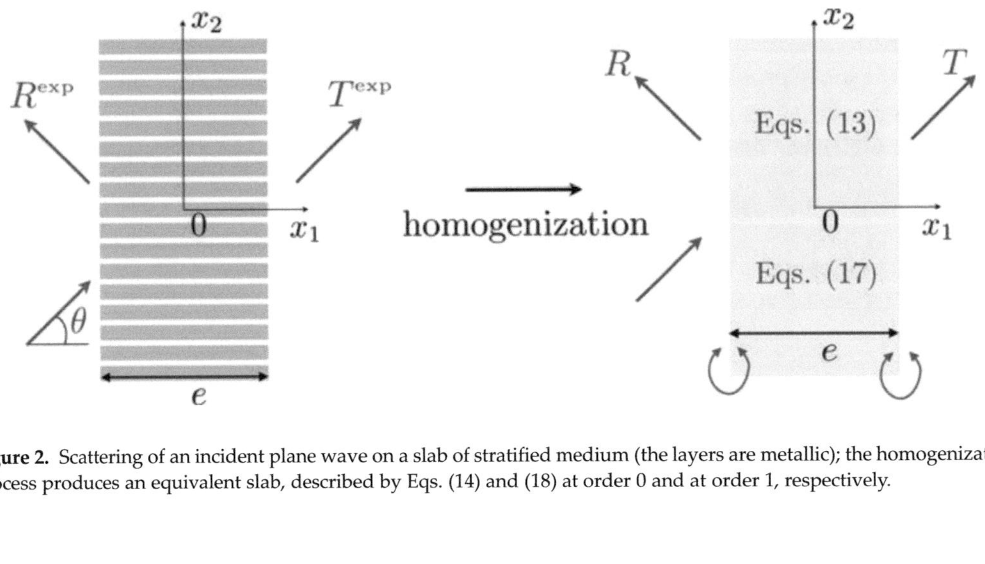

Figure 2. Scattering of an incident plane wave on a slab of stratified medium (the layers are metallic); the homogenization process produces an equivalent slab, described by Eqs. (14) and (18) at order 0 and at order 1, respectively.

shall see that the leading order homogenization may become unsatisfactory for small thicknesses of the array and going up to the homogenization at order 1 is necessary. Thus, the derivation of the scattering coefficients in these homogenized problems is presented first, afterwards comparisons with those measured experimentally are presented.

3.1. The scattering coefficients in the homogenized problems

Let us start with the first-order homogenization, for which the explicit solution of Eq. (14) read

$$\begin{cases} x_1 < -e/2, & H(\mathbf{x}) = \left[e^{ik\cos\ \theta(x_1+e/2)} + Re^{-ik\cos\ \theta(x_1+e/2)}\right]e^{ik\sin\theta x_2}, \\ |x_1| < e/2, & H(\mathbf{x}) = \left[ae^{ikx_1} + be^{-ikx_1}\right]e^{ik\sin\theta x_2}, \\ x_1 > e/2, & H(\mathbf{x}) = Te^{ik\cos\ \theta(x_1-e/2)+ik\sin\theta x_2}. \end{cases} \tag{14}$$

We used the dispersion relations coming from the wave equations $\Delta H + k^2 H = 0$ in the air and $\partial^2_{x_1^2} H + k^2 H$ in the homogenized stratified slab. Next, applying the relations of continuity applying at $x_1 = \pm e/2$, we get the scattering coefficients

$$\begin{cases} R = -\frac{z_1 z_2^* e^{ike} - z_1 z_2 e^{-ike}}{z_1^2 e^{ike} - z_2^2 e^{-ike}}, \\ T = \frac{|z_1|^2 - |z_2|^2}{z_1^{*2} e^{ike} - z_2^2 e^{-ike}}, \end{cases} \tag{15}$$

where z_i^* denotes the complex conjugate of z_i, $i = 1, 2$. In fact, for the leading order homogenization, (z_1, z_2) are real with

$$z_1 \equiv \left(1 - \frac{\cos\theta}{\varphi}\right), \quad z_2 \equiv \left(1 + \frac{\cos\theta}{\varphi}\right), \tag{16}$$

and $\cos\theta/\varphi$ is the effective impedance mismatch between the two media. As previously said, the leading order homogenization will fail and the next order homogenization is required. This homogenization at order 1 has been considered in [7] and it has been shown that the same equation in the bulk (see Eq. (14)) is obtained, but instead of the continuities of H and $\mathbf{C}.\mathbf{n}$ as boundaries conditions, jump conditions are obtained. Specifically, the homogenization at order 1 read

Homogenization at order 1

$$\begin{cases} \mathrm{div}\boldsymbol{C} + k^2\varphi H = 0, \quad \mathrm{C} = \begin{pmatrix} \varphi & 0 \\ 0 & 0 \end{pmatrix} \nabla H, \\ [\![H]\!] = \frac{h\mathcal{B}}{2}(\mathrm{C}^- + \boldsymbol{C}^+).\mathbf{n} \text{ and } [\![\boldsymbol{C}]\!].\mathbf{n} = \frac{h\mathcal{C}}{2}\left(\frac{\partial^2 H^-}{\partial x_2^2} + \frac{\partial^2 H^+}{\partial x_2^2}\right). \end{cases} \tag{17}$$

In the above equations, it appears that both H and $\mathbf{C}.\mathbf{n}$ are now discontinuous (the obtained conditions are jump conditions at the boundaries of the stratified medium). This is why $H^\pm$ (same for $\mathbf{C}$) are defined, as the limit values of H at the boundary. In this case, the expressions of (R, T) in Eq. (16) are still valid, but we get

$$\begin{cases} z_1 \equiv \left(1 - \frac{\cos\theta}{\varphi}\right) + ikh\left(\mathcal{B}\cos\theta + \mathcal{C}\frac{\sin^2\theta}{\varphi}\right) - (kh)^2\sin^2\theta\frac{\mathcal{BC}}{4}\left(1 + \frac{\cos\theta}{\varphi}\right), \\ z_2 \equiv \left(1 + \frac{\cos\theta}{\varphi}\right) - ikh\left(\mathcal{B}\cos\theta - \mathcal{C}\frac{\sin^2\theta}{\varphi}\right) + (kh)^2\sin^2\theta\frac{\mathcal{BC}}{4}\left(1 + \frac{\cos\theta}{\varphi}\right). \end{cases} \tag{18}$$

The parameters $\mathcal{B}$ and $\mathcal{C}$, that we could call boundary parameters, depend only on the filling fraction of air in the layered medium and they are given by

$$\mathcal{B} = -\frac{1}{\pi}\log\sin\left(\frac{\pi\varphi}{2}\right), \quad \mathcal{C} \simeq \frac{\pi}{16}\varphi^2. \tag{19}$$

In principle, $(H, \mathbf{C})$ in the homogenization at order 0, Eq. (14), approximate the solution of the actual problem up to $O(\eta)$ and $(H, \mathbf{C})$ in the homogenization at order 1, Eq. (18), approximate the solution of the actual problem up to $O(\eta^2)$. Thus, we could expect that the difference between both remains incidental. We will see that it is not the case. The reason is that the jump conditions obtained at order 1 encapsulate the effect of the boundary layers at the entrance and at the exit of the stratified slab and these effects may become dominant compared to the effect of the propagation in the bulk of the slab. This is what we inspect below.

3.2. Experimental measurements of the scattering coefficients for varying slab thicknesses e

To test the ability of the leading order homogenization to capture the scattering properties of a metallic array, we realized six arrays of different thicknesses $e = 30$ μm and 0.25, 1, 4, 14 and 20 mm. Otherwise, $h = 6$ mm and $\ell = 5$ mm for the six arrays. We performed the measurements of the transmission coefficients T^{exp} using two X-band frequency horn antennas at both extremities of an electromagnetic chamber (**Figure 3**). The arrays were amounted on a plate able to rotate in order to realize varying incidence angles θ (and the incidence wave is polarized in TM polarization).

We start by reporting in **Figure 4** the spectra $|T^{\text{exp}}|^2$ measured for the two arrays of thickness $e = 30$ μm and $e = 20$ mm (right panels). For frequencies in the range [8,12] GHz, the thinnest array realizes $ke \in [5; 7.5] \cdot 10^{-3}$ and the thickest array $ke \in [3; 5]$. On the two left panels, the corresponding spectra $|T|^2$ given by Eq. (16) using the leading order homogenization ((z_1, z_2) given by Eq. (17)) and using the homogenization at order 1 ((z_1, z_2) given by Eq. (19)) are reported for comparison. For $e = 30$ μm, the leading order homogenization predict $|T| \simeq 1$ in the whole ranges of frequencies and incidence angles. This is clearly not the case for $|T|$ given by the homogenization at order 1 and this latter appears to be in good agreement with $|T^{\text{exp}}|$ (we get $|T - T^{\text{exp}}|/|T^{\text{exp}}| \lceil 3\%$ averaged over all f and θ).

For the thicker array, the transmission predicted by the homogenization at order 0 is closer to the measured transmission; in this case, although the spectra obtained using the homogenization at order 1 reproduces better the form of the measured spectra, the relative errors are in both cases about 15% (these highest discrepancies are due to highest relative errors for transmissions close to 0).

http://dx.doi.org/10.5772/66035

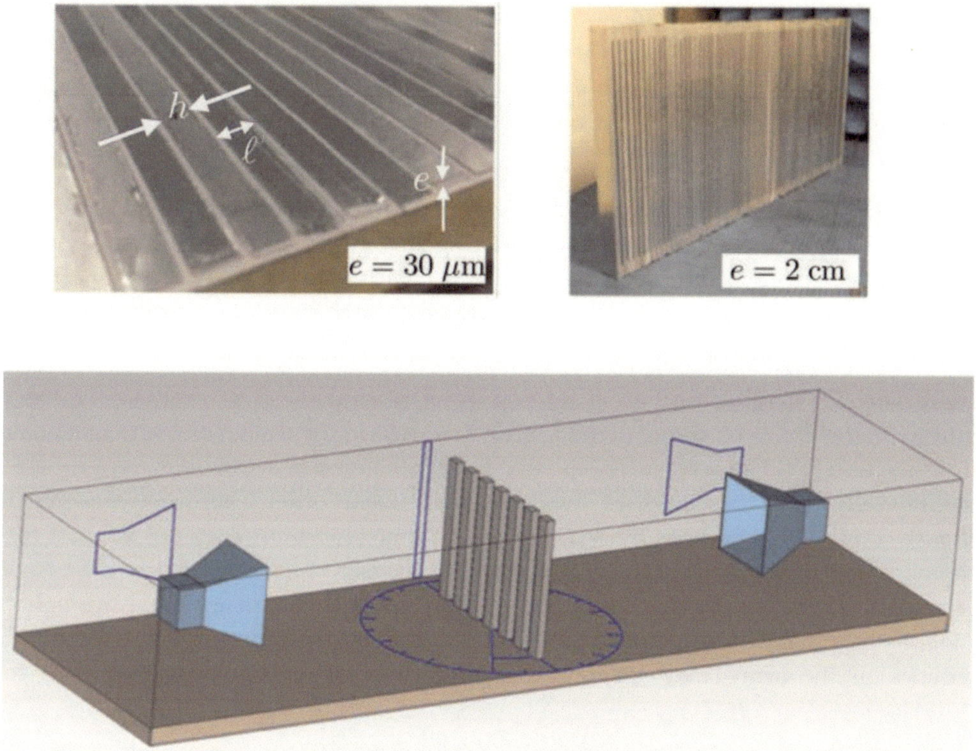

Figure 3. On the top, examples of two metallic arrays; on the bottom, schema of the measurement technique (see text).

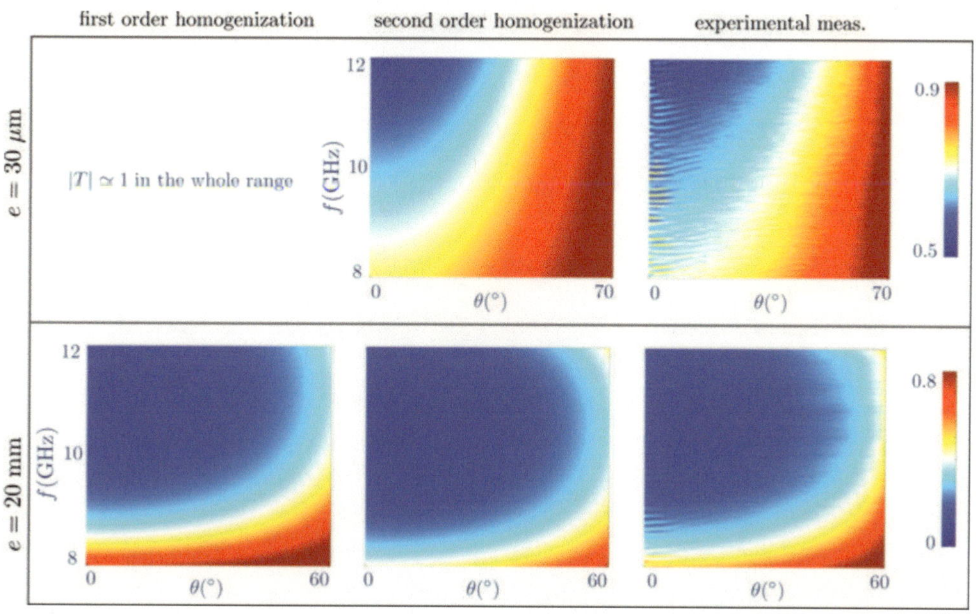

Figure 4. Spectra of transmission as a function of the frequency $f \in [8;12]$ GHz and the incidence angle θ. The left and central panels refer to the results coming from the homogenizations at the order 0 and at the order 1, to be compared with the measured transmission $|T^{\exp}|^2$ (right panel). Spectra are given for the thinner and thicker slabs (see text).

Next, we inspect the variations of the transmission coefficient as a function of *ke* for the six arrays, **Figure 5**. Each plot corresponds to a fixed frequency (f = 8, 10 and 12 GHz) and we reported $|T^{\exp}|^2$ in blue symbols (each blue point corresponds to one of the six arrays) and $|T|^2$ coming from the homogenization at order 0 (dotted gray lines) and coming from homogenization at order 1 (plain gray lines).

It is visible that thin arrays are not correctly described by the homogenization at order 0. More specifically, it largely underestimates the scattering strength of thin arrays (small e produces systematic large errors in the prediction) and it becomes accurate only when $ke > 1$. To the contrary, the homogenization at order 1 is able to describe the scattering strength of thin and thick arrays.

In conclusion, the homogenization at leading order is valid for $ke < 1$ and $kh < 1$ and thus helpful to predict the behavior of metallic arrays as used in the design of many metamaterial devices (see also [6]). However, care has been taken when using arrays with vanishing thicknesses, typically $ke < 1$. In such cases, the homogenization at order 1 has to be considered. As previously said, this is because the leading order homogenization does not account for the scattering effects of the wave at the entrance and at the exit of the array (these are boundary layer effects) which are correctly accounted for in the homogenization at order 1 through the—boundary—parameters ($\mathcal{B}$,$\mathcal{C}$). For even thinner array, the boundary layer effects at both extremities of the array may interact and another homogenization strategy has to be

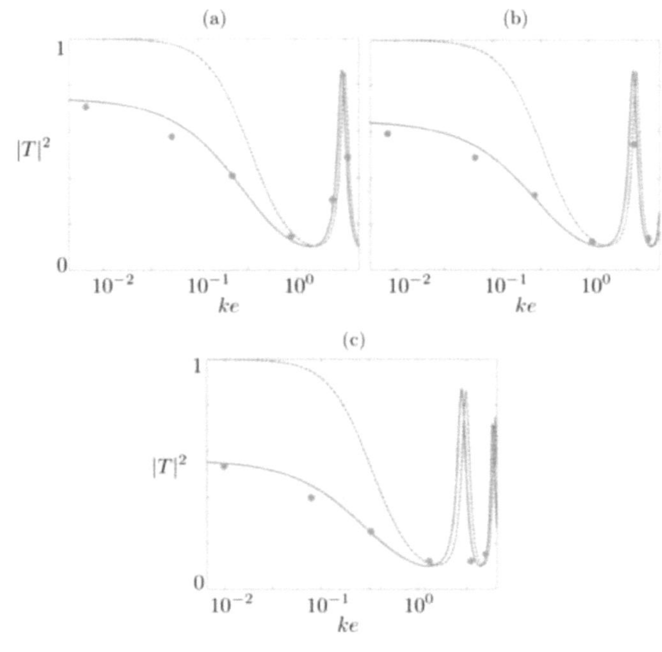

Figure 5. Transmission coefficients as a function of e; $|T^{\exp}|^2$ measured for the six arrays (blue symbols), $|T|^2$ coming from the homogenization at order 0 (dotted gray lines) and coming from the homogenization at order 1 (plain gray lines), (a) $f = 8$ GHz, (b) $f = 10$ GHz and (c) $f = 12$ GHz.

considered; it is called interface homogenization [7] and a practical application of this interface homogenization has been proposed for metallo-dielectric device [5].

4. Use of the results coming for the first-order homogenization for metamaterials with complex geometry

In this section, we investigate another configuration of wave propagation where predictions provided by the homogenization are useful. We consider the ability of periodic structures to present a band structure, with frequency ranges where the wave propagation is forbidden (band gaps) and frequency ranges where the propagation of guided waves is allowed (pass-bands). We study a structure based on such principle. First, the dispersion relation of guided waves within a periodic media (made of metallic plates or metallic rods) is discussed and it is shown that the leading order homogenization is able to reproduce the real dispersion relation. Next, a system of waveguides is thought, based on the band structures of the periodic media, in the waveguide and in the surrounding medium. The application of demultiplexing RF signals is proposed and experimentally validated.

4.1. Dispersion relation of waves guided in a periodic medium

Let us start with the derivation of the wave guided in a 2D stratified medium (the famous "spoof plasmon," **Figure 6a**). In the homogenized problem, Eq. (14), this wave corresponds to the solution of the homogeneous problem (solution in the absence of source), for which the solution reads

$$\begin{cases} 0 < x_1 < e, & H(\mathbf{x}) = \dfrac{\cos kx_1}{\cos ke}\, e^{\mathrm{i}\beta x_2}, \\ x_1 > e, & H(\mathbf{x}) = e^{-\sqrt{\beta^2-k^2}(x_1-e)}\, e^{\mathrm{i}\beta x_2}. \end{cases} \tag{20}$$

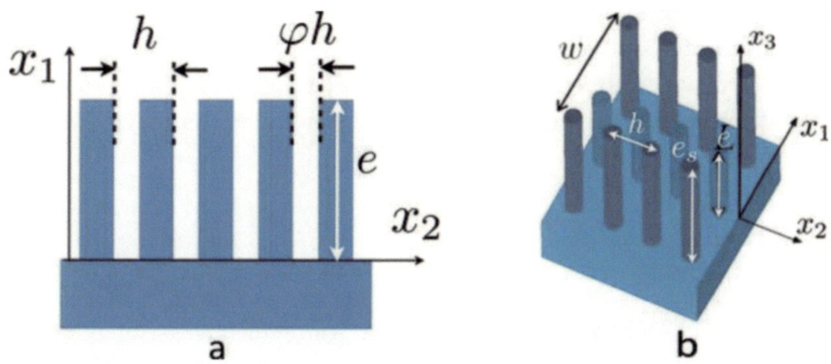

Figure 6. (a) 2D geometry of metallic layers between a ground plane and the air, spoof plasmons can propagate according to Eq. (22), (b) 3D geometry of a structured waveguide of width w, guided modes can propagate according to Eq. (27).

In the above expression, the field H for $0 < x_1 < e$ has been written in order to satisfy the Neumann boundary condition on the ground plane ($\partial_{x_1} H(0,x_2) = 0$) and the continuity of $H(e,x_2)$. It is easy to see that applying the second relation of continuity, namely $\varphi \partial_{x_1} H(e^-,x_2) = \partial_{x_1} H(e^+,x_2)$ coming from Eq. (14), we obtain the dispersion relation of the guided waves

$$\beta^2 = k^2 \left[1 + \varphi^2 \tan^2 ke\right], \quad \tan ke > 0, \tag{21}$$

with φ the filling fraction of air and e the length of the layers in the stratified medium. This dispersion relation has been established previously using approximate modal method [2] and it is easily obtained by considering the equivalent homogenized problem.

It is worth noting that such guided wave propagates in the homogenized medium described by the wave equation

$$\frac{\partial^2 H}{\partial x_1^2} + k^2 H = 0, \tag{22}$$

(according to Eq. (14)) and basically, the wave equation (23) tells us that the wave inside the grooves can only propagate along one direction (the x_1-direction). While in principle the upper frequency f_c^+ of the band gap is obtained for $ke = \pi/2$, from Eq. (22), it is in practice limited by the first Brillouin zone $\beta = \pi/h$, whence

$$f_c^+ \sqrt{1 + \varphi^2 \tan^2 \frac{2\pi e f_c^+}{c}} = \frac{c}{2h}, \quad f_c^+ < \frac{c}{4e}. \tag{23}$$

Now, we want to go toward a 3D structuration (**Figure 6b**), where rods are considered, with radius r, periodic spacing h along x_2 and x_3 and height e. The extension of the homogenization results (Eq. (14)) to three dimensions is easy and we find that a periodic structuration of rods produce an equivalent transverse isotropic medium, with the axis of anisotropy along e_1 (the two other directions are equivalent, now φ the volume fraction of air in the rods, whence $\varphi = 1 - \pi r^2/a^2$). Let us consider that these rods forms a waveguide surrounded by a set of higher rods with height e_s (the surrounding rods form the surrounding medium SM) and imagine that we work in a frequency range such that

$$f > \frac{c}{4e_s}, \tag{24}$$

that is in the band gap of the surrounding medium, from which $H \simeq 0$ (for $x_3 < 0$ and $x_3 > w$). Thus, it sounds reasonable to impose $H = 0$ at $x_3 = 0,w$, as boundary conditions for the field in the waveguide with the rods of height e. Looking for the existence of guided wave in the waveguide, we extend the homogenization result to this 3D configuration assuming a solution of the form

http://dx.doi.org/10.5772/66035

$$\begin{cases} 0 < x_1 < e, & H(\mathbf{x}) = \dfrac{\cos kx_1}{\cos ke}\, e^{i\beta x_2} \sin\left(\dfrac{\pi x_3}{w}\right), \\ x_1 > e, & H(\mathbf{x}) = e^{-\sqrt{\beta^2-k^2}(x_1-e)}\, e^{i\beta x_2} \sin\left(\dfrac{\pi x_3}{w}\right). \end{cases} \tag{25}$$

In the simple form thought above, we added heuristically dependence in the x_3 direction which accounts for Dirichlet boundary conditions at $x_3 = 0,w$, when working in the band gap of the surrounding medium.

Otherwise, Eq. (26) accounts for the continuity of H at the interface with the air and for the Neumann boundary condition at the ground plane; as previously, the additional condition $\varphi\partial_z H(L^-) = \partial_z H(L^+)$ yields the new dispersion relation

$$\beta^2 = k^2\left[1 + \varphi^2 \tan^2 ke\right] - \left(\frac{\pi}{w}\right)^2. \tag{26}$$

In Eq. (27), the band gap above f_c^+ still exists and it does not significantly differ from Eq. (24) (it is sufficient to replace $c/(2h)$ in the right-hand side term of Eq. (24) by $c\sqrt{1 + (h/w)^2}/(2h)$); but now, a new band gap has appeared below the cut-off frequency f_c^- (given from Eq. (27) for $k \tan ke = \pi/(w\varphi)$), whence

$$f_c^- \tan \frac{2\pi f_c^-}{c} = \frac{c}{2w\varphi}. \tag{27}$$

The existence of the resulting finite pass band operating in the frequency range $[f_c^-, f_c^+]$ is the key to realize filtering; in the following, we denote

$$\text{FR(waveguide)} = [f_c^-, f_c^+] \tag{28}$$

this frequency range and we refer to w_n, with n an integer, a waveguide obtained when n lines of rods have been shortened (with resulting height e) with respect to the rods of height e_s forming the surrounding medium (SM); the w_n waveguide has a width $w - (n+1)h$.

4.2. Experimental validation of the homogenized dispersion relation

To begin with, we validate experimentally the existence of the pass band and check the validity of our predictions of the FR with bounds f_c^+, f_c^- in Eqs. (24) and (28) and of the associated wavenumbers Eq. (27). Structures containing a waveguide w_1 ($w = 2h$) and a waveguide w_3 ($w = 4h$) have been realized. The surrounding medium is in both cases made of rods with $h = 7$ mm, $r = 3$ mm and $e_s = 30$ mm. The shortened rods have $e = 17$ mm.

First, we report measurements of the electric field in the range [2.1– 4] GHz (**Figure 7** for w_1). This has been done in a semianechoic chamber using an Agilent 8722ES network analyzer; an S-band coaxial-to-waveguide transition has been used as an excitation source and an electric near-field probe mounted on a motorized two-dimensional scanning system has been used to measure the field distribution at a distance of about 1 mm above the structure. The wave

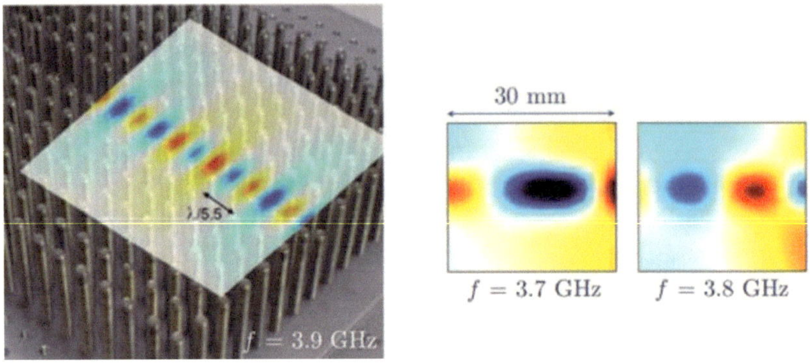

Figure 7. Wavefields of the guided wave in the w_1 waveguide at f = 3.7, 3.8 and 3.9 GHz.

f (GHz)	3.7	3.8	3.9
measured λ_{GW} (mm)	42	30	21
$\lambda_{GW} = 2\pi/\beta$ from Eq. (26)	43	28	20

Table 1. Wavelength λ_{GW} of the guided wave in the w_1 waveguide, measured experimentally and given by Eq. (27).

guided within the waveguide w_1 is visible at f = 3.7, 3.8 and 3.9 GHz, as expected from the frequency range FR(w_1) = [3.6–4] GHz (see **Table 1**) .

To go further, we report the transmission in the w_1 and w_3 waveguides as a function of the frequency (**Figure 8**). This has been done by placing a second coaxial-to-waveguide transition at the end of the waveguide and by implementing a normalization to the free air transmission between the emitter and the receptor. In both cases, the existence of a finite pass band is confirmed (the waveguide is called colored) and the observed bounds of the FR are in good agreement with our predictions (the theoretical dispersion relations are reported, with FR(w_3) = [3.1–4]). One can notice here the importance of the attenuation for the spoof-plasmon modes providing the smallest wavelengths. These wavelengths correspond to the highest frequencies in the transmission bands. This phenomenon, known in plasmonics, happens due to intrinsic losses in the considered materials for the high-wave-vector components.

4.3. Application to the design of a demultiplexer

This validation being performed, a multichannel demultiplexer is easy to design; the principle of the demultiplexing is shown in **Figure 5**. A main waveguide, called white guide, is built in order that the FR(white) covers the working frequency range; this is done by choosing: (i) w large enough to produce a small enough f_c^- (see Eq. (28)) and (ii) e small enough to produce a large enough f_c^+ (see Eq. (24)). By setting $w = 10h = 70$ mm and e = 15 mm, we expect FR(white) [2.5–4.5] GHz.

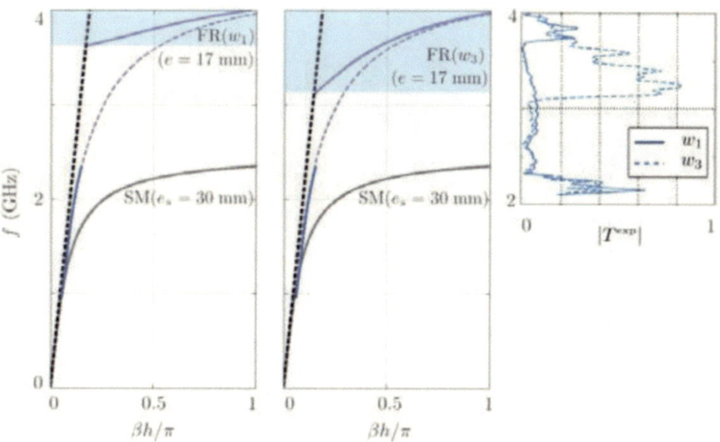

Figure 8. Theoretical dispersion relations of the guided wave in the w_1 (left panel) and w_3 (center panel), given by the plain blue curves, given by Eq. (27). The dispersion relation of the surrounding medium is indicated in plain gray lines, defining a band gap for $f > 2.35$ GHz. Dotted blue lines show the usual dispersion relations of spoof plasmons, Eq. (22), for the periodic array of rods alone (with $e = 17$ mm). Right panel: experimental measurement of the transmission for the w_1 and w_3 waveguides in the range [2.1–4] GHz.

Next, three-colored waveguides (red, green and blue) are thought in order to support guided mode propagation in three different frequency ranges with no overlapping (see **Figure 9**). Again from Eqs. (23) and (27), thin FR are obtained for small w and we set $w = 2h = 14$ mm. By choosing $L = 17$, 19 and 21 mm, we expect this condition to be fulfilled with

$$\begin{cases} \text{FR(red)} = [3 - 3.3] \quad \text{FR(green)} = [3.3 - 3.6] \\ \text{FR(blue)} = [3.6 - 4.0]. \end{cases} \tag{29}$$

The efficiency of the demultiplexer has been tested experimentally and it is illustrated in **Figure 10**. The frequency selection of the colored channel are visible, with the red channel

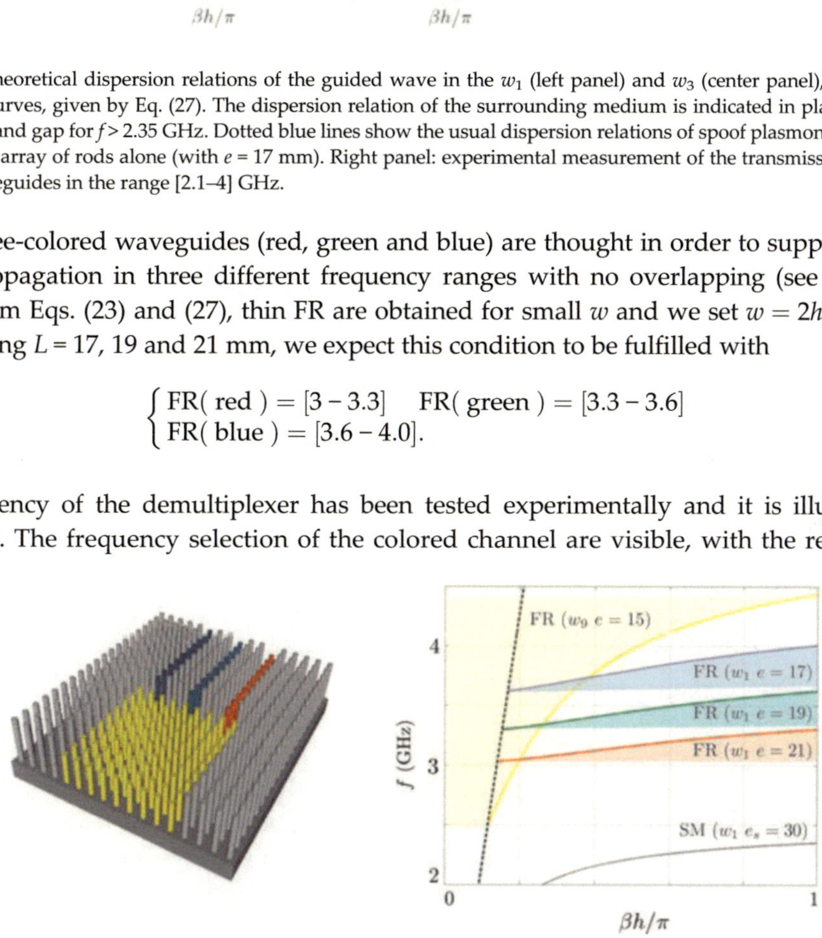

Figure 9. (a) Design of the demultiplexer, the large channel supports guided waves in the whole working frequency range [2.5–4.5] GHz, while each colored channel supports guided waves in a limited frequency range, with no overlapping of the three ranges. (b) Corresponding dispersion relations given by the homogenization, Eq. (27). The gray line shows the dispersion relation of the surrounding medium (all channels work in the band gap of the surrounding medium); the dotted black line shows the light line.

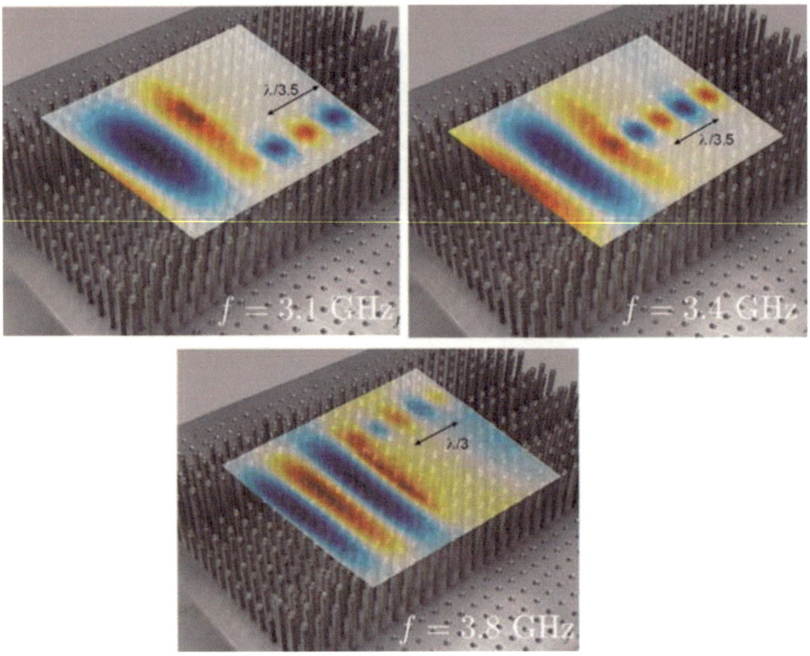

Figure 10. Electric fields scanned above the structure at 3 frequencies chosen respectively in the red (f = 3.1 GHz), in the green (f = 3.4 GHz) and in the blue (f = 3.8 GHz) frequency range $[f_{c-};f_{c+}]$.

being active for $f = 3.1$ GHz, the green channel for $f = 3.4$ GHz and the blue channel for $f = 3.8$. Also as expected, the white channel is active for the three considered frequencies.

5. Conclusion

The design of devices made of subwavelength periodic structure can be helped using the various theoretical predictions provided by homogenization approaches. Among the different homogenization technique, the one presented in this chapter has the advantage to be developed within a rigorous mathematical framework and it yields predictions in a deductive way, that is without any adjustable parameters. We illustrate the ability of such techniques for the scattering properties of metamaterial devices (underlying the limit of the simplest homogenization) and for the ability of certain metamaterial devices to support guided waves of "spoof plasmon" type.

As indicated throughout this chapter, alternative forms of homogenizations can be used, which are more adapted to thin devices. These theoretical tools can be used in order to realize the control of light propagation in a desired way.

Author details

Abdelwaheb Ourir[1]*, Yao Gao[1], Agnès Maurel[1] and Jean-Jacques Marigo[2]

*Address all correspondence to: a.ourir@espci.fr

1 Institut Langevin, CNRS, ESPCI, Paris, France

2 Solid Mechanics Laboratory, CNRS, Polytechnic Institute, Palaiseau, Paris, France

References

[1] D. F. Sievenpiper, High-impedance electromagnetic surfaces, Ph.D. thesis, University of California, Los Angeles (1999).

[2] J. B. Pendry, L. Martn-Moreno, F. J. Garcia-Vidal, Mimicking surface plasmons with structured surfaces, Science. 305 (5685) (2004) 847–848. arXiv:http://www.sciencemag.org/content/305/5685/847.full.pdf, doi:10.1126/science.1098999. URL http://www.sciencemag.org/content/305/5685/847.abstract

[3] M. Farhat, S. Guenneau, S. Enoch, Ultrabroadband elastic cloaking in thin plates, Phys. Rev. Lett. 103 (2009) 024301. doi:10.1103/PhysRevLett.103.024301. URL http://link.aps.org/doi/10.1103/PhysRevLett.103.024301

[4] C. P. Berraquero, A. Maurel, P. Petitjeans, V. Pagneux, Experimental realization of a water-wave metamaterial shifter, Phys. Rev. E. 88 (2013) 051002. doi:10.1103/PhysRevE.88.051002. URL http://link.aps.org/doi/10.1103/PhysRevE.88.051002

[5] A. Maurel, J. J. Marigo, A. Ourir, Homogenization of ultrathin metallo-dielectric structures leading to transmission conditions at an equivalent interface, J. Opt. Soc. Am. B. 33 (5) (2016) 947–956. doi:10.1364/JOSAB.33.000947. URL http://josab.osa.org/abstract.cfm?URI=josab-33-5-947

[6] J. F. Mercier, M. L. Cordero, S. Félix, A. Ourir, A. Maurel, Classical homogenization to analyse the dispersion relations of spoof plasmons with geometrical and compositional effects, Proceedings of the Royal Society of London A: Mathematical, Physical and Engineering Sciences 471 (2182). arXiv:http://rspa.royalsocietypublishing.org/content/471/2182/20150472.full.pdf, doi:10.1098/rspa.2015.0472. URL http://rspa.royalsociety-publishing.org/content/471/2182/20150472

[7] J. J. Marigo, A. Maurel, Homogenization models for thin rigid structured surfaces and films, J. Acoust. Soc. Am. 140 (1) (2016) 260–273. doi:http://dx.doi.org/10.1121/1.4954756. URL http://scitation.aip.org/content/asa/journal/jasa/140/1/10.1121/1.4954756

Chapter 8

Compact Metamaterials Induced Circuits and Functional Devices

He-Xiu Xu, Guang-Ming Wang, Tong Cai,
Qing Peng and Ya Qiang Zhuang

Additional information is available at the end of the chapter

http://dx.doi.org/10.5772/65603

Abstract

In recent years, we have witnessed a rapid expansion of using metamaterials to manipulate light or electromagnetic (EM) wave in a subwavelength scale. Specially, metamaterials have a strict limitation on element dimension from effective medium theory with respect to photonic crystals and other planar structures such as frequency selective surface (FSS). In this chapter, we review our effort in exploring physics and working mechanisms for element miniaturization along with the resulting effects on element EM response. Based on these results, we afford some guidelines on how to design and employ these compact meta-atoms in engineering functional devices with high performances. We found that some specific types of planar fractal or meandered structures are particularly suitable to achieve element miniaturization. In what follows, we review our effort in Section 1 to explore novel theory and hybrid method in designing broadband and dual band planar devices. By using single or double such compact composite right-/left-handed (CRLH) atom, we show that many microwave/RF circuits, i.e., balun, rat-race coupler, power divider and diplexer, can be further reduced while without inducing much transmission loss from two perspectives of lumped and distributed CRLH TLs. In Section 2, we show that a more compact LH atom can be engineered by combining a fractal ring and a meandered thin line. Numerical and experimental results demonstrate that a subwavelength focusing is achieved in terms of smooth outgoing field and higher imaging resolution. Section 3 is devoted to a clocking device from the new concept of superscatterer illusions. To realize the required material parameters, we propose a new mechanism by combining both electric and magnetic particles in a composite meta-atom. Such deep subwavelength particles enable exact manipulation of material parameters and thus facilitate desirable illusion performances of a proof-of-concept sample constructed by 6408 gradually varying meta-atoms. Finally, we summarize our results in the last section.

Keywords: metamaterials, fractal, electrically small, effective medium, broadband, CRLH TL, microwave/RF circuits, super lens, cloaking

1. Miniaturized CRLH atoms for compact microwave/RF circuits

Compact metamaterial element exhibits versatile features and merits over its conventional counterparts. One of the most important and direct features is that it enables more compact microwave/RF circuits. In this section, we will show this first benefit through a set of compact metamaterial transmission lines (TL), i.e., compact composite right/left-handed (CRLH) TLs. For a comprehensive study, two types of compact CRLH TLs, namely, lumped and distributed CRLH TLs, are investigated according to the realization manner. Of particular relevance and importance is the established design guideline, which paves the way for any microwave and millimeter wave integrated circuits and devices with high performances.

1.1. CAD design method for fractal lumped CRLH TL

To begin with, we first derive a general design mythology for the design of any compact CRLH element with fractal or meandered sections. **Figure 1** depicts the CAD design flowchart of such CRLH TLs. For analysis convenience, the general circuit model (CM) of any CRLH TL is reproduced in **Figure 1(a)**. The CRLH TL, with characteristic impedance of Z_c at angular frequency ω and element number of N, consists of left-handed (LH) contribution associated with L_L, C_L, and right-handed (RH) part with L_R, C_R. This CM topology can be easily converted to *T*-type circuit with $L_R/2$ and $2C_L$. In practical realization, the RH part is designed by microstrip line (ML) in meandered topology, whereas the chip components based on surface mount technology (SMT) are utilized to realize the LH part of CRLH atom. As such, the circuit

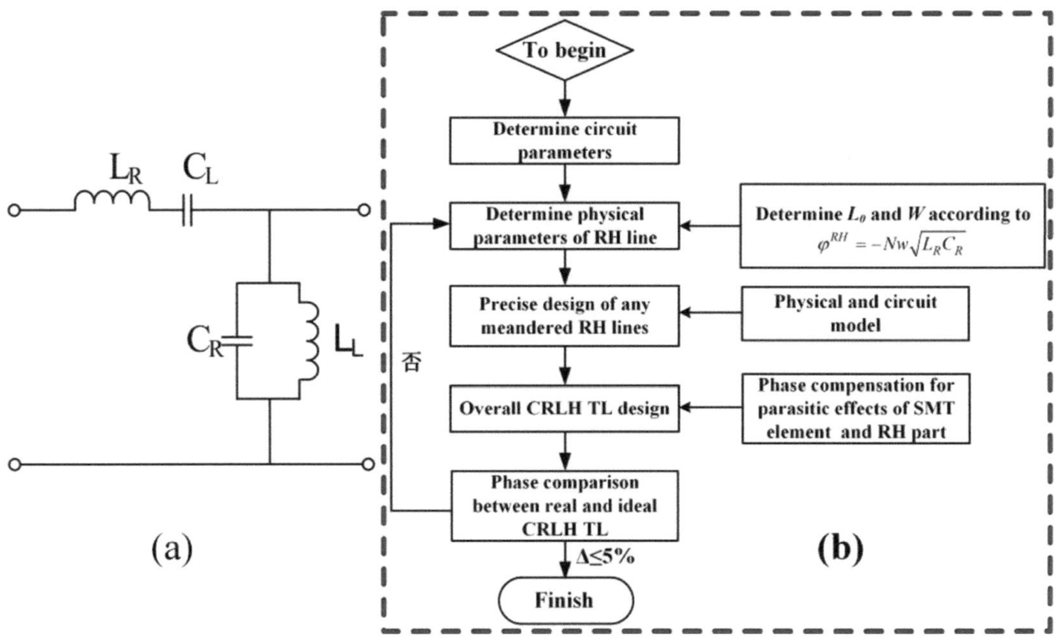

Figure 1. (a) Circuit topology and (b) CAD design flowchart of fractal or meandered CRLH TL.

http://dx.doi.org/10.5772/65603

miniaturization and weakened susceptibility of meandering effects on the characteristic impedance are simultaneously engineered by taking advantages of both space-filling fractals/ meanderings and CRLH technology. The design procedures, which mainly concentrate on phase, are described as follows.

(1) Determine and derive the circuit parameters of any CRLH TL according to the required phase response at operation frequency f_0 based on circuit theory. Here, the desired phase response is synthesized based on the realized functionality.

(2) Given the computed values of LR and CR, the required electrical length of straight RH line (ML) for CRLH TL is determined by

$$\varphi(\omega_0) = -N\omega_0\sqrt{L_R C_R} \tag{1}$$

Therefore, the physical length L_0 and width w are directly obtained referring to the $\varphi(\omega_0)$ and characteristic impedance in circuit simulation package of Ansoft Serenade.

(3) Design any fractal ML with specified phase shift. Since many chamfered bends are generated in iterative meandered process, the nonnegligible discontinuity reactance results in phase-shifting effect of RH line which should be properly evaluated and compensated. Here, a phase-equalizing method is developed. A slight physical length Δb corresponding to the phase shift generated by each bend is applied to model the effect. Consequently, the L_0 should be added by Δb per bend for compensation. The value of Δb in millimeter is determined by [1]:

$$\Delta b = \frac{19.2\pi h}{\sqrt{\varepsilon_{eff}} Z_C}\left[2-\left({}^{f_0 h}/_{0.4Z_C}\right)^2\right] \tag{2}$$

Here, ε_{eff}, h are effective dielectric constant and height of the substrate in millimeter, respectively.

(4) Design the overall CRLH TL by taking into account the parasitic inductances and capacitances of SMT elements and soldering pad. This can be implemented by a close-loop precise optimization through dynamically comparing the phase response of real and ideal CRLH TLs. Here, the parasitic effects are evaluated through a direct comparison of measured transmission response of MLs with and without soldered SMT elements.

1.2. Broadband and dual band balun

1.2.1. Broadband balun using fully artificial fractal-shaped CRLH TL

In this subsection, a compact balun with bandwidth enhancement is realized based on the design methodology established in Section 1.1.

Balun, a three-port network, was commonly used to transfer the input unbalanced signal to two output balanced ones with 180° phase difference. It has been extensively studied for active amplifiers, balanced mixers, passive filter [2] and even antennas [3]. Several types of baluns are available among the open literature, such as the Marchand balun [4–6], power-divider balun [7, 8], branch-line balun [9] and metamaterial balun [10]. Highly integrated circuits with stable communication quality strongly require wideband and miniaturized RF/microwave devices. In this regard, several strategies have been proposed for miniaturization, e.g., by integrating coupled transmission line (TL) [4], using low temperature cofired ceramic technique [5] or fractal/meandered branches [9–11]. Although baluns achieved significant miniaturization in [4, 5], the multilayer design suffered thick profile, complicated structure and high-cost fabrication. As to the bandwidth enhancement of baluns, we have also witnessed several approaches such as using broadband Schiffman phase shifter [3], slot-coupled microstrip lines [6], substrate integrated waveguide [8], phase-adjusting CRLH TL [7] and metamaterial TL [10]. Nevertheless, these baluns occupy a large circuit area. The lack of techniques regarding simultaneous bandwidth and miniaturization make the design of compact broadband balun a pressing task. Here, a compact broadband balun is proposed using fully artificial fractal-shaped CRLH TL [12] and we will begin with the theorem of CRLH TL for broadband objective.

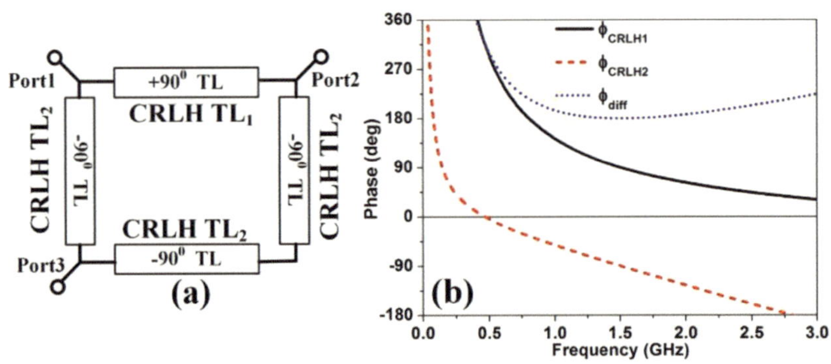

Figure 2. (a) Scheme topology of proposed balun and (b) phase response of dual branches. The derived circuit parameters are: L_{R1} = 0.62 nH, C_{R1} = 0.125 pF, L_{L1} = 8.63 nH, C_{L1} = 1.73 pF, L_{R2} = 6.5 nH, C_{R2} = 1.3 pF, L_{L2} = 89.9 nH, and C_{L2} = 18 pF.

Figure 2 shows the scheme of proposed balun along with the phase response of corresponding dual CRLH branches. Different from branch-line balun with two –90° and two –180° branches, our balun contains one +90° and three –90° TL branches. The characteristic impedance of +90° branch is denoted as Z_{c1} and that of –90° branch as Z_{c2}. To engineer a good impedance match, Z_{c1} and Z_{c2} are related with port impedance Z_0 as [9]

$$Z_{C1} = \frac{Z_{C2}}{\sqrt{2}Z_{C2} - Z_0} Z_0 \tag{3}$$

For simplicity, we select $Z_{C1} = Z_{C2} = \sqrt{2}Z_0$ and Z_0 = 50 Ω. The above four TL branches with specified phase response tailored at a given band can be easily realized through CRLH TL. For convenience, we denote the +90° CRLH branch as CRLH TL_1 while the –90° one as CRLH TL_2. The phase shifts $\varphi_{\mathrm{CRLH\ TL1}}$ and $\varphi_{\mathrm{CRLH\ TL2}}$ are explicit functions of the angular frequency ω:

$$\varphi_{\mathrm{CRLH.TL}}(\omega) = -N[\omega\sqrt{L_R C_R} - \frac{1}{\omega\sqrt{L_L C_L}}], \tag{4}$$

Here, N is the number of adopted CRLH atoms and is chosen as 2 in this work. For good impedance match, the following conditions are compulsorily satisfied:

$$\sqrt{\frac{L_{R1}}{C_{R1}}} = \sqrt{\frac{L_{L1}}{C_{L1}}} = \sqrt{\frac{L_{R2}}{C_{R2}}} = \sqrt{\frac{L_{L2}}{C_{L2}}} = \sqrt{2}Z_0 \tag{5}$$

To maximize the bandwidth around f_0 = 1.5 GHz, identical slope of phase response between CRLH TL_1 and CRLH TL_2 must be fulfilled, which indicates that the phase difference φ_{diff} is minimum at angular frequency ω_0:

$$\varphi_{\mathrm{diff}} = -N[\omega\sqrt{L_{R1}C_{R1}} - \frac{1}{\omega\sqrt{L_{L1}C_{L1}}}] + N[\omega\sqrt{L_{R2}C_{R2}} - \frac{1}{\omega\sqrt{L_{L2}C_{L2}}}] \tag{6}$$

To mathematically guarantee above requirement, the first-order derivative of φ_{diff} with respect to ω should be zero whereas the second-order one should be positive, this yields

$$N[\sqrt{L_{R2}C_{R2}} - \sqrt{L_{R1}C_{R1}} + \frac{1}{\omega_0^2\sqrt{L_{L2}C_{L2}}} - \frac{1}{\omega_0^2\sqrt{L_{L1}C_{L1}}}] = 0 \tag{7}$$

$$\frac{2}{\omega_0^3\sqrt{L_{L1}C_{L1}}} - \frac{2}{\omega_0^3\sqrt{L_{L2}C_{L2}}} > 0 \tag{8}$$

The second derivative ensures the extremum of Eq. (5) to be a minimum. We have seven individual equations from Eqs. (4), (5) and (7) but have eight unknowns. Therefore, it is impossible to exclusively determine a group of solution. This additional degree of freedom can be utilized for other purpose in design of the broadband balun with fully artificial TLs.

Here, C_{L2} is first chosen as a marketable chip capacitor with discrete capacitance of 18 pF. Then, residual elements are obtained by solving above equations conducted in mathematical software Matlab. Since many groups of solution (typically 16) would exist due to the square root equations, we rule out all solutions with negative values and followed by a proof-test process to select the physically meaningful one. To conceptually validate the proposal, we adopted the lumped elements derived by this method to calculate the phase response of CRLH TL_1 and CRLH TL_2, see **Figure 2(b)**. As is shown, the slope of both transmission phases is almost the same over a wide frequency region in vicinity of 1.5 GHz. Therefore, the phase difference is almost with a flat response around 180° across a wide band.

For verification, we designed, fabricated and measured a proof-of-concept sample on F4B substrate with thickness of h = 1 mm, dielectric constant of ε_r = 2.65 and loss tangent of tan δ = 0.001. The basic structure we used is trapezoidal Koch curve (T-Koch), see **Figure 3(a)**, which is determined by its iteration factor (IF = 1/4), Hausdorff dimension (ln5/ln4) and iteration order (IO). It should be highlighted that the T-Koch is very beneficial to load SMT capacitors at its horizontal fractal segments. Since the ML of CRLH TL_1 is too short to be constructed in fractal,

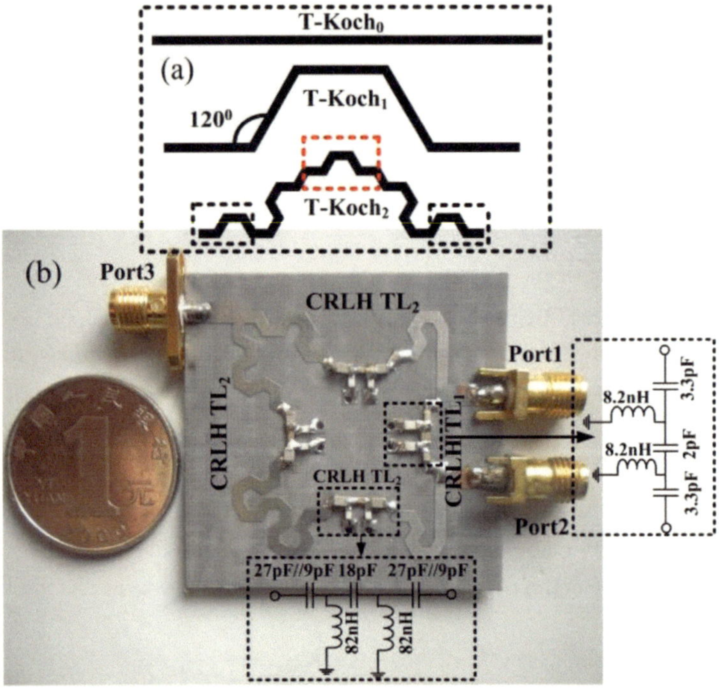

Figure 3. (a) T-Koch curves with zeroth, first and second iteration order and (b) fabricated prototype of the developed balun.

we only considered changing the orientation of several chamfered bends outlined in the black dashed of the T-$Koch_2$ to maximize the miniaturization. The straight length (L_n) and the number of fractal segments (P_n) after n iterations are $L_n = (4/5)^n L_0$ and $P_n = 5^n$, respectively, where L_0 is the initial length. Since T-Koch with higher IO contributes less to the miniaturization while inversely more to fractal segments, the applied T-Koch curve is limited to IO = 2. To load three chip capacitors, five segments outlined in the red dashed of T-$Koch_2$ are removed. Using the established methodology, the ML length of CRLH TL_1 and CRLH TL_2 is eventually obtained as 3.6 and 40.9 mm, respectively. In the fabricated prototype shown in **Figure 3(b)**, two *T*-type cells with chip capacitors and inductors of 0805 and 0603 packages provided by Murata Manufacturing Company Ltd. are cascaded to build the LH part of two-cell CRLH TL. For easy impedance match and enhanced transmission performance, the CRLH TL is terminated in two capacitors of $2C_L$. The size of the balun is only 29 × 30.5 mm^2, corresponding to 24.5% of π × 33.9 × 33.9 mm^2 that conventional rat-race balun occupies and to 56.4% of the 32 × 49 mm^2 that CRLH balun without fractal perturbation occupies.

The performance of developed balun is characterized through the dynamic links and solver of planar EM and circuit cosimulation in MOM-based Ansoft Designer and is measured by the *Anritsu ME7808A* network analyzer. **Figure 4** shows the simulated and measured *S*-parameters along with amplitude and phase imbalance, which are in good consistency. Slightly larger amplitude imbalance in measurements is mainly attributable to the nonideal SMT chip

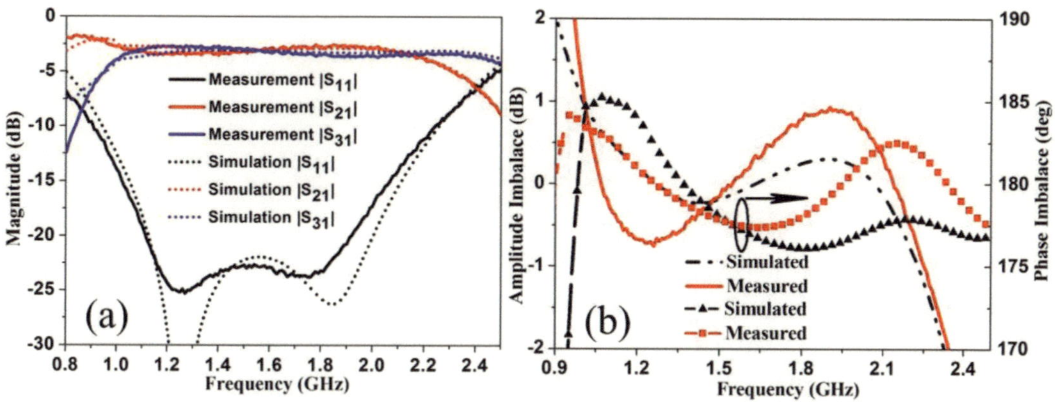

Figure 4. (a) Simulated and measured *S*-parameters of the proposed balun and (b) simulated and measured amplitude and phase imbalance.

components with inductance or capacitance variation of ±10% and is partially to the tolerances inherent in fabrications.

Measurement results reveal that return loss $|S_{11}|$ is better than –10 dB from 1 to 2.25 GHz (a bandwidth of 83.3%), over which the amplitude imbalance varies within 1 dB and the phase differences ranges from 177.3° to 183.4° (180° ± 3.4°). Moreover, the insertion loss $|S_{21}|$ varies between –2.9 and –3.6 dB while $|S_{31}|$ between –2.8 and –3.8 dB. The bandwidth of balun using bare T-Koch fractal is obtained only 10% which is far insufficient relative to proposed design using hybrid technique. To our best knowledge, the proposed balun exhibits comparable performance in broadband and compact size among the available data using double-layer printed circuit board.

To sum up, the insertion loss mainly induced by fractal bends is moderate within the operation band. The hybrid technology of fractal and CRLH TL does not pose severe penalty on device performances but allows additional degree of freedom in developing devices with high integration, promising an elegant alternative for compact and multifunctional devices with high performances.

1.2.2. Dual band rat-race coupler

Rat-race coupler (RRC), see the circuit topology shown in **Figure 5(a)**, is one type of 180° hybrids and one of the most important microwave passive devices. The four-port lossless reciprocal network consists of three –90° branches and one –270° branch with characteristic impedance of $\sqrt{2}Z_0$, where Z_0 is the port impedance 50 Ω. A RRC commonly exhibits two functionalities, namely, a power divider and a power combiner. As a power divider, it can work in two cases of in-phase and 180° out-of-phase state. In the former case, a signal inputted in port 1 divides equally into ports 2 and 3 with identical phase and port 4 is isolated, whereas in the latter case a signal injected into port 4 separates evenly in ports 2 and 3 with 180° phase difference and port 1 is isolated. As a power combiner, two signals with 0°/180° phase difference are input simultaneously at ports 2 and 3 and the sum of signals will be formed at port 1/

port 4. The RRC with four pairs of transmission channel and two pairs of isolation channel (ports 1 and 4, ports 2 and 3) can be modeled by the following scatting matrix [S].

$$[S] = \frac{j}{\sqrt{2}} \begin{bmatrix} 0 & 1 & 1 & 0 \\ 1 & 0 & 0 & -1 \\ 1 & 0 & 0 & 1 \\ 0 & -1 & 1 & 0 \end{bmatrix} \tag{9}$$

Multiband components with miniaturized dimensions have aroused a wide of interest since they enable low cost, high reliability and integrity. Up to date, numerous approaches have been developed for compact RRCs, e.g., using capacitor [13] and periodic slow-wave loading [14], using periodic stepped-impedance ring resonators [15], T-shaped photonic bandgap (PBG) structures [16] and fractal strategy [17–19]. Although above RRCs feature compact, the lack of dual band (DB) performances deserves further improvements. To date, much fewer literatures were devoted to DB applications, e.g., using tri-section branch-line [20], stepped-impedance-stub units [21] and two T-shape open-stub units [22]. However, the design and realization were tedious and typically featured large circuit size. The lack of literature concerning both the DB performance and size reduction makes the design of a compact DB RRC a pressing task.

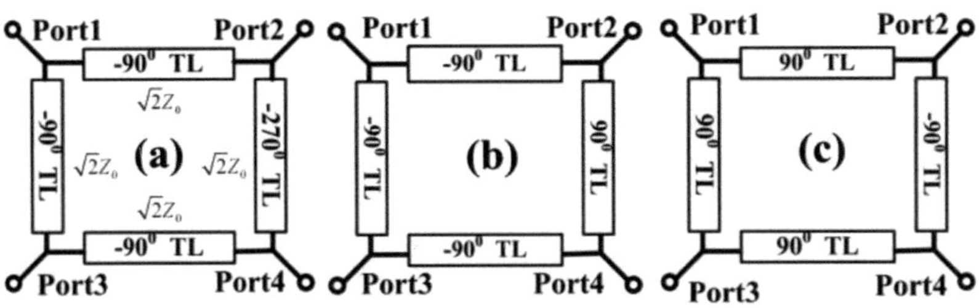

Figure 5. Circuit schematics of the RRCs: (a) conventional one; derivative ones with (b) one +90° branch, and (c) three +90° branches.

In this work, we proposed a novel DB RRC based on the hybrid approach of fractals and CRLH TLs [23]. A new scheme for DB design is proposed by combining two circuit topologies with different phased branches. We noticed that another two RRCs with topologies shown in **Figure 5(b)** and **(c)** also exhibit the same functionality. To develop a DB RRC, we consider combing two of them in one circuit board. Among all three possible combinations, the two networks shown in **Figure 5(a)** and **(c)** are examined as the exclusive solution for DB performance. Two types of CRLH branches with specified phases at two arbitrary frequencies are necessary to integrate these networks. This distinguishes our design from any previous DB synthesis for other devices [24, 25] which only required one set of CRLH branch. Consequently, DB RRC design is less direct and more complicated than any other DB device design, giving rise to the rarely reported work. In this particular design, the operation frequencies are designed at f_L = 0.75 GHz and f_H = 1.8 GHz, respectively.

In what follows, we begin with the dispersion of CRLH TL to briefly derive the fundamental DB theory. At f_L, the RRC works at the state in **Figure 5(c)** with three +90° branches and one –90° branch, whereas at f_H it operates at the state in **Figure 5(a)** with three –90° branches and one –270° branch. For convenience, we denote the CRLH branch with +90° at f_L while –90° at f_H as CRLH TL_1, whereas CRLH TL_2 is the CRLH branch with –90° at f_L while –270° at f_H. Consequently, the DB RRC is formed by three CRLH TL_1 and one CRLH TL_2. Here, φ_L and φ_H are the required phases at f_L and f_H, respectively:

$$\varphi^{\text{CRLH}}(\omega = \omega_L = 2\pi f_L) = \varphi_L \tag{10a}$$

$$\varphi^{\text{CRLH}}(\omega = \omega_H = 2\pi f_H) = \varphi_H \tag{10b}$$

Combining Eqs. (4), (5) and (10), we can readily obtain explicit expressions of four circuit parameters as [25]

$$L_R = \frac{Z_c[\varphi_L(f_L/f_H) - \varphi_H]}{2\pi N f_H[1-(f_L/f_H)^2]}, \tag{11a}$$

$$C_R = \frac{\varphi_L(f_L/f_H) - \varphi_H}{2\pi N f_H Z_c[1-(f_L/f_H)^2]}, \tag{11b}$$

$$L_L = \frac{N Z_c[1-(f_L/f_H)^2]}{2\pi f_L[\varphi_L - \varphi_H(f_L/f_H)]}, \tag{11c}$$

$$C_L = \frac{N[1-(f_L/f_H)^2]}{2\pi f_L Z_c[\varphi_L - \varphi_H(f_L/f_H)]}. \tag{11d}$$

To preserve L_R and C_R positive, the following condition should be satisfied:

$$\varphi_L f_L \geq \varphi_H f_H \tag{12}$$

Following Eq. (12), the possible solution by combining circuit topologies shown in **Figure 5(b)** and **(c)** can ruled out for DB synthesis. Given determined circuit parameters, see **Table 1**, we can readily design the final RRC layout using the approach described in Section 1.1. The

TL type		L_L (nH)	C_L (pF)	$2C_L$ (pF)	L_R (nH)	C_R (pF)	ϕ^{RH} (deg)	P	W
CRLH TL_1	TC	11.1	2.2	4.4	8.42	1.68	–64.3	49.1	1.5
	PU	12	2	4.7					
CRLH TL_2	TC	63.1	12.6	25.2	15.3	3.1	–117.2	89.6	1.5
	PU	56 + 6.8	12	12 + 12					

Note: TC means theoretically computed and PU means practically used, P and W are length and width of the ML in mm.

Table 1. Detailed circuit parameters and dimensions of ML of the designed CRLH TLs.

designed RRC is built on F4B substrate with ε_r = 2.65, h = 1 mm and tan δ = 0.001 based on standard printed circuit board (PCB) fabrication process.

Figure 6 shows the finally engineered layout of the fractal-shaped RRC. Again, we consider realizing the RH and LH part of the 70.7 Ω CRLH branches by MLs and SMT chip elements, respectively. In the former case, the MLs are configured in Koch shape of *IF* = 1/4 and *IO* = 2 to facilitate a super compact size. The middle fractal sections are removed to load SMT capacitors and the right-angle bends are replaced by chamfered ones to minimize current discontinuity. In the latter case, two cascaded T-networks depicted in **Figure 6(c)** are adopted. Since the necessary space accommodating CRLH TL_1 is much smaller than that of CRLH TL_2, we changed orientations of several chamfered bends of CRLH TL_2 to facilitate a super compact circuit. As appreciated from **Figure 6(a)** and **(b)**, the proposed DB RRC occupies a square area of 52.2 × 39.4 mm^2 and is only 10.2% of 150 × 135 mm^2 that its conventional circular counterpart occupies. Therefore, our hybrid technology shrinks the circuit by 89.8%.

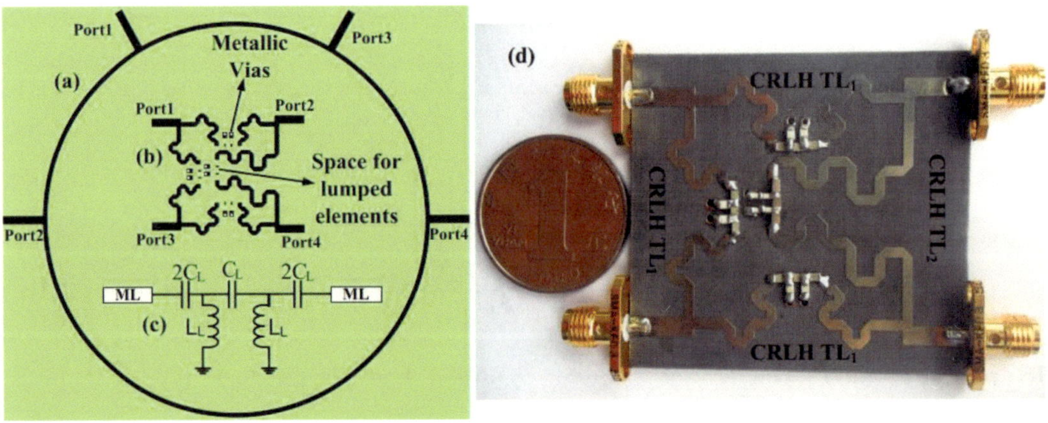

Figure 6. Circuit configuration of the RRCs: (a) conventional design, (b) novel design, (c) circuit topology for LH part of two-cell CRLH TLs, and (d) fabricated prototype.

For characterization, the eventually designed DB RRC is analyzed through the dynamic links and solver of planar EM and circuit cosimulation in Ansoft Designer. For verification, the fabricated RRC sample, see **Figure 6(d)** is measured by *Anritsu ME7808A* vector network analyzer. The SMT chip capacitors and inductors with 0805 and 0603 packages are adopted in prototype fabrication. **Figures 7** and **8** portray the *S*-parameters for in-phase and out-of-phase operation, respectively. In both cases, a reasonable agreement of results between simulation and measurement is observed across the entire frequency band of interest, confirming the effectiveness of our design. The DB performance occurs clearly around 0.75 and 1.8 GHz. For in-phase operation at 0.75 GHz, the measured $|S_{11}|$ is 24.2 dB, $|S_{21}|$ and $|S_{31}|$ are 3.4 and 3.1 dB and $|S_{41}|$ is 28.3 dB, whereas at 1.8 GHz, the measured $|S_{11}|$ is 19.9 dB, $|S_{21}|$ and $|S_{31}|$ are 3.2 and 3.5 dB and $|S_{41}|$ is 28.5 dB. The out-of-phase performance at 0.75 and 1.8 GHz is similar to the in-phase case and is not discussed for brevity of contents. The slight frequency

http://dx.doi.org/10.5772/65603

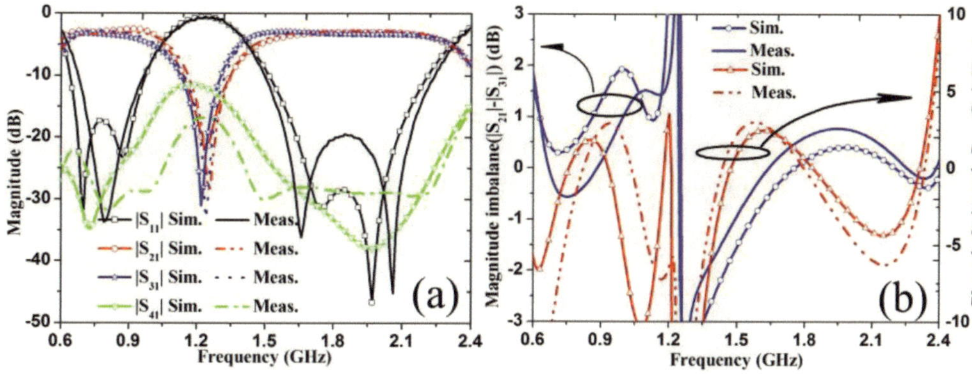

Figure 7. Comparison of *S* parameters between simulation and measurement for in-phase operation: (a) magnitude and (b) output magnitude imbalance and phase imbalance.

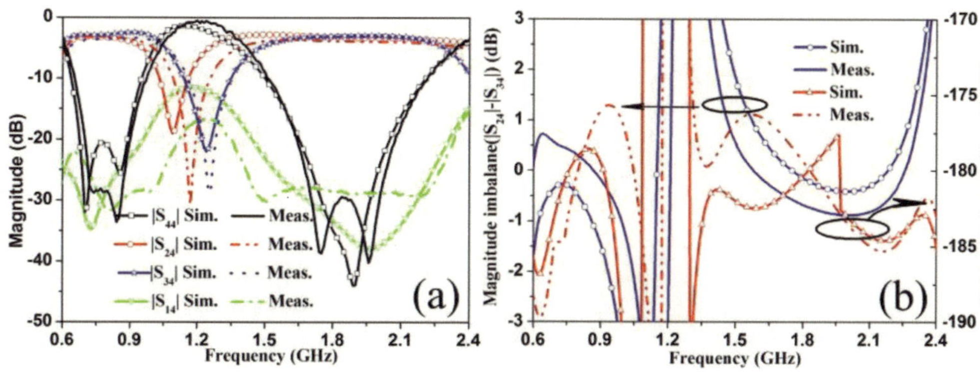

Figure 8. Comparison of *S* parameters between simulation and measurement for 180° out-of-phase operation: (a) magnitude and (b) output magnitude imbalance and phase imbalance.

shift of the first band toward higher frequencies in the measurement case is the same as that discussed in Section 1.2.1. Nevertheless, the discrepancy is within an acceptable range.

Tables 2 and **3** detail the simulated and measured results for both in-phase and out-of-phase operation. In the former case, measured results indicate that $|S_{11}| > 15$ dB, $|S_{21}|$ and $|S_{31}| < 5$ dB, $|S_{41}| > 20$ dB and magnitude and phase imbalances varying within 1 dB and 5° are obtained from 0.73 to 0.92 GHz. Therefore, a fractional bandwidth (FBW) of 190 MHz (a relative bandwidth of 25.3%) is achieved around f_L, whereas a FBW of 26.7% is obtained from 1.56 to 2.04 GHz around f_H. In the latter case, a FBW of 32% characterized by $|S_{44}| > 15$ dB, $|S_{24}|$ and $|S_{34}| < 5$ dB, $|S_{14}| > 20$ dB and magnitude and phase imbalances varying within 1 dB and 5° is acquired from 0.69 to 0.93 GHz around f_L while a FBW of 28.3% from 1.59 to 2.1 GHz is achieved around f_H. In general, the developed DB RRC demonstrated with a modest operation bandwidth and excellent in-band performances in terms of low insertion loss and return loss should be highlighted.

		RL	IL	Isolation	MI	PI
CF_1	Sim.	18.2	2.9 and 3.3	33.1	0.35	–0.3
	Meas.	24.2	3.4 and 3.1	28.3	–0.4	–4.2
FBW_1	Sim.	0.66–0.94	0.61–0.99	0.57–0.97	0.62–0.87	0.66–0.99
	Meas.	0.7–0.92	0.64–1.01	0.54–1.1	0.63–1	0.73–1.1
CF_2	Sim.	29.3	3.1 and 3.2	33.3	0.15	0.5
	Meas.	19.9	3.2 and 3.5	28.5	0.6	–0.15
FBW_2	Sim.	1.58–2.14	1.48–2.29	1.45–2.31	1.57–2.44	1.4–2.36
	Meas.	1.56–2.16	1.46–2.3	1.11–2.33	1.5–2.4	1.37–2.04

Table 2. Detailed in-phase performances of the proposed RRC.

		RL	IL	Isolation	MI	PI
CF_1	Sim.	21.6	3.27 and 2.97	33.1	–0.29	–0.29
	Meas.	28.4	3.25 and 3.7	28.4	0.43	–3.3
FBW_1	Sim.	0.66–0.91	0.62–0.96	0.56–0.96	0.62–0.86	0.66–0.96
	M	0.68–0.93	0.63–1	0.54–1.12	0.58–0.99	0.69–1.09
CF_2	Sim.	34.8	3.2 and 3.1	33.3	–0.15	–0.34
	M	31.7	3.89 and 3.22	28.57	–0.67	0.6
FBW_2	Sim.	1.58–2.11	1.49–2.29	1.46–2.31	1.57–2.28	1.34–2.39
	M	1.59–2.13	1.46–2.28	1.36–2.33	1.47–2.31	1.31–2.1

Note: RL/IL is return/insertion loss in dB, while MI and PI are magnitude imbalance in dB and phase imbalance in degree. FBW_1 and FBW_2 are bandwidth around f_L and f_H in GHz characterized by $|S_{11}|/|S_{44}| \geq 15$ dB, $|S_{21}|$ and $|S_{31}| \leq 5$ dB or $|S_{24}|$ and $|S_{34}| \leq 5$ dB, $|S_{41}| \geq 20$ dB, $|MI| \leq 1$ dB, and $|PI| \leq 5°$.

Table 3. Detailed out-of-phase performances of the proposed RRC.

To sum up, a super compact CRLH RRC is successfully engineered with good DB performances based on proposed DB strategy and hybrid approach. The RRC features low insertion loss, modest operation bandwidth and good isolation between output ports. The 89.8% size reduction is believed to be one of the best miniaturizations in the open literature which should be highlighted.

1.3. Analysis and characterization of novel distributed CRLH atoms

Although lumped-element CRLH TL exhibits the merit of easy design and high degree of freedom, it is rigorously restricted to low-frequency operation due to self-resonant effects of chip components at high frequency. In this subsection, we will introduce a set of distributed resonant-type CRLH TLs made of novel complementary split ring resonators (CSRRs). Three types of CSRRs are involved in terms of compactness, namely complementary single split ring resonator (CSSRR) pair (CSSRRP), Koch-shaped CSSRRP (K-CSSRRP) and Koch-extended CSSRRP (K-ECSSRRP) [26, 27]. They are investigated in depth based on TL theory and electromagnetic (EM) response characterization, aiming to illustrate the novel working mechanism for miniaturization. Of particular irreverence is the dual-shunt branch circuit theory established for a new set of CRLH TLs.

1.3.1. CSSRRP-loaded CRLH atom

The topology of the first CSRRs-evolved CRLH element is shown in **Figure 9(a)**. As is shown, the CSSRRP etched on the ground is composed of two CSSRRs with face-to-face splits beneath a capacitive gap. When the element operates in LH band where the backward propagation is supported, the CSSRRP responds to the time varying axial electric field and affords a negative permittivity. In the circuit model (losses have been excluded) shown in **Figure 9(c)**, the CSSRRP is described by means of a parallel resonant tank formed by L_p, C_p and C_k. C is contributed by the electrical coupling between the series gap and the CSSRRP and the line capacitance. The residual circuit elements share the same physical meaning with those of conventional CSRRs [28]. Here, we introduce additional capacitance C_k to model the interaction between two CSSRRs. Therefore, our proposal allows additional degree of freedom for design and the circuit model can be treated as a special case of a CSSRR-loaded CRLH atom shown in **Figure 9(b)** when C_k is null. The correctness of our model will be validated by full-wave S-parameters calculated through the commercial MOM (moment of method)-based package Ansoft Designer. For characterization, the F4B substrate with a thickness of $h = 0.8$ mm and a dielectric constant of $\varepsilon_r = 2.65$ is considered for all simulations and power divider design.

Figure 10 depicts the frequency response of a CSSRRP- and CSSRR-loaded CRLH TLs obtained from EM full-wave simulation and electrical simulation (circuit model) in Ansoft

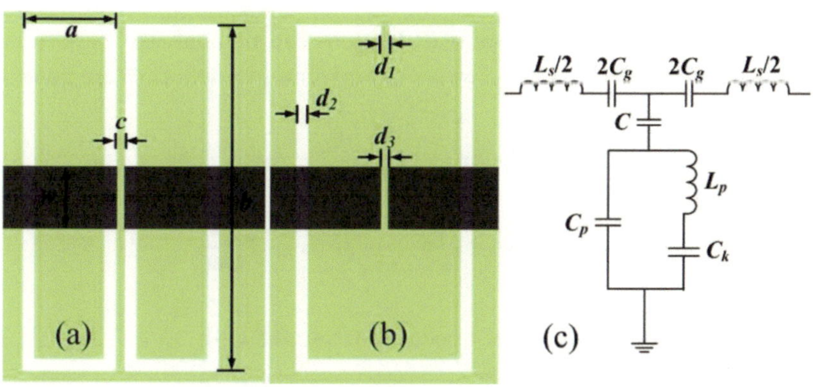

Figure 9. Topology of the: (a) CSSRRP and (b) CSSRR loaded CRLH TLs along with (c) the T-type equivalent circuit model. The capacitive gap is etched on the conductor strip (depicted in dark grey), beneath which the CSSRRP and CSSRR (depicted in white) are etched on the ground (depicted in green). Here, the CSSRRP and CSSRR occupy the same size and have $d_1 = 0.3$ mm, $d_3 = 0.2$ mm, $d_2 = 0.3$ mm, $b = 9.6$ mm, and $w = 1.8$ mm. For CSSRRP, the metallic spacing between two CSSRR is $c = 0.3$ mm and $a = 2.4$ mm.

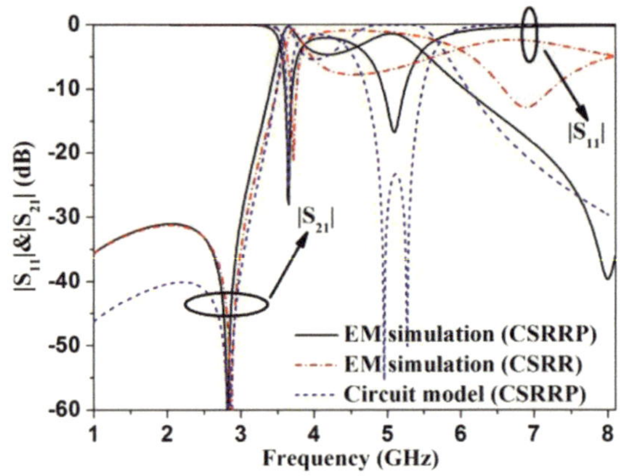

Figure 10. Frequency response of the CSSRRP-loaded and CSSRR-loaded CRLH TL obtained from full-wave EM simulation and electrical simulation. The lumped elements are extracted as L_s = 17.1 nH, L_p = 2.97 nH, C_g = 0.067 pF, C = 363.6 pF, C_p = 1.03 pF, and C_k = 1.06 pF.

Serenade. As is shown, the S-parameters of proposed CRLH atom in both cases are in excellent agreement, illustrating the rationality of the model. Moreover, our meta-atom exhibits slightly lower resonance (LH band) and lower transmission zero below relative to CSSRR-loaded one, enabling a more compact subwavelength particle. Moreover, our atom exhibits lower in-band insertion loss for both the LH and RH band (centered at 5.09 GHz). Most importantly, the frequency interval between the LH band and RH band is much narrower than that of the CSSRR-loaded CRLH atom whose RH band is observed at 6.88 GHz. Therefore, it is easier to realize the balanced condition using proposed meta-atom. The sharp transmission zero above the RH band is more suitable for in-band selectivity and out-of-band harmonics suppression. The identification of aforementioned LH and RH bands will be examined through effective constitutive parameters.

The exotic EM behavior of the proposed CRLH atom can be analyzed based on equivalent circuit model inspired by Bloch theory. The series and shunt impedances are given by

$$Z_s(j\omega) = \frac{1-\omega^2 L_s C_g}{j2\omega C_g}, \tag{13}$$

$$Z_p(j\omega) = \frac{(1-\omega^2 L_p C_k)}{j\omega(C_p + C_k)-j\omega^3 L_p C_p C_k} + \frac{1}{j\omega C}. \tag{14}$$

The shunt admittance can be obtained immediately from Eq. (14):

$$Y_p(j\omega) = 1/Z_p(j\omega) = \frac{j\omega C[C_p + C_k-\omega^2 L_p C_p C_k]}{C_p + C_k + C-\omega^2 L_p C_k(C_p + C)} \tag{15}$$

As crucial parameters for circuit design, the electrical length φ and characteristic impedance Z_β of a CRLH TL are determined by

http://dx.doi.org/10.5772/65603

$$\cos\varphi = \cos(\beta l) = 1 + \frac{Z_s(j\omega)}{Z_p(j\omega)}, \tag{16}$$

$$Z_\beta = \sqrt{Z_s(j\omega)[Z_s(j\omega) + 2Z_p(j\omega)]} \tag{17}$$

Thus, by insertion of Eqs. (13) and (14) into Eqs. (16) and (17), the propagation constant β and Z_β are obtained. The EM wave propagation is allowed in the region where both β and Z_β are real and positive values. In what follows, we will discuss the limits of the LH and RH band and finally derive the balanced condition for broadband operation. Similar to CSRRs-loaded CRLH atom [28], the lower limit of the RH band f_{RH}^L corresponds to the resonances of series branch:

$$f_{RH}^L = f_{se} = 1/2\pi\sqrt{L_s C_g} \tag{18}$$

In a similar manner, the transmission zero f_T occurs below the lower limit of LH band f_{LH}^L when the shunt branch resonates, namely Eq. (14) is null or Eq. (15) takes an extreme value:

$$f_T = f_{sh} = \frac{1}{2\pi}\sqrt{\frac{C_p + C_k + C}{(L_p C_k (C_p + C)}} \tag{19}$$

When the parallel tank of CSSRRP resonates, we obtain the upper limit of the LH band f_{LH}^H by forcing Eq. (15) to be zero:

$$f_{LH}^H = f_p = \frac{1}{2\pi}\sqrt{\frac{C_p + C_k}{L_p C_k C_p}} \tag{20}$$

Here, Eqs. (18) and (20) are reasonable on assumption that $f_p \leq f_{se}$. If $f_p > f_{se}$, f_{RH}^L and f_{LH}^H interchange their value. Under balanced condition $f_{LH}^H = f_{RH}^L = f_0$, the LH band transits to the RH one continuously without a gap. We can obtain f_{LH}^L and the upper limit of RH band f_{RH}^H when Eq. (17) is null. The analytic expressions for f_{LH}^L and f_{RH}^H are tedious and hence are not reproduced here. However, they are revealed in **Figure 11** through series impedance, shunt admittance and characteristic impedance calculated using the extracted circuit parameters shown in the caption of **Figure 10**.

Referring to **Figure 11**, Y_p is null at 4.03 GHz (f_{LH}^H) and takes extreme value at 2.83 GHz (f_T) which coincide well with f_T = 2.82 GHz obtained in **Figure 10**. Moreover, Z_{se} is observed as null at 4.71 GHz (f_{RH}^L) and therefore a stop band emerges in the unbalanced case as $f_{LH}^H \neq f_{RH}^L$. In addition, Z_β takes zero value at 3.61 (f_{LH}^L) and 5.59 GHz (f_{RH}^H). Within 3.61~4.03 GHz, Z_β/β takes positive/negative real number which reveals a LH transmission, whereas within 4.03–4.71 GHz both Z_β and β take pure imaginary value which represents loss and a stop band. Finally, both Z_β and β are positive real numbers within 4.71–5.59 GHz, indicating a RH transmission. In general, the results in **Figure 11** coincide well with those in **Figure 10**.

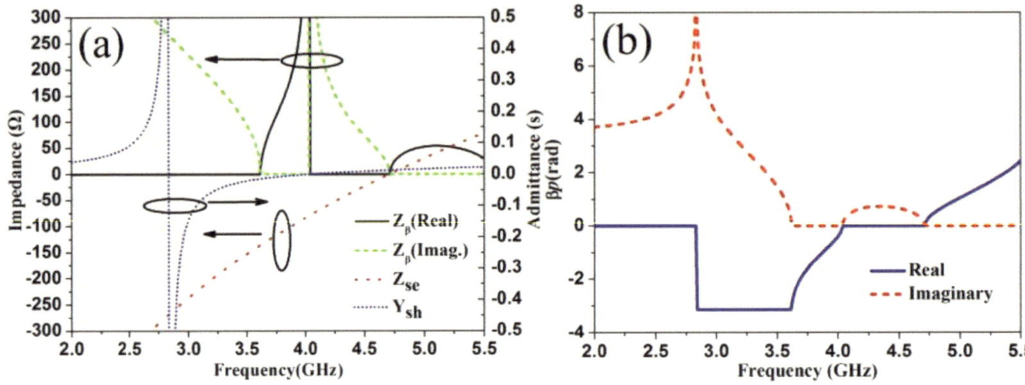

Figure 11. Representation of: (a) series impedance, shunt admittance, and characteristic impedance versus frequency and (b) propagation constant of the CSSRRP-loaded unbalanced CRLH TL in Figure 9.

Figure 12 illustrates the extracted effective constitutive parameters of the CSSRRP-loaded CRLH atom. Negative refractive index and propagation constant can be clearly observed within 2.74–3.94 GHz. The LH band that allows signals to be transmitted freely is observed within 3.56–3.94 GHz, where the imaginary parts of refractive index, permittivity and permeability approximate to zero. However, the imaginary parts associated with electric or magnetic loss are considerable within 2.74–3.56 GHz where the EM wave propagation is still not allowed. A further inspection reveals that an obvious electric resonance occurs around 3.6 GHz with negative permittivity. The vanished electric resonance for the meta-atom without CSSRRP illustrates that the new slot gives rise to the negative permittivity. Since the effective permeability is negative across the entire band, the simultaneous negative LH band is solely dependent on the negative permittivity band (3.56– 3.94 GHz). The retrieved β is consistent with φ obtained from Bloch analysis except for the magnitude discrepancy which is inherently introduced by the multiplication factor l in Eq. (4). In general, the constitutive EM parameters interpreted all exhibited EM phenomena.

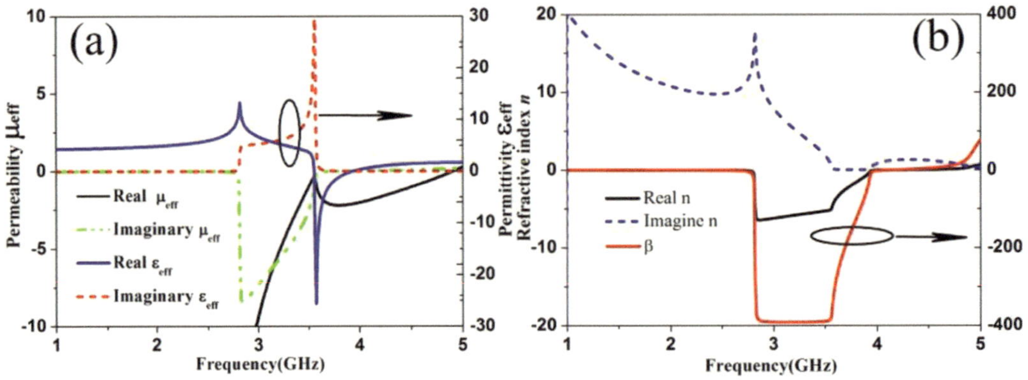

Figure 12. The effective constitutive EM parameters of the CSSRRP-loaded CRLH TL: (a) effective permittivity and permeability and (b) refractive index n and propagation constant β.

Our further numerical results (not included here for brevity) show that LH band shifts downwards when either width a or length b increases. The increase of a and b enhances both C_P and C. However, the in-band insertion loss increases drastically and the LH band may even vanish. The influence of the series gap and microstrip line is analogous to that of conventional CSRRs [28] and thus is not discussed. Most importantly, it is easier to engineer a balanced condition when $2a \approx b$ is guaranteed. In this case, the CRLH meta-atom may exhibit an inherent balanced point. Nevertheless, such balanced point is nonexclusive which will be validated below.

1.3.2. KCSSRRP-loaded CRLH atom with balanced condition

Since the LH band of above CRLH atom is narrow, it is preferred for narrow-band filter applications. Moreover, the potential for further miniaturization of CSSRRP-loaded CRLH atom is still available. Here, we will improve the in-band bandwidth and demonstrate the zero-phase behavior by engineering a miniaturized balanced CRLH TL.

As discussed earlier, the position of RH band is much influenced by a and b. As such, we construct the slot of CSSRRP in Koch curve for other demand and further miniaturization. The new K-CSSRRP shares the same operation mechanism with CSSRRP and thus exhibits similar LH characteristic. The Koch slot of zig-zag boundary significantly extends the current path in ground plane, namely, increases the electrical wavelength in terms of an enhanced C_p. To facilitate a balanced condition ($f_{LH}^{H} = f_{RH}^{L} = f_0$), a low-impedance patch is introduced in the upper conductor line to increase L_s and C_g and thus shifts the RH band downwards. Moreover, it is advisable to maintain the C as a small value to preserve the LH characteristic with low loss.

In the first design example (see **Figure 13**), the K-CSSRRP is with partial fractal boundary only by constructing the two outmost vertical sides as Koch curves of third order. **Figure 14** depicts

Figure 13. Layout of K-CSSRRP-loaded zero phase-shift CRLH TL operating at Satellite DMB band. The characteristic impedance corresponds to 50 Ω and detailed geometrical parameters are $a = 4$ mm, $b = 12$ mm, $c = 0.3$ mm, $d_1 = d_2 = d_3 = 0.3$ mm, $m = 1.7$ mm and $h = 8$ mm. The width of microstrip line is 0.83 mm. The IF of Koch curve is 1/3.

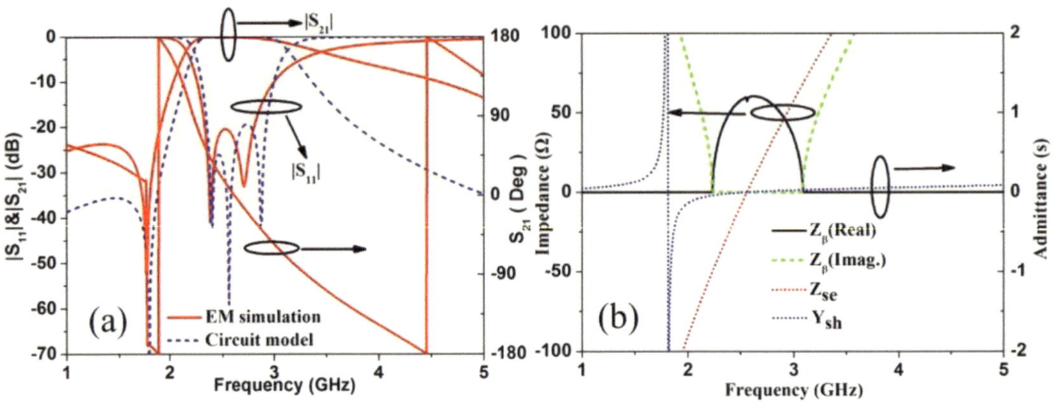

Figure 14. (a) *S*-parameters and (b) representation of series impedance, shunt admittance, and characteristic impedance of the K-CSSRRP-loaded balanced zero phase-shift CRLH atom. The extracted lumped elements are: $L_s = 22.5$ nH, $L_p = 2.36$ nH, $C_g = 0.172$ pF, $C = 369.4$ pF, $C_p = 3.217$ pF, and $C_k = 3.34$ pF.

the simulated *S*-parameters and the representation of series impedance, shunt admittance and characteristic impedance. From **Figure 14(a)**, a reasonable agreement of results is clearly observed between circuit and full-wave EM simulations. The exact zero phase response is obtained within the Satellite DMB band centered at 2.63 GHz. The balanced condition can be further validated through theoretical analysis results shown in **Figure 14(b)**, where the shunt admittance and series impedance intersect at 2.56 GHz. At this frequency, the LH band transforms to the RH band continuously without a band gap.

In the second design example (see **Figure 15**), the four-vertical and four-horizontal slots are constructed as Koch curves of quasi-second iteration order to roughly guarantee $2a \approx b$ in a square configuration with full fractal boundary. In this regard, a balanced CRLH TL with bandwidth enhancement will be achieved as discussed above. To accommodate all fractal

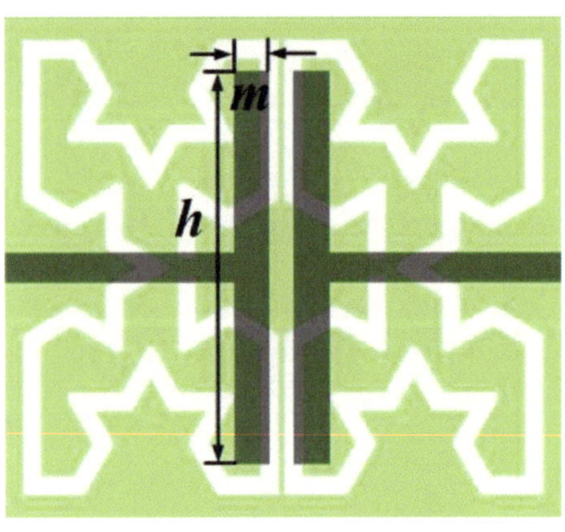

Figure 15. K-CSSRRP-loaded phase-shift CRLH atom with characteristic impedance of 70.7 Ω operated at WiMAX band. The geometrical parameters (in mm) are: $a = 4$, $b = 7.2$, $c = 0.1$, $d_1 = d_2 = 0.3$, $d_3 = 0.4$, $m = 0.6$, $h = 6.5$. The width of microstrip line is 0.5 mm and the iteration factor of Koch slot is 1/3.

http://dx.doi.org/10.5772/65603

sections in a highly compact region for further size reduction, some Koch sections formed in the first-order iteration process are removed. The lumped-element circuit parameters are also extracted for theoretical analysis, see caption of **Figure 16**.

Figure 16 portrays the corresponding frequency response, along with representation of series impedance, shunt admittance and characteristic impedance. From **Figure 16(a)**, it is obvious that *S*-parameters obtained from circuit and EM simulations are in excellent agreement over the entire frequency range, further confirming the correctness of proposed model. The return loss is better than –10 dB from 3.17 to 4.13 GHz and all EM phenomena reveal a balanced pass band without stop band interruption. Moreover, the zero phase shift occurs within WiMAX band centered at 3.5 GHz. From **Figure 16(b)**, Z_β takes a positive real number within the pass band of 3.11–4.22 GHz and the balanced point occurs at 3.52 GHz where the series impedance and shunt admittance intersect on the frequency axis. The transmission zero is always observed below the LH band no matter in balanced or unbalanced case. The fractal perturbation in CSSRRP results in a significant lower operation band whose center frequency reduces from 5.01 (without fractal geometry) to 3.5 GHz. Thus, a frequency reduction of 30.2% is obtained.

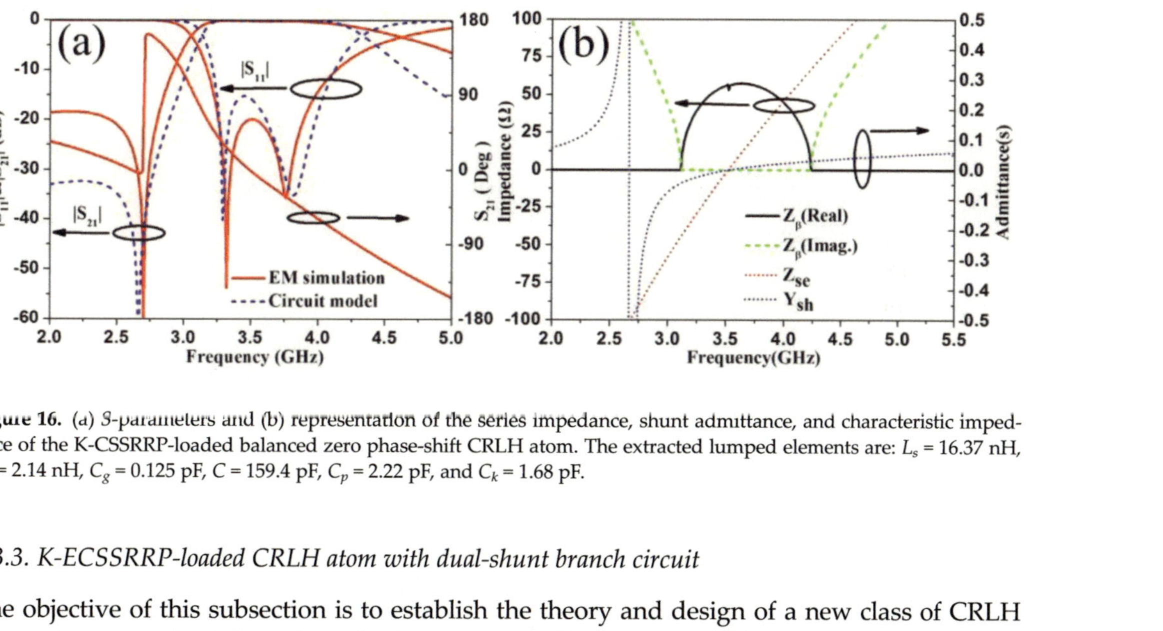

Figure 16. (a) *S*-parameters and (b) representation of the series impedance, shunt admittance, and characteristic impedance of the K-CSSRRP-loaded balanced zero phase-shift CRLH atom. The extracted lumped elements are: L_s = 16.37 nH, L_p = 2.14 nH, C_g = 0.125 pF, C = 159.4 pF, C_p = 2.22 pF, and C_k = 1.68 pF.

1.3.3. K-ECSSRRP-loaded CRLH atom with dual-shunt branch circuit

The objective of this subsection is to establish the theory and design of a new class of CRLH atoms that own dual-shunt branch circuit and enable compact RF/microwave devices with harmonic suppression. As a result, it is unnecessary to cascade a chain of CSRRs for harmonic suppression, which in turn leads to super compact devices.

Figure 17 shows novel dual-shunt brunch circuit model and the layout of novel CRLH atom evolved from previous CSSRRP-loaded CRLH structure. Here, previous CRLH atom, see **Figure 17(a)** is reproduced for comparison convenience. The major difference of novel CRLH atom lies in the K-ECSSRRP etched in the ground, see **Figure 17(b)**. It is constructed by expanding each end-point of CSSRRP with an inner smaller slot made of three second-order Koch curves and one first-order Koch curve.

Similar to CSSRRP [26], the K-ECSSRRP beneath the capacitive gap excites current along the zig-zag boundary and generates two effects in response to the time varying axial electric field, i.e., the electric excitation to external CSSRRP and that to inner Koch slot. The electric excitation to inner slot is independent of external CSSRRP and thus also provides a shunt branch in the circuit model like CSSRRP. To guarantee effective excitation, the center area of the complementary resonator should be unoccupied. Both excitations contribute to the negative permittivity. In the CM shown in **Figure 17(c)** and **(d)**, losses are irrespective of for analysis convenience. L_s represents the line inductance, C_g models the gap capacitance, CSSRRP is described by means of a parallel resonant tank formed by L_{p1}, C_{p1} and C_{k1}, C_1 contains the line capacitance and the electrical coupling through the gap to CSSRRP. In like manner, the shunt brunch formed by C_2, L_{p2}, C_{p2} and C_{k2} models the corresponding effect of the Koch slot. Consequently, novel CRLH atom enhances considerably the design flexibility.

To begin with, we perform numerical circuit analysis to identify the transmission zeros, cutoff frequencies, LH characteristic and balanced condition. Assume a two-port CRLH TL with N cells, the transmission behavior can be readily predicted using the [*ABCD*] matrix method. Here, the *ABCD* parameters can be easily obtained from *S*-parameters. Provided a [*ABCD*] matrix for a single-cell CTLH atom with known input and output voltage and current, the transmission characteristic of *N*-cell CRLH TL can be straightly achieved by cascading *N* two-port networks:

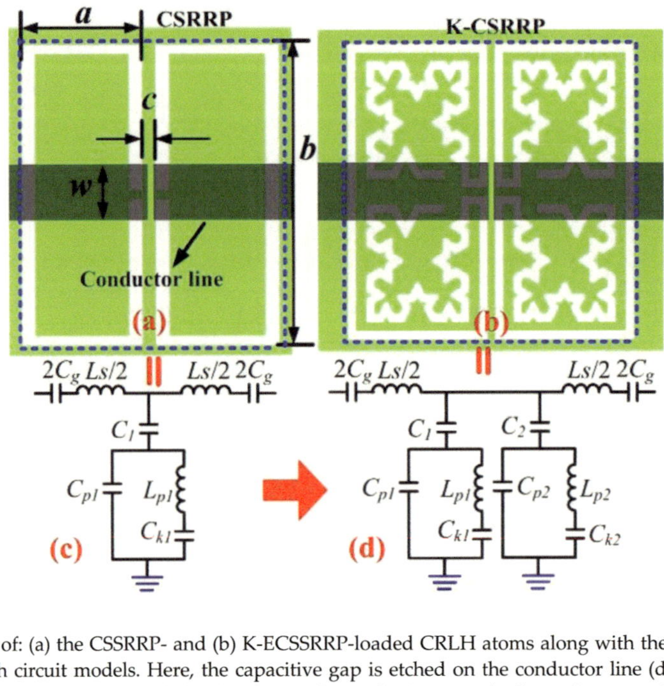

Figure 17. Topology of: (a) the CSSRRP- and (b) K-ECSSRRP-loaded CRLH atoms along with their (c) single-shunt and (d) dual-shunt branch circuit models. Here, the capacitive gap is etched on the conductor line (depicted in lights grey) beneath which, the CSSRRP and K-ECSSRRP (depicted in white) are etched on the ground (depicted in green). Both CRLH atoms have identical slot width 0.3 mm and gap width of 0.2 mm, the residual geometrical parameters are $a = 5.3$ mm, $b = 10.3$ mm, $c = 0.3$ mm, and $w = 2$ mm.

http://dx.doi.org/10.5772/65603

$$\begin{bmatrix} V_{\text{out}} \\ I_{\text{out}} \end{bmatrix} = \begin{bmatrix} A_N & B_N \\ C_N & D_N \end{bmatrix} \begin{bmatrix} V_{\text{in}} \\ I_{\text{in}} \end{bmatrix} = \prod_{k=1}^{N} \begin{bmatrix} A & B \\ C & D \end{bmatrix}^N \begin{bmatrix} V_{\text{in}} \\ I_{\text{in}} \end{bmatrix} \tag{21}$$

Although CRLH atoms are isotropic and periodic in this design for computational and fabrication convenience, it is unnecessary to require this condition in practice. The $[ABCD]$ matrix of a CRLH atom is related with the series impedance Z and shunt admittance Y (Y_1 and Y_2 for each branch) as

$$\begin{bmatrix} A & B \\ C & D \end{bmatrix} = \begin{bmatrix} 1 & Z/2 \\ 0 & 1 \end{bmatrix} \begin{bmatrix} 1 & 0 \\ Y_1+Y_2 & 1 \end{bmatrix} \begin{bmatrix} 1 & Z/2 \\ 0 & 1 \end{bmatrix} = \begin{bmatrix} 1+Z(Y_1+Y_2)/2 & Z[1+Z(Y_1+Y_2)/4] \\ Y_1+Y_2 & 1+Z(Y_1+Y_2)/2 \end{bmatrix} \tag{22}$$

Z is formulated as a function of circuit elements:

$$Z = \frac{2(1-\omega^2 L_s C_g)}{j\omega C_g} \tag{23}$$

The calculation of Y is somewhat tedious and can be attained as

$$Y = Y_1 + Y_2 = \frac{j\omega C_1[C_{p1} + C_{k1} - \omega^2 L_{p1} C_{p1} C_{k1}]}{C_{p1} + C_{k1} + C_1 - \omega^2 L_{p1} C_{k1}(C_{p1} + C_1)} + \frac{j\omega C_2[C_{p2} + C_{k2} - \omega^2 L_{p2} C_{p2} C_{k2}]}{C_{p2} + C_{k2} + C_2 - \omega^2 L_{p2} C_{k2}(C_{p2} + C_2)} \tag{24}$$

Apply the periodic boundary condition to the two-port network, we have $V_{\text{out}} = e^{-i\beta p} V_{\text{in}}$ and $I_{\text{out}} = e^{-i\beta p} I_{\text{in}}$, where p is the physical length of a CRLH atom and β is the propagation constant. Taking this condition together with Eq. (21) yields

$$\begin{bmatrix} A & B \\ C & D \end{bmatrix} \begin{bmatrix} V_{\text{in}} \\ I_{\text{in}} \end{bmatrix} - e^{-i\beta p} \begin{bmatrix} V_{\text{in}} \\ I_{\text{in}} \end{bmatrix} = \begin{bmatrix} A - e^{-i\beta p} & B \\ C & D - e^{-i\beta p} \end{bmatrix} \begin{bmatrix} V_{\text{in}} \\ I_{\text{in}} \end{bmatrix} = 0 \tag{25}$$

The determinant of Eq. (25) should be zero to guarantee a nontrivial solution, which yields [25]

$$AD - (A + D)e^{-i\beta p} + e^{-2i\beta p} - BC = 0 \tag{26}$$

Insertion of Eq. (22) into Eq. (26), we can obtain the dispersion relation associated with Z, Y_1 and Y_2 as

$$\cos(\beta p) = 1 + Z(Y_1 + Y_2)/2. \tag{27}$$

Take the condition of Brillouin zone ($\beta p = \pi$), we immediately have

$$Z(Y_1 + Y_2) = -4 \tag{28}$$

The lower cutoff of LH band $\omega_{LH}^{L} = 2\pi f_{LH}^{L}$ and the upper cutoff of RH band $\omega_{RH}^{H} = 2\pi f_{RH}^{H}$ can be obtained by solving Eq. (28). In the same time, we obtain the equation of $Z(Y_1 + Y_2) = 0$

when take $\beta = 0$. In this regard, the upper limit of LH band $\omega_{LH}^{H} = 2\pi f_{LH}^{H}$ and lower limit of RH band $\omega_{RH}^{L} = 2\pi f_{RH}^{L}$ are achieved by solving the following equations:

$$\begin{aligned} &C_1[C_{p1} + C_{k1} - \omega^2 L_{p1} C_{p1} C_{k1}][C_{p2} + C_{k2} + C_2 - \omega^2 L_{p2} C_{k2}(C_{p2} + C_2)] \\ &+\omega C_2[C_{p2} + C_{k2} - \omega^2 L_{p2} C_{p2} C_{k2}][C_{p1} + C_{k1} + C_1 - \omega^2 L_{p1} C_{k1}(C_{p1} + C_1)] = 0 \end{aligned} \tag{29}$$

$$1 - \omega^2 L_s C_g = 0 \tag{30}$$

Moreover, the Bloch impedance Z_β is derived from Eq. (25) after solving Eq. (26) for $e^{-i\beta p}$ and taking reciprocal ($AD - BC = 1$) and symmetric ($A = D$) condition in this particular design:

$$Z_\beta = \frac{V_{\text{in}}}{I_{\text{in}}} = \frac{B}{A - e^{-i\beta p}} = \frac{B}{\sqrt{A^2 - 1}} \tag{31}$$

Insertion of Eq. (22) into Eq. (31) yields

$$Z_\beta = \sqrt{Z[1/(Y_1 + Y_2) + Z/4]} \tag{32}$$

The f_{LH}^{L} and f_{RH}^{H} can be achieved by forcing Eq. (32) to be null. Moreover, two transmission zeros of the two shunt branches are determined when the denominator of Eq. (24) is null:

$$f_{T1} = \frac{1}{2\pi}\sqrt{\frac{C_{p1} + C_{k1} + C_1}{(L_{p1} C_{k1}(C_{p1} + C_1)}}, \tag{33}$$

$$f_{T2} = \frac{1}{2\pi}\sqrt{\frac{C_{p2} + C_{k2} + C_2}{(L_{p2} C_{k2}(C_{p2} + C_2)}}. \tag{34}$$

The CRLH TL is rigorously balanced only when $f_{LH}^{H} = f_{RH}^{L}$ which is particular interest for broadband design. In this case, the LH band (backward wave propagation) switches to the RH region (forward wave propagation) without a gap. Otherwise the continuous pass band is perturbed by a stop band. Rather than obtain cut-off frequencies from tedious analytical expressions by solving Eqs. (28), (29) and (32), we will illustrate them by means of representations of these equations, see **Figure 18**. Moreover, these cutoffs can also be ambiguously identified from transmission zeros of CRLH TLs with large number of cells, see **Figure 19**. Note that f_{LH}^{L} and f_{RH}^{H} do not correspond to the attenuation poles that the two shunt branches afford. In a general case, they fulfill the requirement of $f_{T1} \leq f_{LH}^{L} < f_{LH}^{H} \leq f_{RH}^{L} < f_{RH}^{H} \leq f_{T2}$.

For characterization, **Figure 18(a)** illustrates the *S*-parameters of proposed CRLH atoms obtained from full-wave EM simulation and circuit simulation. The low-loss RT/duroid 4300C substrate with h = 0.5 mm and ε_r = 3.38 is utilized as the dielectric board. As is shown, the results obtained from EM simulation and circuit simulation are in desirable agreements. Therefore, the rationality of the circuit model and lumped parameters are ambiguously verified. Slight discrepancy beyond the upper transmission zero is attributable to the higher-order mode (third attenuation pole) which is not considered in the CM. Novel CRLH atom obviously

http://dx.doi.org/10.5772/65603

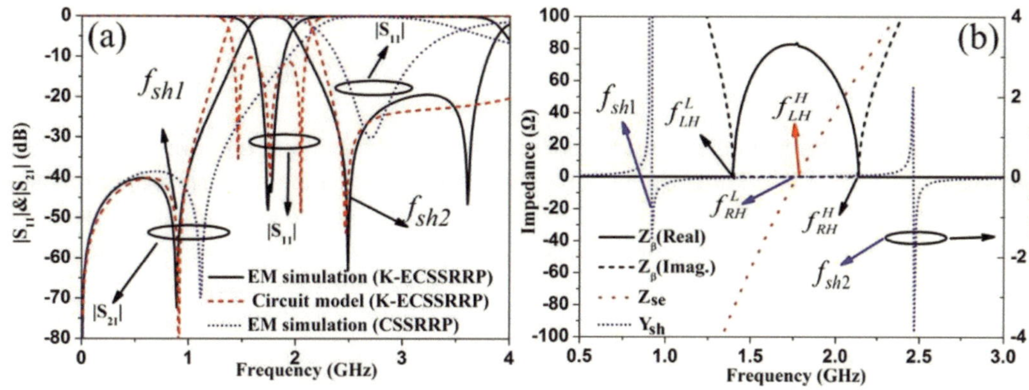

Figure 18. (a) S parameters of proposed CRLH atoms and (b) representation of the series impedance, shunt admittance, and characteristic impedance. The extracted circuit parameters are: $L_s = 31.02$ nH, $C_g = 0.26$ pF, $C_1 = 6.74$ pF, $C_{k1} = 351.4$ pF, $C_{p1} = 0.59$ pF, $L_{p1} = 4.26$ nH, $C_2 = 1.33$ pF, $C_{k2} = 2.16$ pF, $C_{p2} = 0.1$ pF, and $L_{p2} = 4.84$ nH.

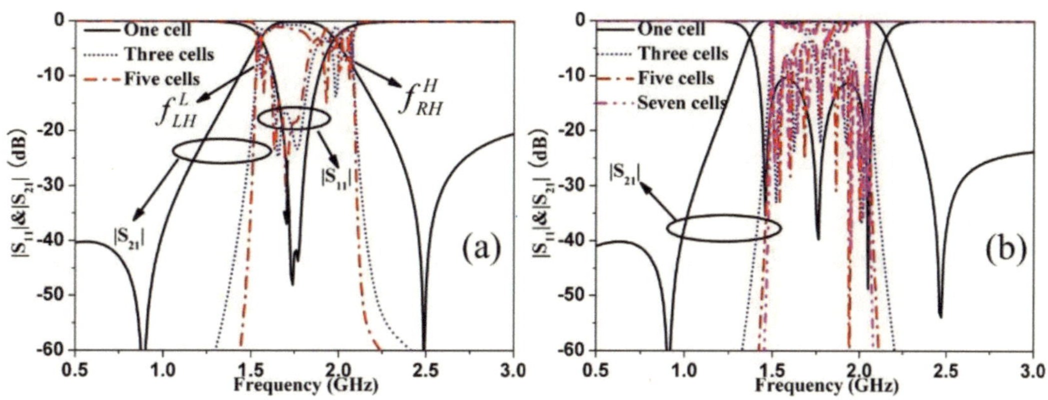

Figure 19. S-parameters of proposed CRLH TLs owning different elements: (a) EM full-wave simulation and (b) CM simulation.

exhibits two transmission zeros around 0.9 and 2.48 GHz in the lower and upper edge of pass band, whereas the upper transmission zero is not evidenced for CSSRRP. In addition, the Koch slot also induces a significant red shift of 33%. From **Figure 18(b)**, two transmission zeros related to resonances of Y_1 and Y_2 are expected in both pass band edges where Z_β takes real numbers. It also reveals that the particle works in balanced condition and the resultant transition occurs at the intersection (1.78 GHz) of Y and Z on the frequency axis.

Figure 19 depicts the S-parameters of proposed CRLH TLs owning different number of cells. The interval between adjacent cells is chosen as 4.6 mm which is a good tradeoff for homogeneity and inter-element coupling. Both EM and CM simulation results indicate that the signal suppression and selectivity at the lower and upper band have been obviously enhanced with increasing number of cells. Very accurate cutoff frequencies can be identified when $N \geq 3$ in EM simulation while $N \geq 7$ in circuit simulation. The $f_{LH}^{L} = 1.52$ and $f_{RH}^{H} = 2.09$ GHz characterized by –10 dB suppression in former case are quantitative very similar to $f_{LH}^{L} = 1.48$ and $\omega_{RH}^{H} = 2.05$ GHz in latter case. The ripples in frequency response spectra are intrinsic effects of finite size and discontinuous network.

It should be emphasized that proposed meta-atom not only exhibits additional attenuation pole in the upper edge band but also has a miniaturized circuit which is on the order of 65% of the CRLH atom using CSSRRP. The extended current path on the ground due to the large compression ratio of Koch curves enhances considerably the LC values of lumped elements in parallel resonant tanks of CM. Further study shows that f_{sh1} is mainly determined by external CSRRP but also affected by inner slot and interspace c between two CSSRRs, however, f_{sh2} is exclusively decided by the inner slot. This affords individual controllability over these two resonant modes.

1.4. Compact microwave device applications

1.4.1. Multiple-way and two-way Wilkinson power dividers

In this subsection, we will demonstrate the possibility of employing novel K-CSSRRP-loaded distributed CRLH TLs in the design of compact three-way fork power divider and a four-way series power divider by substituting zero phase-shift CRLH TLs for conventional 2π TLs. These artificial TLs are qualified for this task because they are readily engineered to work in balanced condition by tuning the circuit parameters. The electrically small size and the possibility to control the phase and impedance over wide ranges further facilitate this task. Although a chain of CRLH elements are more appropriate to be related as MTM TLs, the CRLH TL in current design is implemented using only one element for the sake of compactness.

Since we only have three equations regarding the balanced condition, specified phase φ and characteristic impedance Z_β, it is impossible to uniquely determine the six unknowns. Therefore, several degrees of freedom can be exploited for other requirements, e.g., miniaturization. The design procedure of an impedance transformer with φ and Z_β using distributed CRLH meta-atoms lies in four steps: (1) Select the shape and dimension of CSSRRP to roughly locate the operation frequency within the target band. Given the geometrical parameters of CSSRRP, the elements of L_P, C_P and C_k in the parallel resonant tank are determined. (2) Derive the residual three unknowns through the three available equations via a mathematical software. (3) Roughly synthesize the geometrical parameters of low-impedance patch and gap according to the newly obtained parameters. (4) Optimize the overall layout by taking into account all requirements.

Figure 20(a) depicts the schematic of the three-way fork divider. It is composed of two conventional 2π lines (in-phase power division) with characteristic impedance of Z_0 and three 90° microstrip lines with characteristic impedance of $\sqrt{3}Z_0$ which ensures an impedance match at center frequency. Here, Z_0 is the port reference impedance. We adopt the zero phase-shift CRLH atom shown in **Figure 13** to replace conventional 2π line to reduce the circuit size. Its geometrical parameters shown in the caption are cautiously designed following the design procedure. The host line length of the CRLH atom is only 9.5 mm, representing a significant

http://dx.doi.org/10.5772/65603

size reduction with regard to 76.9 mm of conventional 2π line designed on the same substrate. The host line width of the CRLH atom is deliberately chosen the same as that of 90° microstrip lines for convenience. This is applicable since CRLH atom exhibiting a wide tuning range of Z_β. For verification, the designed power divider is fabricated and measured by an *Anritsu ME7808A* vector network analyzer, see the prototype shown in **Figure 20(b)** and **(c)**. The sample occupying an area of 30 × 43 mm^2 is very compact in size, corresponding to 25.8% of the 30 × 166.4 mm^2 that its conventional counterpart occupies.

To examine the performances, **Figure 21** plots the simulated and measured S-parameters of the power divider. A reasonable agreement of results is observed except for slightly larger insertion loss in the measurement case. From 2.24 to 2.95 GHz, the measured return loss is better than –10 dB and the transmission losses of $|S_{21}|$ and $|S_{41}|$ suggest almost flat response around –4.98 dB. The variation of measured $|S_{31}|$ is slightly larger but within 4.77 ± 1 dB ranging from 2.28 to 2.78 GHz. At the center frequency of 2.63 GHz, an almost equal power split to all three

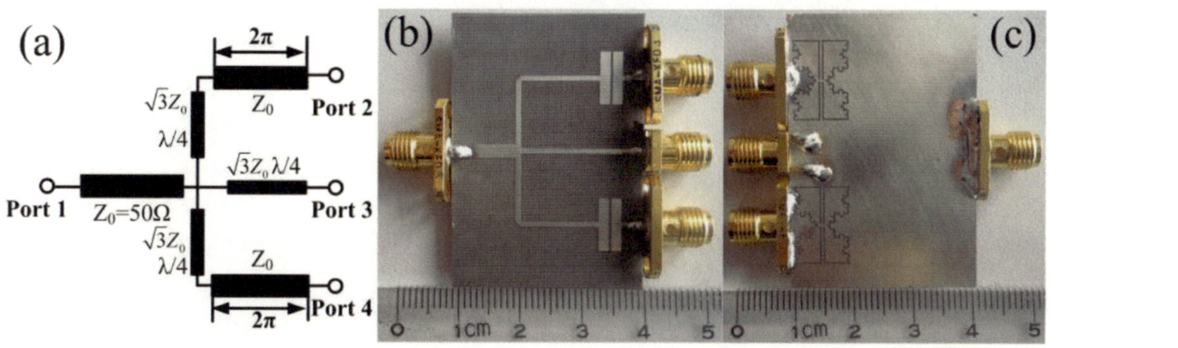

Figure 20. K-CSSRRP-loaded three-way fork power divider: (a) schematic, (b) top view, and (c) bottom view of the fabricated prototype.

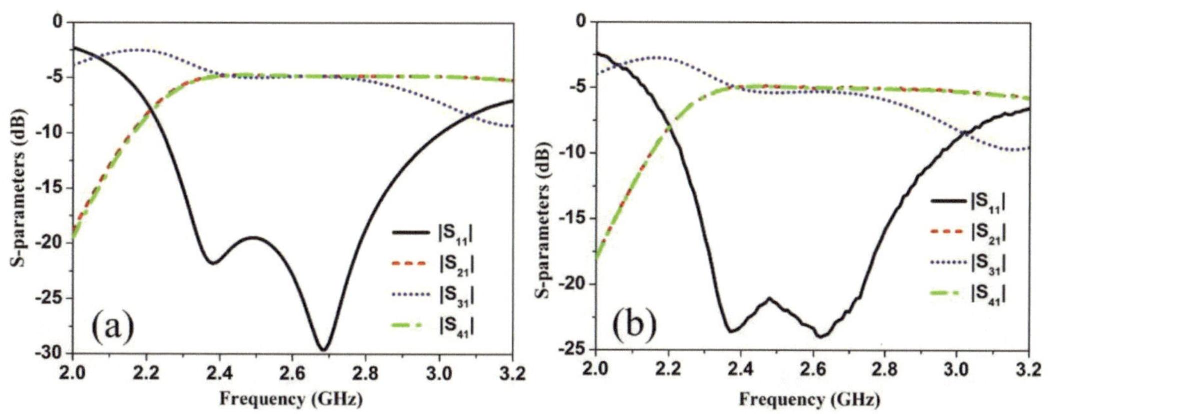

Figure 21. S-parameters of the three-way power divider prototype: (a) full-wave simulation and (b) measurement.

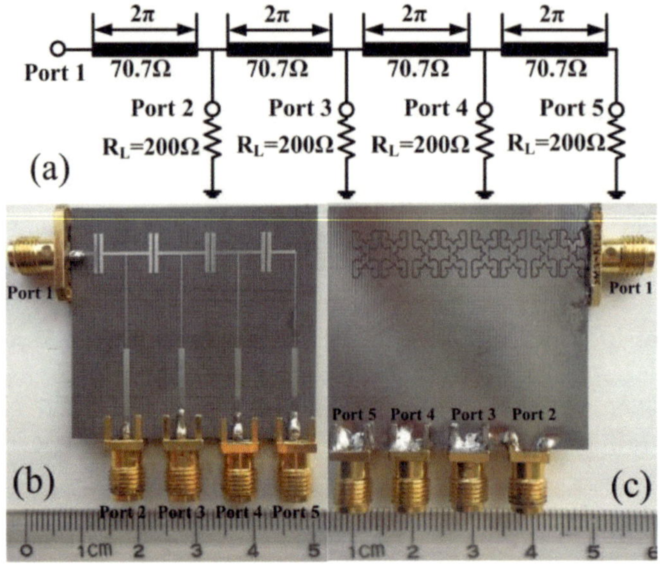

Figure 22. K-CSSRRP-loaded 1:4 series power divider: (a) schematic, (b) top view and (c) bottom view of the fabricated prototype.

output ports is observed. The slight discrepancy is possibly attributable to the radiation loss induced by the complex fractal perturbations on the ground and partially to the soldering pad which is in much proximity to the Koch slot and thus may affect the excitation of Koch slot.

For further application, a 1:4 series power divider is also designed and fabricated using the zero phase-shift CRLH atom shown in **Figure 15**. The corresponding schematic and the fabricated prototype are shown in **Figure 22**. As can be seen, the power divider consists of four series connected 2π lines for in-phase output signals. Therefore, the 1:4 power divider commonly occupies a large area especially at low frequency. Fortunately, such a 2π line can be replaced by a 0° CRLH TL due to the phase advance and the nonlinear dispersion of CRLH TL.

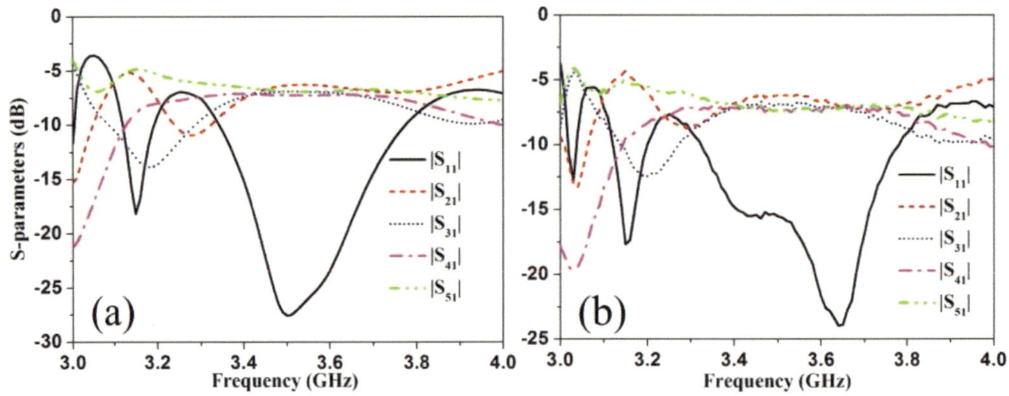

Figure 23. *S*-parameters of the proposed 1:4 series power divider: (a) simulation and (b) measurement.

Again, the zero phase-shift CRLH atom with detailed parameters shown in caption of **Figure 15** is synthesized following the design procedure mentioned above. For impedance matching, the characteristic impedance of 2π line or 0° CRLH TL should be $\sqrt{2}Z_0$ and the input impedance of each power split branch is 200 Ω. As such, two-staged $\lambda/4$ impedance transformers are necessary to afford a broadband impedance transformation from the required 200 Ω load impedance to 50 Ω test impedance. The narrow (15.3 × 0.16 mm^2) and wide (14.8 × 0.99 mm^2) lines function as the 158.1 Ω high-impedance transformer and the 79.1 Ω low-impedance transformer, respectively.

For verification, **Figure 23** portrays the full-wave simulated and measured *S*-parameters of our fabricated 1:4 series power divider. Satisfactory agreement of results is observed across the entire band of interest. Measurement results illustrate that return loss of the input port is better than –10 dB and the transmission power of four ports is near 7 dB. The power variation of each output port is within 5.96–8.8 dB from 3.3 to 3.8 GHz and thus the maximum variation is obtained as 2.8 dB relative to the ideal 6.02 dB. Therefore, our divider successfully fulfills the demand of equal power division. For sharp comparison, we have also fabricated and measured a 1:4 series power divider based on conventional meandered-line technique. Measured results show that the –10 dB return loss bandwidth is only 320 MHz varying from 3.38 to 3.7 GHz. Consequently, the bandwidth of our design has been broadened by 56%. Moreover, the total occupied area of our divider is only 41 × 38 mm^2 corresponding to 42% of the area 43.5 × 61.6 mm^2 that the meandered-line divider occupies.

To sum up, a novel compact resonant-type CRLH atom along with equivalent circuit model is proposed. It features inherent balance condition, additional transmission zero above the RH band and a higher degree of freedom in design over previous CSRRs-loaded counterpart. The fractal perturbation in novel CRLH atom leads to a significant shrinking of LH and RH bands and thus is of particular interest in compact device applications with a super miniaturization

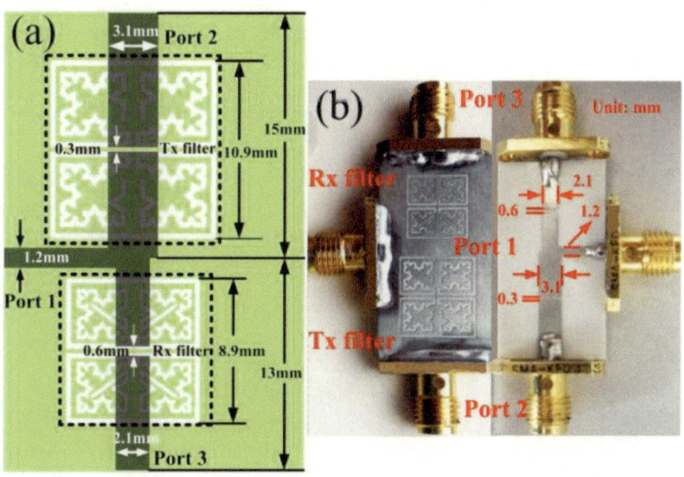

Figure 24. (a) Layout and (b) fabricated prototype of the diplexer sample. The slot width is 0.3 mm for CSSRRP while 0.2 mm for the inner slot, other geometrical parameters are: a = 4.3 mm, b = 8.3 mm, and c = 0.3 mm.

factor. The high-performance of super compact three-way and series 1:4 power divider has qualified it a good candidate.

1.4.2. Harmonic suppressed diplexer

The additional transmission zeros of CRLH atom with dual-shunt branch circuit can be directly applied to design a diplexer with harmonics suppression and enhanced frequency selectivity. A diplexer used to make receiving and transmitting share a common antenna consists of three ports, a receiver (Rx) filter and a transmitter (Tx) filter. Here, we realize both the Rx and Tx filter utilizing single K-ECSSRRP-loaded CRLH atom with specified φ and Z_β following the theory established in Section 1.3.3. The dimensions of CRLH atoms are designed and optimized to locate the Rx filter operation frequencies within GSM band centered at 1.8 GHz while Tx filter at 2.2 GHz. To ease the design, the Tx filter is directly designed from Rx filter by loading inner Koch slot of latter CRLH atom with additional stub to lower the operation frequency. Moreover, both Rx and Tx CRLH atoms are engineered to operate in balanced condition for broadband performance. For verification, the diplexer is fabricated, see **Figure 24** and measured through the *Anritsu ME7808A* vector network analyzer. The prototype occupies only 15.6 × 28 mm^2 corresponding to 0.094 λ_g × 0.168 λ_g, where λ_g is the free-space wavelength at 1.8 GHz. To the best of our knowledge, this is one of the most compact diplexers among available data.

Figure 25 illustrates the simulated and measured results of proposed diplexer. A reasonable agreement of results is observed in both cases. As shown in **Figure 25(a)**, the bandwidth of the Tx filter is measured as 220 MHz (1.61–1.83 GHz), in which the return loss $|S_{11}|$ is better than –10 dB and the insertion loss $|S_{21}|$ and $|S_{31}|$ are better than –1.5 dB. The measured bandwidth is 330 MHz (2.08–2.41 GHz) for Rx filter, where $|S_{11}|$ is better than –10 dB, $|S_{21}|$ and $|S_{31}|$ are less than –1.3 dB. Furthermore, the diplexer exhibits $|S_{11}|$= –12.9 dB, $|S_{21}|$ = –0.94 dB and $|S_{31}|$ = –14.7 dB at f_0 = 1.8 GHz, while $|S_{11}|$ = –28.1 dB, $|S_{21}|$ = –23.8 dB and $|S_{31}|$ = –0.54 dB at 2.2 GHz. The bandwidth for out-of-band suppression has been enhanced evidently which is up to 3.66 GHz characterized by $|S_{21}|$ and $|S_{31}|$ better than –20 dB. From **Figure 25(b)**, decent

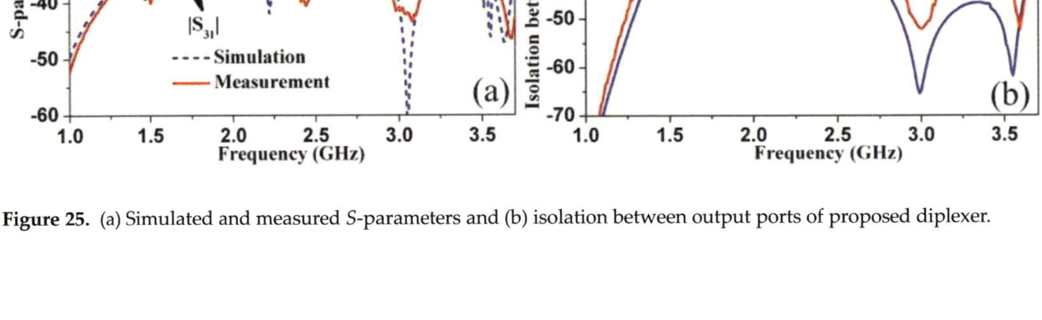

Figure 25. (a) Simulated and measured *S*-parameters and (b) isolation between output ports of proposed diplexer.

http://dx.doi.org/10.5772/65603

isolation between output ports 2 and 3 is observed. Measurement results indicate that the isolation level maintains below –15 dB up to 3.7 GHz. Slight discrepancies between simulation and measurement is attributable to the fabrication tolerances. This is especially true for the complicated CRLH atoms with zig-zag boundary. For validation, a sensitive analysis is carried out for K-ECSSRRP atom with different slot width. Numerical results suggest that the operation frequency shift upward by 1.8% while the transmission and impedance match performances are almost preserved when slot width increases by 0.1 mm. To sum up, the behavior of our diplexer fulfills unambiguously the criterion set by the wireless communication system.

In summary, the theory of dual-shunt branch circuit has been numerically studied and experimentally vilified. Due to the inherent additional transmission zero of dual-shunt branch CRLH atom, the resulting devices can be engineered with enhanced harmonic suppression and selectivity without posing penalty on circuit dimension. Moreover, the compact CRLH atom further reduces the circuit size. The high performances of designed diplexer corroborate our proposal and above statements, promising potentials in transceiver front-ends of mobile and wireless local area network (WLAN) systems.

2. Compact LH atoms for three-dimensional super lens

Three-dimensional (3D) LH-TL super lens with free-space excitation is very fascinating in practical applications since it is unnecessary to embed the sources and fields in the TL network [29, 30]. Here, we propose a 3D super lens with super resolution [31] using fractal perturbed LH TL in printed circuit board (PCB) fabrication process. The distributed LH-TL lens allows geometry scalability and avoids the parasitic RH effects in soldering the lumped elements. Due to the space-filling feature of fractals, the operation frequency in terms of subwavelength resonance is drastically lowered down. Consequently, a compact LH-TL element is engineered with large inductance and capacitance (LC) values in a limited volume.

2.1. Analysis and characterization of fractal LH-TL metamaterial

Figure 26 portrays the schematic and equivalent circuit model of the TL atom and resulting 3D lens. The TL atom with a lattice of $a \times a \times p$ consists by a metallic pattern printed on a dielectric spacer, see **Figure 26(a)**. The metallic pattern is constructed by a Sierpinski ring loaded in middle of its four concaves with four interdigitals each with a capacitance C_i. The adjacent TL atoms are interconnected through four thin meandered arms (3/2 fractal curve of first iteration order) each with an inductance L_m. Consequently, four subrings each with a self-inductance L_s are engineered. The loaded L_m and C_i contribute to the LH behavior, whereas the host medium (a combination of air gap and dielectric medium) modeled by L_0 and C_0 contribute to the RH behavior. The TL atom is isotropic since it exhibits four-fold rotational symmetry. The Sierpinski ring is designed as a fractal curve of second iteration order for the sake of compactness. In such a strategy, the inductance and capacitance (LC values) can be enhanced considerably in a limited volume, enabling the multilayered TL lens to be more appropriately described by an effective medium with homogenized μ_{eff} and ε_{eff}. By periodically arranging the fractal TL atom in *xoz*-plane and then layering the resulting structures along *y*-direction, a

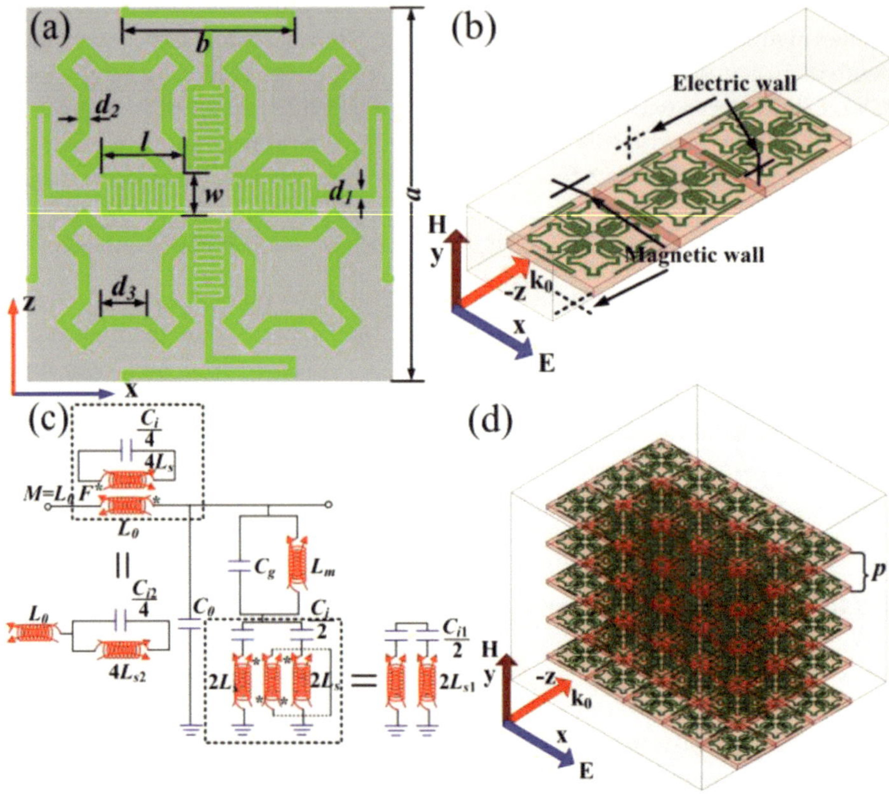

Figure 26. Topology and equivalent CM of the fully-printed LH-TL atom and lens: (a) top view and the geometrical parameter illustration of the TL atom, (b) simulation setup of an infinite slab illuminated by a plane wave of normal incidence, (c) equivalent CM, and (d) perspective view of the volumetric TL lens. The geometrical parameters (in mm) are: $a = 10.6$, $b = 5$, $l = 2.4$, $w = 1.2$, $d_1 = 0.2$, $d_2 = 0.3$, $L = 6.8$, and $d_3 = 1.32$.

3D LH-TL lens can be envisioned, see **Figure 26(d)**. It is shined by an x-polarized plane wave normally incident along z-direction. Such a configuration supports transverse electric (TE) mode with respect to xoz-plane. The computation volume is reduced to a TL atom by assigning electric and magnetic walls to four boundaries along x- and y-direction, respectively, see **Figure 26(b)**. This setup mimics an infinite array in xoy-plane.

In the model shown in **Figure 26(c)**, similar to reference [29], the fractal ring is driven by the y-oriented incident magnetic field, inducing inductive currents flowing along the interdigitals and ring which is modeled by a resonant tank formed by $4L_{s2}$ and $C_{i2}/4$ in series branch. The fractal ring couples magnetically to the host medium through a mutual inductance $M = L_0F$ which along with above resonant tank contribute the negative permeability. Here, F is the area ratio of the ring to the entire TL atom. The electric field impinging to the TL atom generates two effects which both contributing to the negative electric response. The two effects are: the oscillating current through the meandered arm and interdigital fringe modeled by a parallel tank of L_m and C_g in shunt branch; the oscillating current through the interdigitals and ring modeled by a pair of resonant tanks of $C_{i2}/2$ and $2L_{s1}/2$. The mechanism for LH behavior

http://dx.doi.org/10.5772/65603

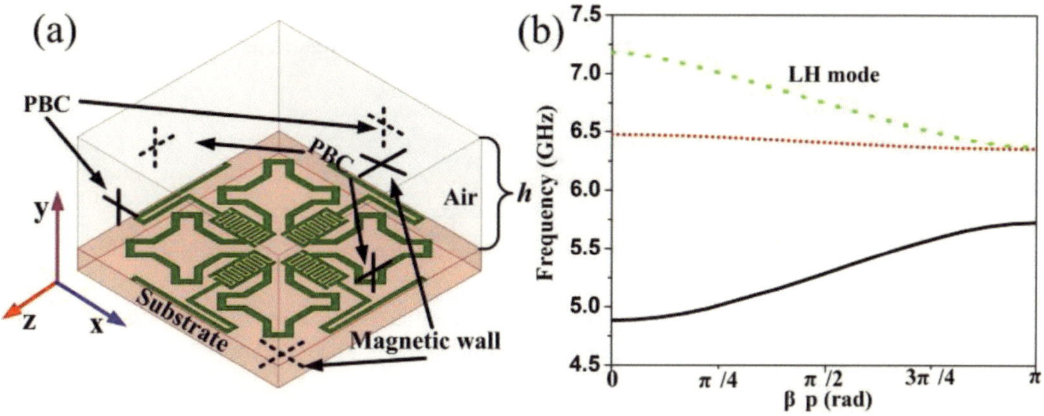

Figure 27. (a) Simulation setup for the eigenmode analysis in HFSS and (b) the dispersion curve of the TL atom for the initial design.

distinguishes from SRR-wire medium in following three aspects. First, the TL atom is physically connected to its adjacent one in coplanar structure, whereas that of SRR-wire medium is with double-layer topology. Second, the ring (magnetic response) is relevant to meandered arm (electric response) in a circuit model, whereas SRR (magnetic response) and cut wire (electric response) are independent. Most importantly, the EM wave permeating the periodically-arranged particles resembles a signal propagating along a TL, enabling low loss of resulting structure.

For characterization, the commercial full-wave finite-element-method (FEM) solver Ansoft HFSS is employed. The fractal patterns are built on 1 mm thick F4B substrate with $\varepsilon_r = 2.65$. The height of the air gap is $h = 5$ mm and thus $p = 6$ mm. The LH behavior can be appreciated from the dispersion data shown in **Figure 27(b)** using eigenmode simulation setup shown in **Figure 27(a)** where four walls along x- and z-directions are assigned as periodic boundary (master/slave in HFSS) while the top-and-bottom walls are assigned as magnetic walls. As expected, the backward wave dispersion (negative slope) is clearly observed from 6.37 to 7.18 GHz with a fractional bandwidth of 12%.

2.2. Design and realization of super lens

Now, we show how to design the basic TL atom and resulting super lens. To lay a solid theoretical method for implementation of $Z = Z_0$ and $n_{\mathrm{eff}} = -1$ which are necessary for super lens, we derive constitutive effective parameters for the 3D layered lens from CM. According to Floquet-Bloch theory, the phase shift φ per cell and the Bloch impedance Z_β can be calculated as

$$\cos\varphi = \cos(\beta p) = 1 + Z_s(j\omega)Y_p(j\omega)/2 \tag{35}$$

$$Z_\beta = \sqrt{Z_s(j\omega)[Z_s(j\omega)/4 + 1/Y_p(j\omega)]} \tag{36}$$

where $Z_s(j\omega)$ and $Y_p(j\omega)$ are the impedance and admittance of the series and shunt branch, respectively and are associated with the effective material parameters as

$$Z_s(j\omega) = j\omega\mu_{\text{eff}}\mu_0 = j\omega L_0 + \frac{1}{\frac{1}{j4\omega L_{s2}} + \frac{j\omega C_{i2}}{4}} \tag{37}$$

$$Y_p(j\omega) = j\omega\varepsilon_{\text{eff}}\varepsilon_0 = j\omega C_0 + \frac{1}{\frac{1}{j\omega C_g + \frac{1}{j\omega L_m}} + \frac{1}{j\omega C_{i1}} + j\omega L_{s1}} \tag{38}$$

Here, L_{r1}, C_{r1}, L_{r2} and C_{r2} are the equivalent inductances and capacitances by taking the coupling effects into consideration and are associated with L_r and C_r as

$$4L_{s1} = L_s - M^2/L_0,\ 4L_{s2} = \frac{M^2}{4L_s},\ C_{i1} = C_i,\ C_{i2} = \frac{4L_s^2 C_i}{M^2} \tag{39}$$

Insertion of Eq. (39) into Eqs. (37) and (38) yields immediately the effective material parameters

$$\mu_{\text{eff}} = 1 + \frac{C_i M^2}{4L_0} \frac{\omega_{r2}^2}{(1 - \omega^2/\omega_{r2}^2)} \tag{40a}$$

$$\varepsilon_{\text{eff}} = 1 + \frac{C_i}{C_0} \frac{(1 - \omega^2/\omega_g^2)}{(1 - \omega^2/\omega_{r1}^2)(1 - \omega^2/\omega_g^2) - \omega^2/\omega_m^2} \tag{40b}$$

where ω_{r1}, ω_{r2}, ω_g and ω_m are

$$\omega_{r1}^2 = \frac{1}{(L_s - M^2/L_0)C_i},\ \omega_{r2}^2 = \frac{1}{L_s C_i},\ \omega_g^2 = \frac{1}{L_m C_g},\ \omega_m^2 = \frac{4}{L_m C_i} \tag{41}$$

The effective permittivity and permeability tensors of the TL lens can be written as

$$\varepsilon(\omega) = \varepsilon_0 \begin{pmatrix} \varepsilon_{\text{eff}} & 0 & 0 \\ 0 & \varepsilon_{\text{avg}} & 0 \\ 0 & 0 & \varepsilon_{\text{eff}} \end{pmatrix},\ \mu(\omega) = \mu_0 \begin{pmatrix} 1 & 0 & 0 \\ 0 & \mu_{\text{eff}} & 0 \\ 0 & 0 & 1 \end{pmatrix}, \tag{42}$$

where ε_{avg} represents the averaged permittivity of the host medium. Insertion of $M = L_0F$ into Eq. (39), we conclude that L_{s1} is inversely proportional to L_0. Therefore, the electric resonance is proportional to L_0. Moreover, Eq. (40) suggests that the air gap plays an important role in affecting the constitutive parameters and the effective impedance while Eq. (41) indicates that L_m, L_s and C_i can be utilized to control the electric and magnetic resonances.

With above fundamentals known, we designed a 3D TL super lens working at f_0 = 5.35 GHz with optimized geometrical parameters shown in caption of **Figure 26**. The commercially available 3 mm-thick FR4 substrate with ε_r = 4.2 is selected in the super-lens design. The air space is selected as h = 9 mm to guarantee good impedance matching of the lens to free space. **Figure 28** depicts the dispersion diagram and S-parameters of designed TL atom which will be utilized as a building block for super lens. From **Figure 28(a)**, the backward LH band ranges

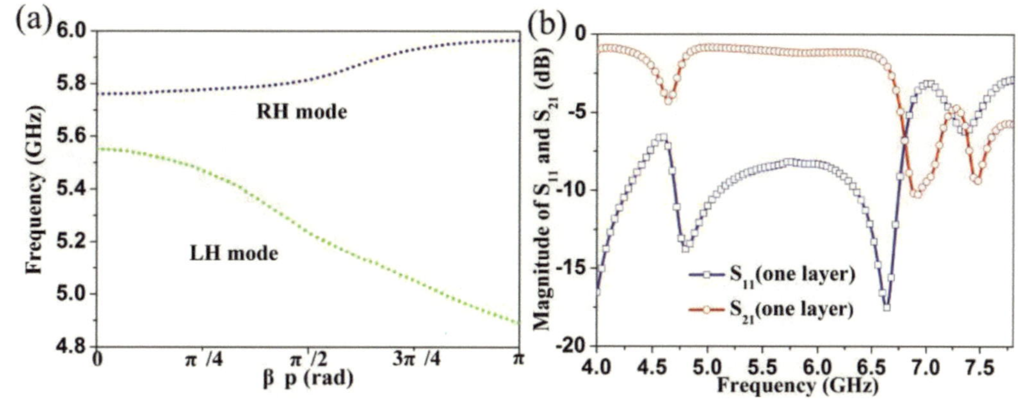

Figure 28. (a) Dispersion diagram and (b) simulated S parameters against frequency of designed TL atom.

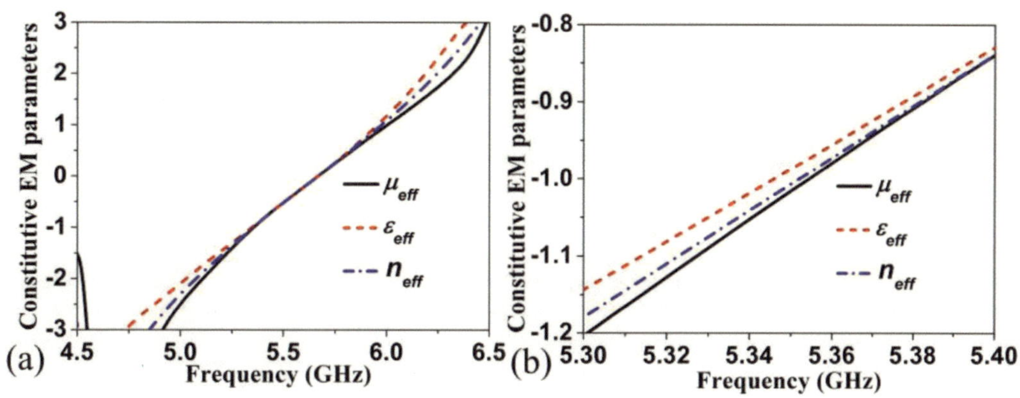

Figure 29. Constitutive EM parameters (real parts) extracted from simulated S-parameters: (a) overall view and (b) zoom-in view around $f_0 = 5.35$ GHz.

from 4.89 to 5.55 GHz, corresponding to a fractional bandwidth of 12.6%. The element size is evaluated as $\lambda_0/5.3 \times \lambda_0/5.3$ at 5.35 GHz which is within the limit of an effective medium ($\lambda_0/4$). The βp is on the order of 70°, yielding a refractive index of $n = -1.028$ according to $n = c\beta/\omega a$. Moreover, the RH band with positive negative index follows the LH band though a narrow interval of 5.55–5.76 GHz. This narrow signal inhibition does not lead to notable stop band in **Figure 28(b)** but induces slightly large insertion loss and weak ripple in S_{11} around 5.74 GHz. It enhances with increased element number in propagation direction and thus an obvious stop band would emerge when more TL atoms are cascaded. The insertion loss retains at a decent level in the pass band.

As shown in **Figure 29**, the effective parameters are retrieved as $\mu_{\text{eff}} = -1.006 - j0.132$, $\varepsilon_{\text{eff}} = -0.995 + j0.049$ and $n = -1.004 - j0.041$ at $f_0 = 5.35$ GHz under time-harmonic progression $e^{j\omega t}$. The refractive index is in excellent agreement with that calculated from dispersion diagram shown in **Figure 28(a)**. Therefore, the condition for Veselago-Pendry lens which require simultaneously good impedance matching ($\mu_{\text{eff}} = \varepsilon_{\text{eff}} = -1$) and $n = -1$ is successfully fulfilled.

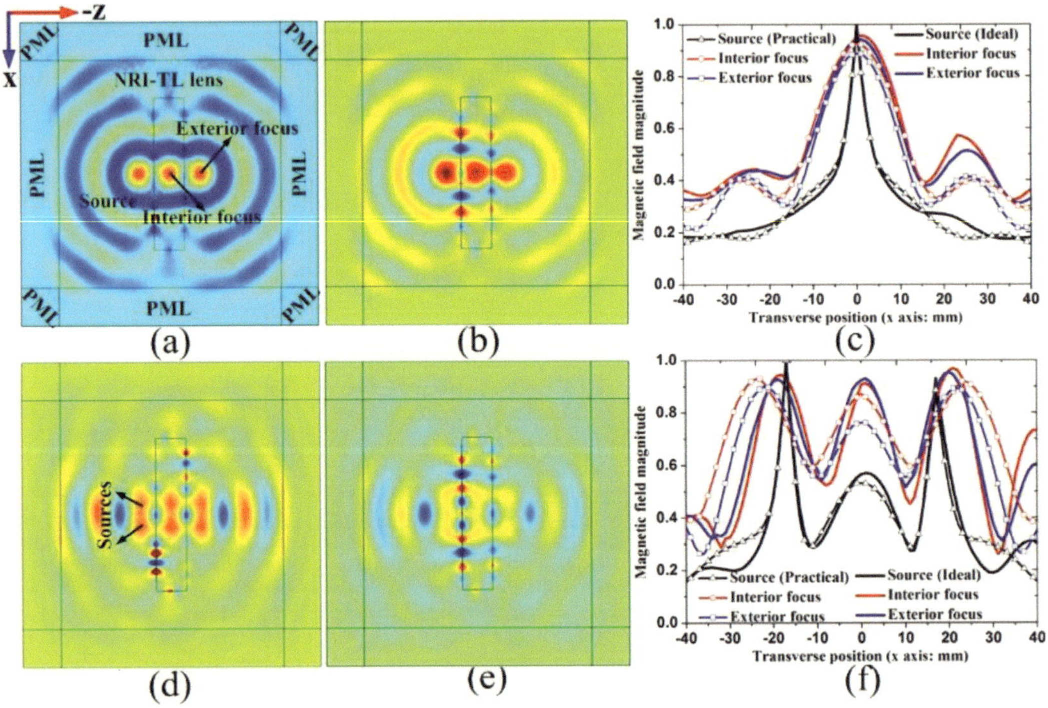

Figure 30. Numerically simulated magnetic field distributions of the designed TL lens. Field distributions for (a and d) ideal lens with $\mu_y = \varepsilon_x = -1$ and (b and e) designed (practical) lens with $\mu_y = -1.006 - j0.132$, and $\varepsilon_x = -0.995 + j0.049$ in (a–c) single-source and (d and e) dual-source case. (c and f) The field curve plotted along x-axis at the center of the lens.

2.2. Numerical and experimental results

To verify the subwavelength focusing, we numerically designed a TL lens using the extracted effective parameters. The simulation is conducted in Comsol Multiphysics™ software package, where a current source oriented along y-axis is placed in front of the TL lens with an overall size of 200 × 40 mm^2 to generate EM wave propagating along z-direction. A sufficiently large PML boundary with a total area of 400 × 400 mm^2 is employed to encompass the lens and the source. **Figure 30** portrays the magnetic field distribution within the entire calculation domain. For comprehensive study, two cases are considered for both designed (practical) lens and ideal lens with $\mu_y = -1$ and $\varepsilon_x = -1$, i.e., single source and dual closely arranged identical sources with an interval of 34 mm.

Obvious focusing behavior is clearly suggested inside (interior focusing) and in front (exterior focusing) of the lens for both designed and ideal lens in either single-source or dual-source case. The evanescent wave amplification is clearly observed along the two-slab interfaces. The large field concentration on interfaces is attributable to plasmonic surface waves at the interface of LH and RH media. The nonideal material parameters away from –1 induce slightly asymmetric fields, discontinuous wave front and lower imaging intensity observed in practical lens due to slight loss and impedance mismatch at the interface of lens and free space. Nevertheless, the loss originated from imaginary parts of μ_{eff} and ε_{eff} do not pose critical

http://dx.doi.org/10.5772/65603

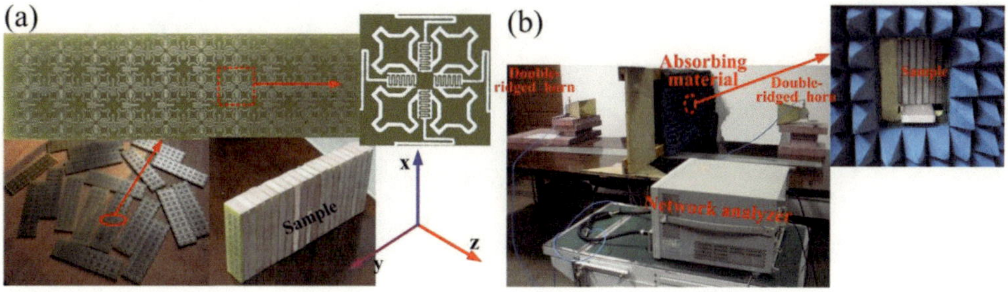

Figure 31. (a) Photograph of the fabricated TL lens and (b) free-space measurement system for *S*-parameters.

influence on focusing. The clearly converged spots for dual-source practical lens give strong support to above statement, which should be emphasized. The size of converged images measured by 3 dB contour (half-power beam width, HPBW) is about 19 mm, corresponding to 0.34 λ_0 at 5.35 GHz. Therefore, the subwavelength imaging is unambiguously validated beyond the diffraction limit with both focused propagating wave and amplified evanescent wave. The spot size characterized by HPBW in the dual-source case is observed slightly narrower than that in single-source case. This is attributable to the wave interference of the two sources.

For verification, the designed TL lens is fabricated and measured, see **Figure 31(a)**. In fabrication, a total of 10 × 3 TL atoms are fabricated on FR4 substrate board using standard PCB technology. To sustain the bulk TL lens with desired air spacer, each substrate board is supported by a plastic foam of identical size with $\varepsilon_{\text{foam}} = 1.2$ and $h = 9$ mm. Then, 20 substrates and 20 plastic foams are alternatively stacked by adhesives and are reinforced by a hot press to form a lens occupying a volume of 106 × 240 × 31.8 mm^3. To guarantee axial magnetic excitation, two small loop antennas each with an inner diameter of 4.6 mm are implemented in shielded-loop configuration to mitigate unbalanced currents.

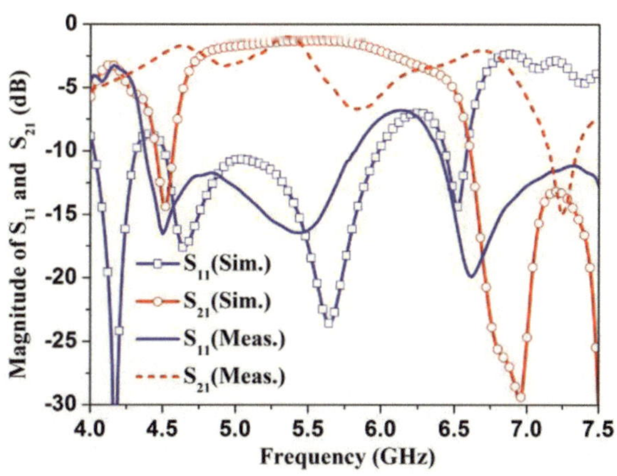

Figure 32. Simulated and measured *S*-parameters of the TL lens.

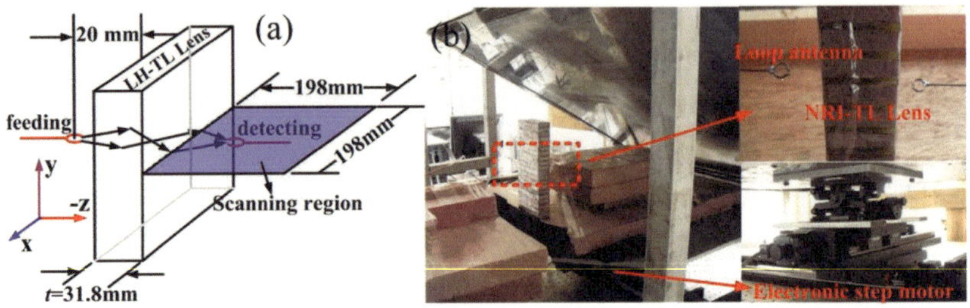

Figure 33. (a) Schematic illustration and (b) near-field measurement system of the free-space focusing apparatus.

The *S*-parameters of the lens are measured utilizing the setup shown in **Figure 31(b)**. Two double-ridged horn antennas functioning as the receiver and transmitter, respectively, are placed at two sides of the sample, each is 90 mm away from the center sample to provide a plane EM wave with required polarization. The sample is placed in a drilled aperture on a sufficiently large wooden board aside which a mass of absorbing materials were used to block diffracted energy. As shown in **Figure 32**, numerical and experimental *S*-parameters coincide reasonably except for a transmission dip around 5.75 GHz and blue shifted signal inhibition band at high frequencies in measurement case. The mutual coupling between adjacent atoms in longitudinal direction gives rise to wider pass band. Moreover, the nonideal foams with $\varepsilon \neq 1$ in experiments and infinite dimension in *xoy*-plane in simulations also give rise to the discrepancy. Nevertheless, these factors play a negligible role in degrading the electrical performances at 5.35 GHz where the focusing performance will be evaluated. The simulated (measured) return loss is observed as –13.1 (–15.9 dB) at this point, implying a good impedance matching to free space.

For dual-source imaging, additional power-division circuit is necessary and wave interference of two closely distributed sources may degrade the focusing quality in terms of weak field intensity at focal point. Here, we only performed experiments for single-source imaging by measuring magnetic field intensity using loop antenna, see setup shown in **Figure 33**. The transmitting loop antenna (20 mm away from the slab), is stationary and illuminates at the front side of the lens (slightly larger than $t/2$) while the detecting loop antenna scans a region of 198 × 198 mm^2 at rear side of the sample by an electronic step motor. **Figure 34** plots the measured magnetic field intensity and phase distribution in steps of 1 mm at rear side (focal region) of the lens at three different frequencies. Here, the field intensity has been normalized to their maximum amplitude. From **Figure 34(a)** and **(b)**, the well-resolved imaging can be evidenced from both localized field intensity and reversal of smooth concave phase fronts near the focal plane at 5.35 and 5.4 GHz. The HPBW indicated by –3 dB intensity (along *y*-axis) is measured about 21.9 mm (0.39 λ_0) which is in reasonable agreement with that obtained in simulations. The slightly deteriorative spot size may be attributable to the adhesives employed in experiments which also introduce additional loss. Therefore, significant resolution improvement is achieved with respect to the 0.55 λ_0 [30], which suggests that the imaging resolution in this approach is less susceptible to the loss than that in other approaches. However, there is none evidence for subwavelength focusing from magnitude or phase plotted in **Figure 34(c)**, where the transverse size of elongated

http://dx.doi.org/10.5772/65603

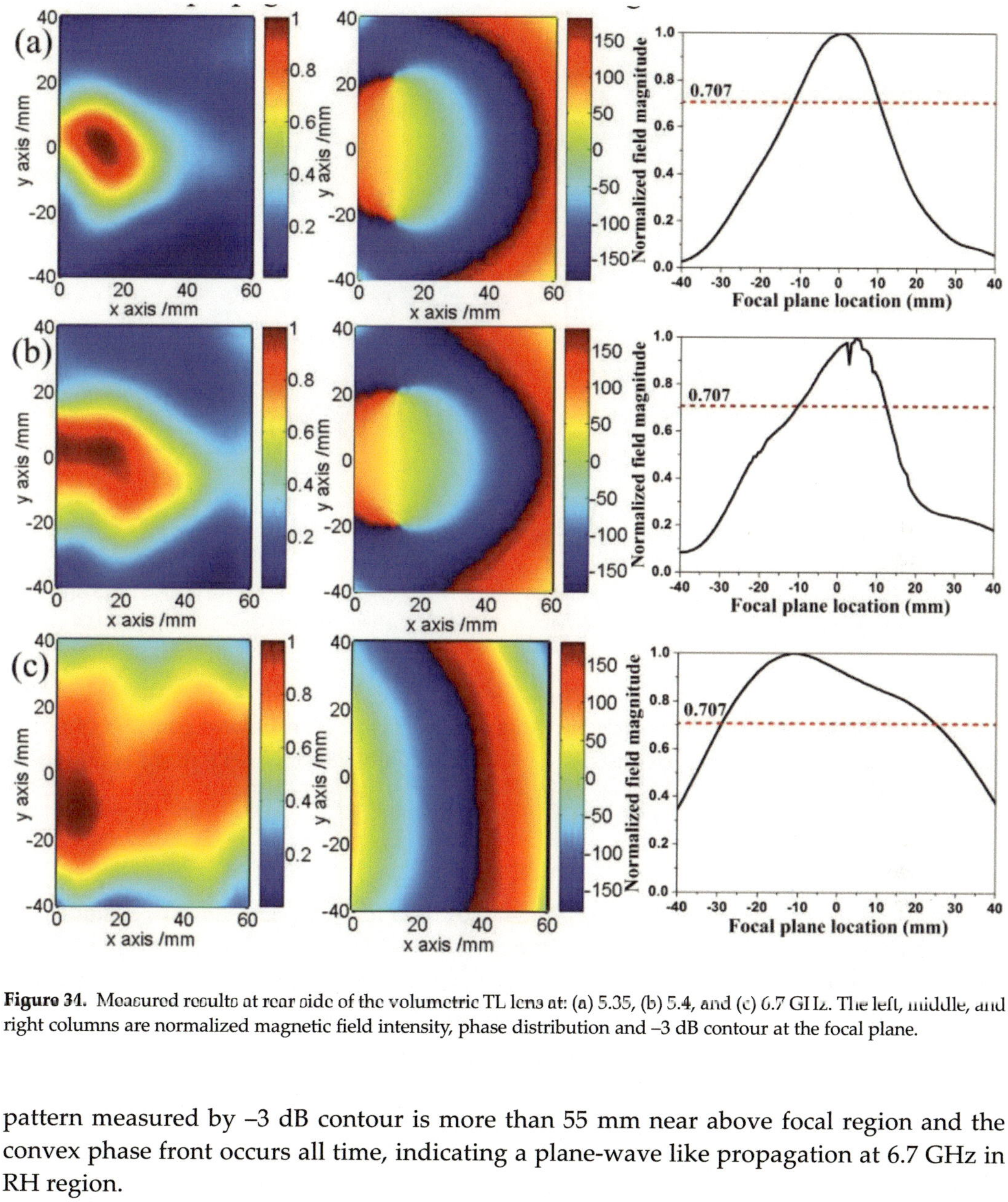

Figure 31. Measured results at rear side of the volumetric TL lens at: (a) 5.35, (b) 5.4, and (c) 6.7 GHz. The left, middle, and right columns are normalized magnetic field intensity, phase distribution and –3 dB contour at the focal plane.

pattern measured by –3 dB contour is more than 55 mm near above focal region and the convex phase front occurs all time, indicating a plane-wave like propagation at 6.7 GHz in RH region.

3. Compact meta-atoms for superscatterer illusions devices

In this section, we conceptually proposed and experimentally demonstrated a superscatterer illusion device [32] with abundant functionality inspired by the concept of magnifying lenses [33, 34] using transformation optics (TO) theory. A new strategy to miniaturize the meta-atom

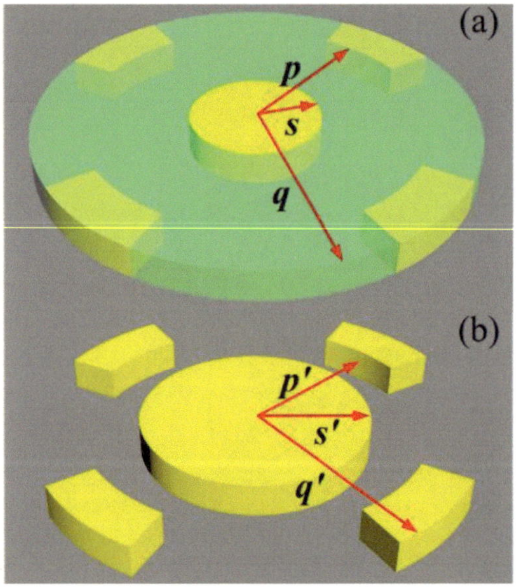

Figure 35. Scheme of the proposed superscatterer illusion device: (a) the original metallic object (yellow) is wrapped by the illusion device (light green) and (b) the virtual enlarged metallic object at the original center with four wing dielectric objects.

is proposed by combining both electric and magnetic particles. Using such a compact building block, a proof-of-concept sample consisted by 6408 gradually varying meta-atoms is designed, fabricated and measured.

3.1. Scheme and theoretical design

The device which tightly wraps an original metallic object (denoted as yellow) in actual space, see **Figure 35(a)**, is a shell (denoted as light green) embedded by four symmetrically located wing objects (denoted as light yellow). It functions to transform the metallic object into multiple isolated ghost objects with isotropic material properties ($\overline{\overline{\varepsilon}}' = \varepsilon_r \overline{I}, \overline{\overline{\mu}}' = \mu_r \overline{I}$) and an enlarged metallic object at center in virtual space, see **Figure 35(b)**. In other words, the functional device enables the radar scattering signature of a metallic object to equal that of an enlarged metallic object and four symmetrically arranged ghost objects under EM wave illumination. The proposed scheme with complex and versatile functionality inherits the merits of both superscatterers and cognitive deception and is not confined to metallic objects but also suitable for dielectric objects.

In actual space, the radius of the object (also the inner radius of the ghost device), the inner and outer wings are denoted as s, p and q (also the outer radius of the device), respectively, whereas they are s', p' and q' in virtual space. To ease the design, we choose $p = p'$, $q = q'$ and assume that the material property and geometrical layout of the four wing objects are the same. The anistropic and inhomogenious material parameters of the shell (region I) and wings (region II) are different and require precise design based on the TO theory. We remark the number of

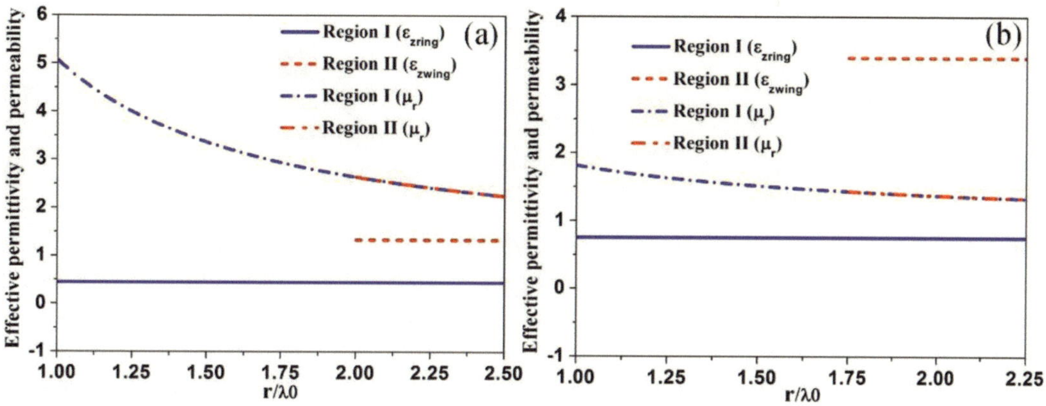

Figure 36. Theoretically calculated material profiles as a function of normalized radius in region I and region II which has been normalized to the free-space working wavelength λ_0 = 30 mm: (a) case 1: $\varepsilon_{\text{virtual}} = 3$, $s = \lambda_0$, $= 1.5\lambda_0$, $p = p' = 2\lambda_0$, and $q = q' = 2.5\lambda_0$ and (b) Case 2: $\varepsilon_{\text{virtual}} = 4.53$, $s = \lambda_0$, $s' = 7/6\lambda_0$, $p = p' = 1.75\lambda_0$, and $q = q' = 2.25\lambda_0$.

wings can be arbitrary engineered provided that the required complex material profiles can be realized by metamaterials and in this work four wings each separated by 90° are selected with four-fold rotational symmetry. Moreover, the material properties of the virtual wing object can also be arbitrarily predesigned. Here, we tailor the wing property only by tuning the dielectric constant $\varepsilon'_{\text{virtual}}$ since most materials in nature are nonmagnetic.

We derived the required material parameters in region I and II through a general transformation $r = f(r')$ in 3D spherical coordinate (r, φ, θ). According to TO theory, the diagonal permittivity tensor $\overline{\overline{\varepsilon}}$ and permeability tensor $\overline{\overline{\mu}}$ in physical space associate with $\overline{\overline{\varepsilon'}}$ and $\overline{\overline{\mu'}}$ in virtual space as $\overline{\overline{\varepsilon}} = \Lambda\overline{\overline{\varepsilon'}}\Lambda^T/|\Lambda|$ and $\overline{\overline{\mu}} = \Lambda\overline{\overline{\mu'}}\Lambda^T/|\Lambda|$. Here, $\Lambda_{ij} = \partial x_i/\partial x'_j$ is the Jacobian matrix of the transformation and $|\Lambda|$ is its determinant. After some derivations, the permittivity tensor with similar form as the permeability tensor is obtained as

$$\overline{\overline{\varepsilon}}(r,\varphi,\theta) = \begin{bmatrix} \dfrac{f'(r')r'^2\varepsilon'_r}{f(r')^2} & 0 & 0 \\ 0 & \dfrac{\varepsilon'_\varphi}{f'(r')} & 0 \\ 0 & 0 & \dfrac{\varepsilon'_\theta}{f'(r')} \end{bmatrix} \tag{43}$$

In this particular design, the proposed superscatterer illusion functionality can be readily realized through following transformation:

$$\begin{pmatrix} r & \varphi & \theta \end{pmatrix} = \begin{pmatrix} k(r'-s') + s & \varphi' & \theta' \end{pmatrix} \tag{44}$$

where $k = (q-s)/(q-s')$ is a constant related to the magnifying factor that can be arbitrarily designed. By substituting Eq. (44) into Eq. (43), we immediately obtain the required permittivity or permeability tensor as

$$\overline{\overline{\varepsilon}}(r,\varphi,\theta)=\begin{bmatrix}\frac{(r-s+ks')^2\varepsilon'_r}{kr^2} & 0 & 0\\ 0 & \frac{\varepsilon'_\varphi}{k} & 0\\ 0 & 0 & \frac{\varepsilon'_\theta}{k}\end{bmatrix} \tag{45}$$

For 2D illusion device in the cylindrical coordinates (r, φ, z), see cross-section shown in **Figure 35**, the full diagonal parameter tensors (μ_r, μ_φ, μ_z) and (ε_r, ε_φ, ε_z) in regions I and II have the same form as those in the spherical 3D case. Since all cases considered in this work are transverse-electric (TE) wave excitation with z-polarization, the full diagonal-parameter tensors for proposed EM functionality can be simplified to three permittivity and permeability components (μ_r, μ_φ, ε_z) for an easy design [35]

$$\begin{pmatrix}\mu_r & \mu_\varphi & \varepsilon_{zwing}\end{pmatrix}=\begin{pmatrix}\frac{f'(r')^2r'^2}{f(r')^2} & 1 & \frac{\varepsilon'_{virtual}}{f'(r')^2}\end{pmatrix} \tag{46a}$$

$$\begin{pmatrix}\mu_r & \mu_\varphi & \varepsilon_{zring}\end{pmatrix}=\begin{pmatrix}\frac{f'(r')^2r'^2}{f(r')^2} & 1 & \frac{1}{f'(r')^2}\end{pmatrix} \tag{46b}$$

where ε_{zring} and ε_{zwing} are z-directed permittivity components in region I and II, respectively.

The superscatterer illusion device is designed to work around 10 GHz in X band. **Figure 36** portrays the required anisotropic and inhomogeneous material profile as a function of radius in both regions. Without loss of generality, two illusion devices with large and smaller magnifying factor of $s'/s = 1.5$ and $s'/s = 1.167$ are considered. From **Figure 36**, it is learned that the radial permeability μ_r exhibits an anomalous negative dispersion in both regions, whereas μ_r has a positive slope for a shrinking device [35]. Moreover, μ_r increases drastically and thus exhibits a sharper slope while ε_z presents an inverse manner with respect to μ_r when s' increases or s and q decrease. Notice that the sensitivity of q on the material property is more mitigative than that of s and s'. In Case 1, the material parameters is ε_{zring}= 0.444 and $2.25 \le \mu_r \le 5.06$ in region I ($\lambda_0 \le r \le 2.5\ \lambda_0$) while they are ε_{zwing}= 1.333 and $2.25 \le \mu_r \le 2.64$ in region II ($2\lambda_0 \le r \le 2.5\ \lambda_0$). In Case 2, they are ε_{zring} = 0.75 and $1.33 \le \mu_r \le 1.812$ in region I ($\lambda_0 \le r \le 2.25\ \lambda_0$) while ε_{zwing}= 3.34 and $1.33 \le \mu_r \le 1.435$ in region II ($1.75\ \lambda_0 \le r \le 2.25\ \lambda_0$). In practice, there is no size limitation on our superscatterer device which can be engineered substantially larger to deceive any sized object. However, it is engineered relatively small in this design for experimental convenience due to scanning capability limit of available near-field measurement system.

3.2. Metamaterial design and device fabrication

The above required gradient material profiles in both regions can be realized using compact meta-atoms. Since it is easy to fully cover a small scope of permeability using only one type of subwavelength elements, we consider implementing the superscatterer illusion in Case 2 with more relaxed slope of μ_r. **Figure 37** depicts the topology and fabricated prototype of designed illusion device. Here, a new strategy is proposed to simultaneously fulfill the less-than-unity

http://dx.doi.org/10.5772/65603

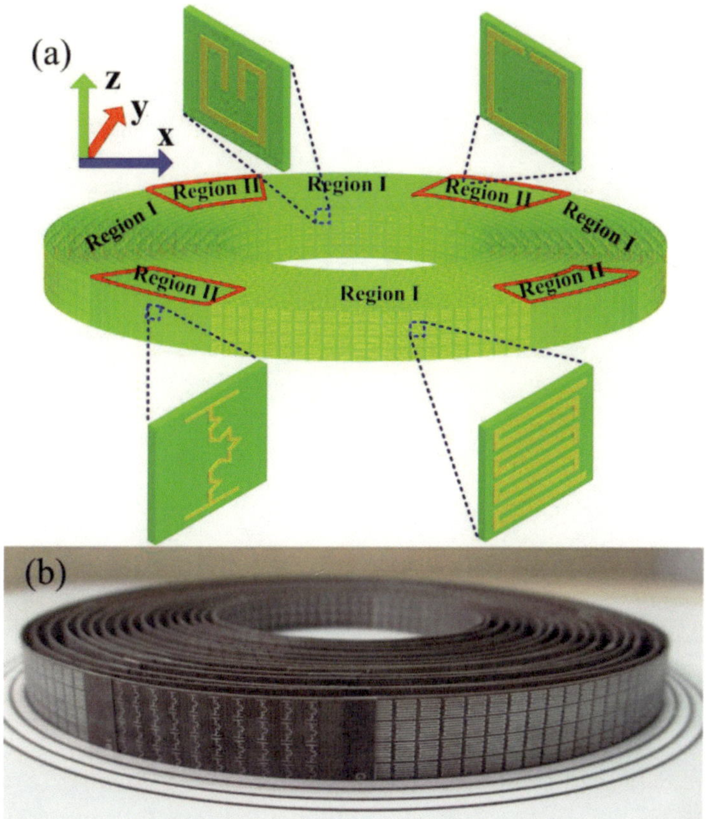

Figure 37. (a) Topology and (b) fabricated prototype of the proposed superscatterer illusion device, which is composed of four types of subwavelength electric and magnetic particles.

permittivity ε_z while the larger-than-unity permeability μ_r. Two kinds of meta-atoms each composed of an electric resonator and a magnetic inclusion are designed on both sides of a 0.2 mm-thick flexible F4B board with ε_r= 2.65, see **Figure 37(a)**. Through a proper design, the working frequency can be precisely engineered above the electric plasma frequency ω_p while below the magnetic resonance. In region I, the meta-atom consists of a meander line electric resonator and a split-ring resonator (SRR) with two concave arms; whereas in region II the meta-atom consists by SRR and Koch-shaped cut-wire resonator. Two factors give rise to the meta-atom miniaturization within a limited volume: meandered or fractal strategy for electric resonator design and layered coupling between layered electric and magnetic particles. In this regard, the volumetric devices made of these meta-atoms are more feasible to be described by an effective medium and thus the collective EM response is less affected by the parasitic diffractions, enabling an exact design due to good approximation of an infinite array to a finite bulk medium. Moreover, the low loss and relaxed material parameter dispersion enable desirable performance and working bandwidth of the superscatterer illusion device since the meta-atoms operate at nonresonance region.

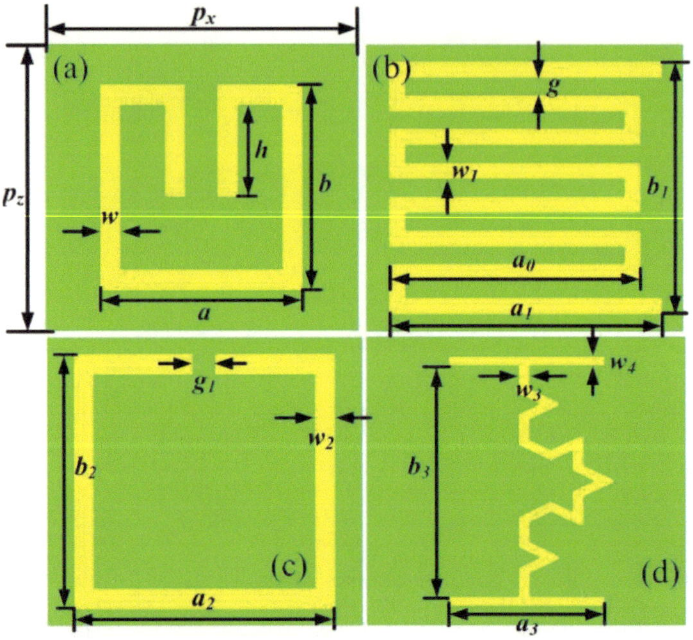

Figure 38. The illustration of physical parameters of four electric and magnetic particles. "(a)SRR with two concave arms. (b) Meanderline electric resonator." behind the word "particles.(c) conventional SRR. (d)Koch-shaped cut-wire resonator."

In the fabricated prototype, the superscatterer illusion device is discretized into a total of 6408 grids (meta-atoms) each occupying an area of $p_r \times p_\varphi \times p_z = 2.5 \times 3 \times 2.7\ \text{mm}^3$ ($0.083\ \lambda_0 \times 0.1\ \lambda_0 \times 0.09\ \lambda_0$). This size level goes beyond the shrinking illusion device [35] in terms of effective medium and homogenization. The prototype is synthesized by 16 concentric layers each owning one or two particles of different geometrical parameters and composed of four meta-atoms along *z*-direction, corresponding to a height of 10.8 mm ($0.36\ \lambda_0$). The 2.5 mm interval between neighboring layers is guaranteed by adhering the 16 flexible boards to 0.2 mm-wide concentric circular rabbets carved on a hard-foam board by the LPKF milling machine. The shell extends from the 1st to 16th layer and contains a sum of 5288 elements, whereas the four-wing objects extend from the 10th to 16th layer and consist by 1120 particles. A total of 22 groups of different geometrical parameters of meta-atoms, see **Figure 38** and **Table 4**, arc finely optimized in Ansoft HFSS to generate the required permittivity and permeability.

In the simulation setup shown in **Figure 39(a)**, *z*-polarized EM wave is incident parallel to the dielectric board with magnetic fields penetrating through the magnetic inclusions along the *x*-direction (radial direction in cylindrical coordinates). To mimic an infinite array, four transverse walls are assigned master/slave boundary in HFSS while the two walls along propagation direction are assigned the Floquet ports. From **Figure 39(b)** and **(c)**, it is learned that the operation frequency of both meta-atoms locates precisely above ω_p of the electric resonator while below resonance of the magnetic inclusion. The material parameters are

	Region I			Region II					
Number of Layers	H [mm]	ε_z	μ_r	w_3 [mm]	a_3 [mm]	g_1 [mm]	a_2 [mm]	ε_z	μ_r
1	0.9	0.7328	1.807						
2	0.87	0.7459	1.745						
3	0.835	0.7362	1.679						
4	0.8	0.7451	1.621						
5	0.77	0.7383	1.583						
6	0.74	0.7376	1.548						
7	0.71	0.7331	1.515						
8	0.68	0.7387	1.485						
9	0.655	0.7395	1.461						
10	0.633	0.746	1.429	0.108	1.548	0.11	2.5	3.395	1.441
11	0.61	0.7382	1.417	0.108	1.548	0.13	2.5	3.405	1.416
12	0.59	0.7418	1.395	0.105	1.4	0.2	2.6	3.396	1.394
13	0.56	0.7391	1.371	0.105	1.47	0.15	2.5	3.392	1.378
14	0.54	0.7402	1.36	0.102	1.462	0.17	2.5	3.397	1.358
15	0.51	0.745	1.342	0.111	1.48	0.2	2.5	3.405	1.34
16	0.49	0.7413	1.33	0.105	1.435	0.2	2.5	3.393	1.334

Note: The other geometrical parameters are given as follows: p_z = 2.7mm, p_x = 3 mm, $a = b$ = 1.95 mm, w = 0.2 mm, a_1 = 2.607mm, $b_1 = a_0$ = 2.394 mm, $w_1 = g$ = 0.16 mm, b_2 = 2.4 mm, w_2 = 0.2 mm, b_3 = 2.16 mm, and w_4 = 0.09 mm.

Table 4. The elaborate geometrical parameters of the electric and magnetic meta-atoms utilized in region I and II.

$\mu_r = 1.807 + i0.057$ and $\varepsilon_z = 0.733 + i0.09$ for the first layer in region I and are $\mu_r = 1.441 + i0.032$ and $\varepsilon_z = 3.395 + i0.08$ for the 10th layer in region II. Due to small ε_z required in region I while large ε_z in region II, the working frequency is designed immediately above ω_p in region I while far away above ω_p in region II. In all cases, the imaginary parts of permittivity and permeability are found to be near zero while the real parts of them change slowly near 10 GHz, indicating a low loss and a relatively relaxed material dispersion. Moreover, the combined electric and magnetic resonators have lowered ω_p from 12 GHz (meta-atom without SRR) to 9.4 GHz, further facilitating electrically smaller particles which should advance a step toward the homogenization of bulk composites.

3.3. Numerical and experimental results

Now, we carry out full-wave simulations in COMSOL Multiphysics to demonstrate the functionality of proposed superscatterer illusion device. For comprehensive study and not lose generality, the superscatterer illusion for a dielectric object is also studied. As shown in **Figure 40**, a line source (a monopole probe in experiment) is located 50 mm (1.67 λ_0) away

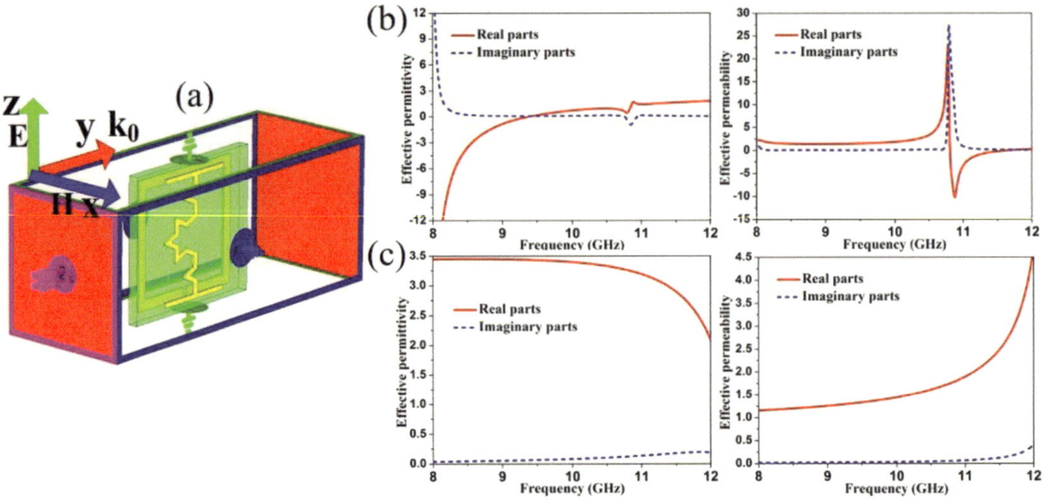

Figure 39. (a) Simulation setup for the meta-atom with periodic boundary conditions in HFSS. The retrieved constitutive parameters for (b) meta-atoms of the first layer in region I, and (c) those of the tenth layer in region II.

from the left edge of the illusion device in Case 1, whereas it is 57.5 mm (1.92 λ_0) in Case 2. Notice that the geometrical parameters of the devices in **Figure 40(a–d)** are exactly the same except for different number of wing objects. In all cases, the scattering patterns in actual and virtual spaces are in good consistency and equivalence, indicating that the superscatterer illusions are efficiently generated for both dielectric and metallic objects. Our device enables the camouflage of the radar image of a small cylinder to radar images of a large cylinder and two/four standalone objects. Moreover, the similar field patterns of the illusion devices with two and four wings indicate the robustness of the method in creating multiple isolated radar ghost images. Most importantly, the small signature of the original object is transformed to a larger one with arbitrary scaling ratios, enabling a superscatterer.

To further quantitatively evaluate the superscatterer illusion performance, we compare the simulation results of scattered electric fields in actual and virtual spaces within the region $-2\lambda_0 \leq y \leq 2\lambda_0$ at the rear side of the shell, see **Figure 41(a)**, where the field intensities along the line $x = 8.27\lambda_0$ in both spaces range from 0.014 to 0.182 V/m and have good consistency. As additional analysis, we also numerically discuss the sensitivity of the illusion performance to μ_r (real part) and loss (imaginary part of μ_r) along the line $x = 8.67\lambda_0$ in actual space. When it varies from μ_r to $\mu_r + i0.01\mu_r$ and finally to $\mu_r + i0.02\mu_r$, the wavefronts do not suffer substantial changes except for gradually decreased field intensities (e.g. the maximum values decrease from 0.158 to 0.131 V/m), see **Figure 41(b)**. Similarly, the scattered fields do not suffer significant change except for varied peak intensity when the real part of μ_r varies from μ_r to 1.25 μ_r, see **Figure 41(c)**. Therefore, the superscatterer illusion performance of our devices is insensitive to material loss and should exhibit a desirable operation bandwidth within the acceptable tolerances of material parameters.

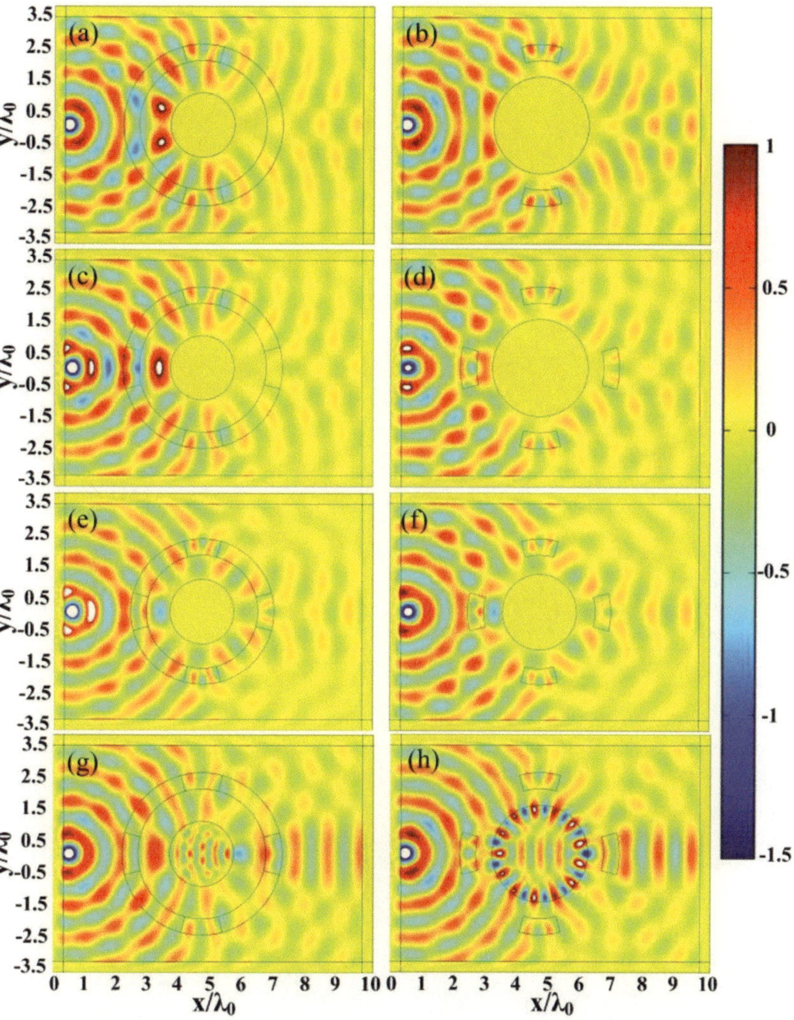

Figure 40. Simulated snapshots of electric-field distributions at 10 GHz in actual space (the left column) and virtual space (the right column). A metallic cylinder is wrapped by simplified-parameter illusion devices for (a) two and (c) four dielectric wings in Case 1, and (e) four dielectric wings in Case 2. An enlarged metallic cylinder with (b) two dielectric wings in Case 1, and four dielectric wings in (d) Case 1 and (f) Case 2. (g) A dielectric cylinder wrapped by an illusion device with four dielectric wings and simplified parameter in Case 1. (h) An enlarged dielectric cylinder with four dielectric wings in Case 1.

To validate the superscatterer illusion functionality, we measured 2D electric-field mappings in near-field parallel-plate waveguide measurement system (**Figure 42**). A monopole probe fixed inside the planar waveguide excites the sample over a discrete cluster of frequencies. The impinging EM wave interacts with thousands of particles and thus affords desired scattering signatures. Due to large size of the sample, the simulated and measured electric fields in horizontal plane are mainly recorded at rear side of the device ($7\lambda_0 \leq x \leq 10.2\lambda_0$ and $-2\lambda_0 \leq y \leq 2\lambda_0$) at 10.1, 10.15, 10.21, 10.25 and 10.3 GHz, respectively, see **Figure 43**.

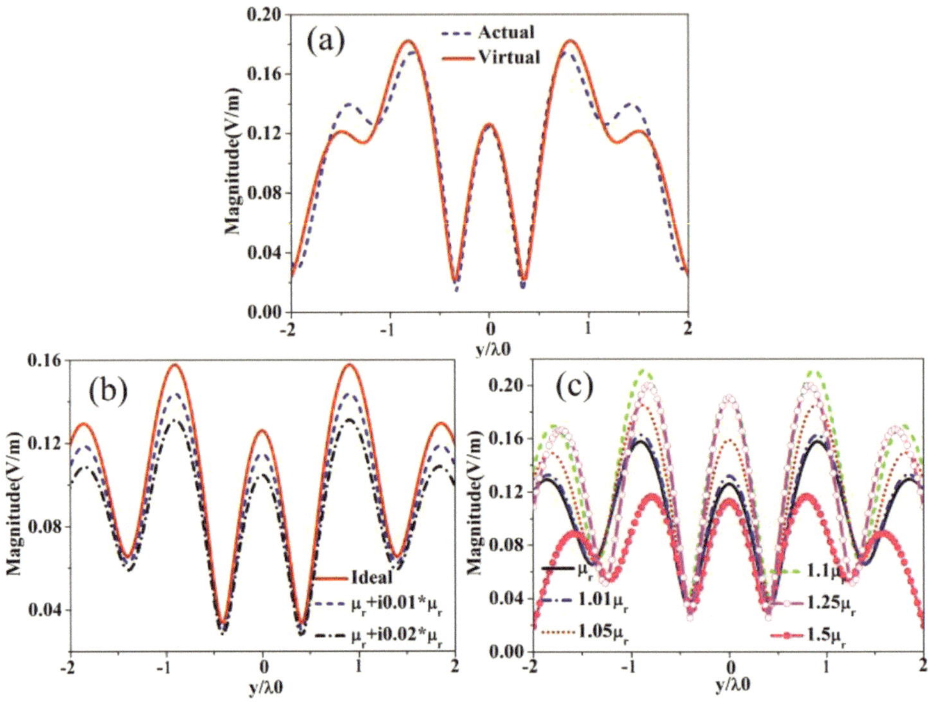

Figure 41. (a) Scattered electric fields in actual and virtual spaces along the line $x = 8.27\lambda_0$. Scattered electric fields in actual space along the line $x = 8.67\lambda_0$ for (b) different loss and (c) different real parts of μ_r.

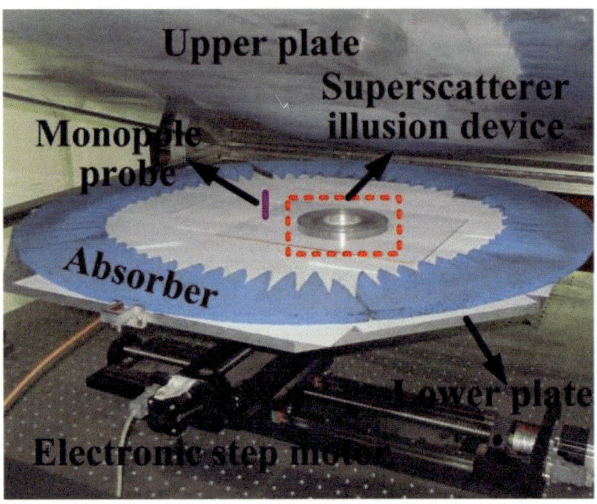

Figure 42. Experimental setup for 2D electric-field mapping in near-field parallel-plate waveguide measurement system.

As shown in **Figure 43(a)** and **(d)**, the simulated and measured scattering patterns are in reasonable agreement. The consistency can be further inspected from electric-field intensities shown in **Figure 43(g)**, where the fields at 10.1 GHz are normalized to the maximum along the black dashed line $x = 9.57\lambda_0$. In all simulated and measured cases shown in **Figure 43(a–f)**, the

http://dx.doi.org/10.5772/65603

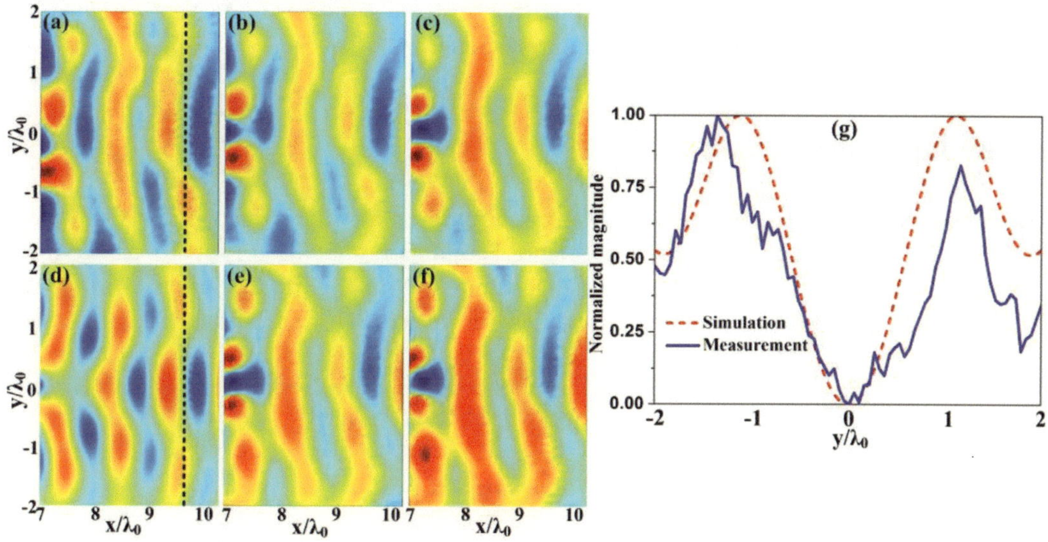

Figure 43. Comparison of simulated and measured electric-field distributions. (d) Simulated and (a–c, e, and f) measured results at (a and d) 10.1 GHz, (b) 10.15 GHz, (c) 10.21 GHz, (e) 10.25 GHz, and (f) 10.3 GHz. (g) Normalized electric-field intensity along the black dashed line $x = 9.57\lambda_0$.

uneven wavefronts with convex curves at both ends and concave curves in the interspace, along with a remarkable perturbation in the first outgoing wavefront from the illusion device are clearly observed, illustrating the effectiveness and robustness of the design. Hence, the superscatterer illusion functionality, i.e., the transformation of the radar signature of an object to the equivalence of an enlarged object and four isolate dielectric objects is unambiguously validated over a bandwidth of more than 200 MHz. The shifted operation frequency and slight distortion of electric fields in measurements are possibly attributable to the tolerance inherent in fabrications and assembling, the utilized adhesives to connect meta-atoms at the interfaces of different regions and the misalignment of particles along radial direction.

4. Conclusions

In summary, we have reviewed in this chapter our recent effort in synthesizing electrically small meta-atoms from effective medium perspective and utilizing compact meta-atoms to design microwave circuits and functional devices. Several strategies have been proposed for such a purpose and the mechanisms have been studied in depth. The advantages of compact meta-atoms can be classified in two categories. First, it can significantly reduce the circuit size without posing penalty on device performances. Second, it brings about additional degree of freedom for device design and broadband deep out-of-band signal inhibition which can be employed for harmonic suppression. Third, it enables manipulation of precise material parameters and smooth outgoing field which is preferable for functional devices with high performances and new physics demonstration with high-quality phenomena. Moreover, the precise-material parameters will improve the success rate of correct design. Our compact approach lays a platform and gives a promising alternative for both engineers and scientists to realize their devices or demonstrate their find using metamaterials.

Acknowledgements

This work was supported by National Natural Science Foundation China under Grant Nos. 61501499 and 61372034 and also Natural Science Foundation of Shaanxi Province under Grant 2016JQ6001. The authors deliver their special gratitude to Prof. Tie Jun Cui and his meta-group for the guidance, discussions and help afforded in the work of 3D super lens and illusion device.

Author details

He-Xiu Xu*, Guang-Ming Wang, Tong Cai, Qing Peng and Ya Qiang Zhuang

*Address all correspondence to: hxxuellen@gmail.com

Microwave Laboratory, Air Force Engineering University, Changle, Xi'an, China

References

[1] Bahl I, "Lumped elements for RF and microwave circuits," Boston: Artech House, Ch. 14, 2003, pp. 462–465.

[2] E.-Y. Jung and H.-Y. Hwang, "A balun-BPF using a dual mode ring resonator," *IEEE Microw. Wireless Compon. Lett.*, vol. 17, no. 9, pp. 652–654, Sep. 2007.

[3] Y.-X. Guo, K.-W. Khoo and L. C. Ong, "Wideband circularly polarized patch antenna using broadband baluns," *IEEE Trans. Antennas Propag.*, vol. 56, no. 2, pp. 319–326, Feb. 2008.

[4] T.-G. Ma and Y.-T. Cheng, "A miniaturized multilayered Marchand balun using coupled artificial transmission lines," *IEEE Microw. Wireless Compon. Lett.*, vol. 19, no. 7, pp. 446–448, Jul. 2009.

[5] Y.-X. Guo, Z. Y. Zhang, L. C. Ong and M. Y. W. Chia, "A novel LTCC miniaturized dualband balun," *IEEE Microw. Wireless Compon. Lett.*, vol. 16, no. 3, pp. 143–145, Mar. 2006.

[6] C.-H. Tseng and Y.-C. Hsiao, "A new broadband Marchand balun using slot-coupled microstrip lines," *IEEE Microw. Wireless Compon. Lett.*, vol. 20, no. 3, pp. 157–159, Mar. 2010.

[7] C.-H. Tseng and C.-L. Chang, "Wide-band balun using composite right/left-handed transmission line," *Electron. Lett.*, vol. 43, no. 21, pp. 1154–1155, Oct. 2007.

[8] Z.-Y. Zhang and K.Wu, "A broadband substrate integrated waveguide planar balun," *IEEE Microw. Wireless Compon. Lett.*, vol. 17, no. 12, pp. 843–845, Dec. 2007.

[9] J.-L. Li, S.-W. Qu and Q. Xue, "Miniaturised branch-line balun with bandwidth enhancement," *Electron. Lett.*, vol. 43, no. 17, pp. 931–932, Aug. 2007.

[10] C. Liu and W. Menzel, "Broadband via-free microstrip balun using metamaterial transmission lines," *IEEE Microw. Wireless Compon. Lett.*, vol. 18, no. 7, pp. 437–439, Jul. 2008.

[11] C.-C. Chen and C.-K. C. Tzuang, "Synthetic quasi-TEM meandered transmission lines for compacted microwave integrated circuits," *IEEE Trans. Microw. Theory Tech.*, vol. 52, no. 6, pp. 1637–1647, Jun. 2004.

[12] H.-X. Xu, G.-M. Wang, X. Chen, et al. "Broadband balun using fully artificial fractal-shaped composite right/left handed transmission line," *IEEE Microw. Wireless Compon. Lett.*, vol. 22, pp. 16–18, 2012.

[13] W. Shao, J. He and B.-Z. Wang, "Compact rat-race ring coupler with capacitor loading," *Microwave Opt. Technol. Lett.*, vol. 52, no. 1, pp. 7–9, 2010.

[14] J. Wang, B.-Z. Wang, Y.-X. Guo, L. C. Ong and S. Xiao, "Compact slow-wave microstrip rat-race ring coupler," *Electron. Lett.*, vol. 43, no. 2, pp. 111–113, 2007.

[15] J.-T. Kuo, J.-S. Wu and Y.-C. Chiou, "Miniaturized rat race coupler with suppression of spurious passband," *IEEE Microw. Wireless Compon. Lett.*, vol. 17, no. 1, pp. 46–48, 2007.

[16] S. Opozda, P. Kurgan and M. Kitlinski, "A compact seven-section rat-race hybrid coupler incorporating PBG cells," *Microwave Opt. Technol. Lett.*, vol. 51, no. 12, pp. 2910–2913, 2009.

[17] P. H. Rao, J. Subramanian and M. Kamki, "Miniaturized broadband microstrip fractal rat-race coupler with shunt stubs," *Microwave Opt. Technol. Lett.*, vol. 52, no. 2, pp. 289–292, 2010.

[18] H. Ghali and T. A. Moselhy, "Miniaturized fractal rat-race, branch-line and coupled-line hybrids," *IEEE Trans Microwave Theory Tech.*, vol. 52, no. 11, pp. 2513–2520, 2004.

[19] H. Ghali and T. Moselhy, "Design of fractal rat-race coupler," *IEEE MTT-S Int. Microwave Symp. Dig.*, pp. 324–327, 2004.

[20] K.-K. M. Cheng and F.-L. Wong, "Dual-band rat-race coupler design using tri-section branch-line," *Electron. Lett.*, vol. 43 no. 6, pp. 41–42, 2007.

[21] K.-S. Chin, K.-M. Lin, Y.-H. Wei, T.-H. Tseng and Y.-J. Yang, "Compact dual-band branch-line and rat-race couplers with stepped-impedance-stub lines," *IEEE Trans. Microwave Theory Tech.*, vol. 58, no. 5, pp. 1213–1221, 2010.

[22] M. K. Mandal and S. Sanyal, "Compact dual-band rat-race couplers," *Microwave Opt. Technol. Lett.*, vol. 50, no. 2, pp. 277–281, 2008.

[23] H.-X. Xu, G.-M. Wang, X.-K. Zhang, et al. "Novel compact dual-band rat-race coupler combining fractal geometry and CRLH TLs," *Wireless Pers. Commun.*, vol. 66, no. 4, pp. 855–864, 2012.

[24] P.-L. Chi and T. Itoh, "Miniaturized dual-band directional couplers using composite right/left-handed transmission structures and their applications in beam pattern diversity systems," *IEEE Trans Microwave Theory Tech.*, vol, 57, no. 5, pp. 1207–1215, May 2009.

[25] C. Caloz and T. Itoh, "Electromagnetic metamaterials: transmission line theory and microwave applications: the engineering approach ," Hoboken, New Jersey: Wiley-Interscience, 2006.

[26] H.-X. Xu, G.-M. Wang, C.-X. Zhang, Z.-W. Yu and X. Chen, "Composite right/left-handed transmission line based on complementary single-split ring resonator pair and compact power dividers application using fractal geometry," *IET Microw. Antennas Propag.*, vol. 6, no. 9, pp. 1017–1025, 2012.

[27] H.-X. Xu, G.-M. Wang, Z.-M. Xu, et al. "Dual-shunt branch circuit and harmonic suppressed device application," *Appl. Phys. A*, vol. 108, no. 2, pp. 497–502, 2012.

[28] R. Marques, F. Martin and M. Sorolla. "Metamaterials with negative parameters: theory, design and microwave applications," Hoboken, NJ: Wiley, 2008.

[29] A. K. Iyer and G. V. Eleftheriades, "Volumetric layered transmission-line metamaterial exhibiting a negative refractive index," *J. Opt. Soc. Am. B*, vol. 23, pp. 553–570, 2006.

[30] A. K. Iyer and G. V. Eleftheriades. "A multilayer negative-refractive-index transmission-line (NRI-TL) metamaterial free-space lens at X-Band," *IEEE Trans. Antennas Propag.*, vol. 55, pp. 2746–2753, 2007.

[31] H.-X. Xu, G.-M. Wang, M. Q. Qi, et al. "Three-dimensional super lens composed of fractal left-handed materials," *Adv. Opt. Mater.*, vol. 1, no. 7, pp. 495–502, 2013.

[32] H.-X. Xu, G.-M. Wang, K. Ma and T. J. Cui, "Superscatterer illusions without using complementary media," *Adv. Opt. Mater.*, vol. 2, no. 6, pp. 572–580, 2014.

[33] B. L. Zhang and G. Barbastathis, "Dielectric metamaterial magnifier creating a virtual color image with far-field subwavelength information," *Opt. Express*, vol. 18, p. 11216, 2010.

[34] W. X. Jiang, C.W. Qiu, T. C. Han, Q. Cheng, H. F. Ma, S. Zhuang and T. J. Cui, "Broadband all-dielectric magnifying lens for far-field high-resolution imaging," *Adv. Mater.*, vol. 25, pp. 6963–6968, 2013.

[35] W. X. Jiang, C.-W. Qiu, T. Han, et al. "Creation of ghost illusions using wave dynamics in metamaterials," *Adv. Funct. Mater.*, vol. 23, pp. 4028–4034, 2013.

Chapter 9

Polarization State Manipulation of Electromagnetic Waves with Metamaterials and Its Applications in Nanophotonics

Shuqi Chen, Wenwei Liu, Zhancheng Li,
Hua Cheng and Jianguo Tian

Additional information is available at the end of the chapter

http://dx.doi.org/10.5772/66036

Abstract

Polarization state is an important characteristic of electromagnetic waves. The arbitrary control of the polarization state of such wave has attracted great interest in the scientific community because of the wide range of modern optical applications that such control can afford. Recent advances in metamaterials provide an alternative method of realizing arbitrary manipulation of polarization state of electromagnetic waves in nanoscale via ultrathin, miniaturized, and easily integrable designs. In this chapter, we give a review of recent developments on polarization state manipulation of electromagnetic waves in metamaterials and discuss their applications in nanophotonics, such as polarization converter, wavefront controller, information coding, and so on.

Keywords: polarization manipulation, optical activity, wave plates, wavefront controlling, optical communication

1. Introduction

Harnessing electromagnetic waves for modern nanophotonics applications often involves the control and manipulation of polarization state. The ability to manipulate the polarization state of electromagnetic waves can enable us to control electromagnetic waves for a wide range of applications such as polarization manipulation, wavefront controlling, and optical communication [1–4]. Conventional approaches to manipulate the polarization state of electromagnetic waves employ bulky wave-plates, which are made of birefringent materials composed of crystalline solids and liquid crystals. However, the inherent disadvantages in the size, collimation, and bandwidth of these configurations prevent optical system miniaturization and integration. Thus, realizing polarization state manipulation of electromagnetic waves in nanoscale has become one of the key problems for the development of modern optics and nanophotonics.

Metamaterials are generally composed of subwavelength artificial nanostructures, which can overcome the physical limitations imposed by natural materials and provide exceptional capabilities for manipulating waves with greater precision. Over the past decade, numerous novel optical properties have been demonstrated in this area, such as negative refractive index, super-lenses, cloaking, etc. [5–14]. The soul of metamaterials is the ability to realizing arbitrary manipulation of the electromagnetic waves in multiple parameters (frequency, amplitude, phase, and polarization) with multiple degrees of freedom, which make it possible for people to design devices with optical properties on demand. Recently, metamaterials have been reported to provide a promising pathway toward the realizing of efficient manipulation of polarization state of electromagnetic waves via ultrathin, miniaturized, and easily integrable designs, which open up intriguing possibilities toward the realization of polarization state manipulation of electromagnetic waves in nanoscale and show infinite prospection in nanophotonics applications [4, 5, 15–17].

In this chapter, we will review the development of metamaterials in polarization state manipulation of electromagnetic waves and discuss its applications in nanophotonics, which will provide a guidance for its further designs and applications. The chapter is organized as follows:

In the second section, we begin with a brief introduction of polarization state of electromagnetic waves and give a review of the development process of metamaterials for polarization state manipulation.

In the third section, the fundamental applications of metamaterials in nanophotonics, such as wave plates, polarization converter, opt-isolator, arbitrary generation of vector beam, polarization-based wavefront-control, and so on will be discussed in detail.

In the fourth section, the polarization integrated metamaterials devices, such as tunable polarization controlling devices, photonic spin Hall effect, information coding and optical communication, and polarization-switchable phase holograms, will be discussed in detail.

In the fifth section, we will overlook the whole area of metamaterials-based polarization devices, summarize the main difficulties, possible solutions, and further applications in future.

In the last section, we will give a summary of the chapter.

2. Polarization state manipulation in metamaterials

2.1. Polarization state of electromagnetic waves

The polarization state of electromagnetic waves, which cannot be detected by human eyes, forms an important characteristic of such waves. The polarization state of electromagnetic waves at a fix position is determined by the time course of the electric-field vector $\mathbf{E}(\mathbf{r}, t)$ [18].

For monochromatic electromagnetic waves, the electric-field vector **E**(**r**, *t*) can be divided into two orthogonal components with different amplitude and phases, which can be concisely expressed in the form of **Jones vector** as:

$$\mathbf{E}(\mathbf{r}, t)=\begin{pmatrix}\mathbf{E}_x\\ \mathbf{E}_y\end{pmatrix}=\begin{pmatrix}a_x e^{i(kx-\omega t+\varphi_x)}\\ a_y e^{i(kx-\omega t+\varphi_y)}\end{pmatrix} \tag{1}$$

The endpoint of the vector **E**(**r**, *t*)is determined by the envelope of two orthogonal components A_x=Re($\mathbf{E}_x$) and A_y=Re($\mathbf{E}_y$), which vary in cosinusoid with time and the trajectory of the endpoint can be expressed as:

$$\frac{A_x^2}{a_x^2}+\frac{A_y^2}{a_y^2}-2\cos\varphi\frac{A_x A_y}{a_x a_y}=\sin^2\varphi \tag{2}$$

where $\varphi=\varphi_y-\varphi_x$ is the phase difference. The trajectory of Eq. (**2**) is an ellipse and the properties of it vary with the position for wavefront with different directions at different positions. For the plane wave whose wavefront is parallel transverse plane, the elliptical trajectory stays the same. Thus, the polarization state of plane wave can be described by a single ellipse. Moreover, such polarization ellipse can be determined by its orientation and shape which can be characterized by two angles:

$$\tan 2\Psi=\frac{2r}{1-r^2}\cos\varphi, \tag{3}$$

$$\sin 2\chi=\frac{2r}{1+r^2}\sin\varphi. \tag{4}$$

where r= a_x / a_y is the ratio of the magnitude of two orthogonal electric field components. The angle Ψ determines the orientation of the polarization ellipse while the angle χ determines the ellipticity. Thus, the polarization state of electromagnetic waves can be described by the magnitude ratio $r=a_x / a_y$ and phase difference $\varphi=\varphi_y-\varphi_x$. For a_x (or a_y) equal to zero or the phase difference $\varphi=0$ (or $\varphi=\pi$), the polarization ellipse becomes a straight line, and the electromagnetic waves is linearly polarized (LP). If the magnitude ratio is r=1 and the phase difference is $\varphi=\pm\pi/2$, then the polarization ellipse become a circle and the electromagnetic waves is said to be circularly polarized. For $\varphi=\pi/2$ (electric field rotates in a clockwise direction when viewed from the direction toward which the wave is propagating), the electromagnetic waves is said to be right-handed circularly polarized (RCP). And the case $\varphi=-\pi/2$ corresponds to counter-clockwise rotation and left-handed circularly polarized (LCP). For other condition, the polarization state of electromagnetic waves is said to be elliptically polarized.

2.2. Polarization state manipulation with metamaterials

As indicated above, the polarization state of electromagnetic waves can be described by magnitude ratio r and phase difference φ. Thus, the polarization state manipulation with metamaterials always involves the tailoring of the wave interference at the subwavelength scale by introducing the anisotropic optical resonance mode to effectively manipulate the magnitude and phase of electric components in two orthogonal directions. Thus, the polarization state manipulation of electromagnetic waves in metamaterials always involves structures with two orthogonal resonance including elliptical nanoholes, L-shaped nanoparticles, crossed nanodipoles, nanoslits, and nanorods, as shown in **Figure 1** [19–24].

However, the polarization state manipulative ability of initial complanate metamaterials with the above-mentioned structures is limited because of the limited interaction between electromagnetic waves and structures, thus the efficiency and bandwidth are lower than that required for practical applications. Over the past decade, many efforts have been made in the scientific community to overcome the drawbacks of complanate metamaterials and improve the polarization state manipulation of electromagnetic waves in nanoscale with new types of metamaterials [25, 26]. As one of the solutions, sandwich-like metamaterials constructed with anisotropic resonators, dielectric layer, and highly reflective metallic film have been proposed to provide an alternate way to realize effective polarization state manipulation in reflection mode as shown in **Figure 2a** and **b** [27, 28]. The near-field interference in this type of designs improves the interaction between electromagnetic waves and metamaterials effectively, thus realizing

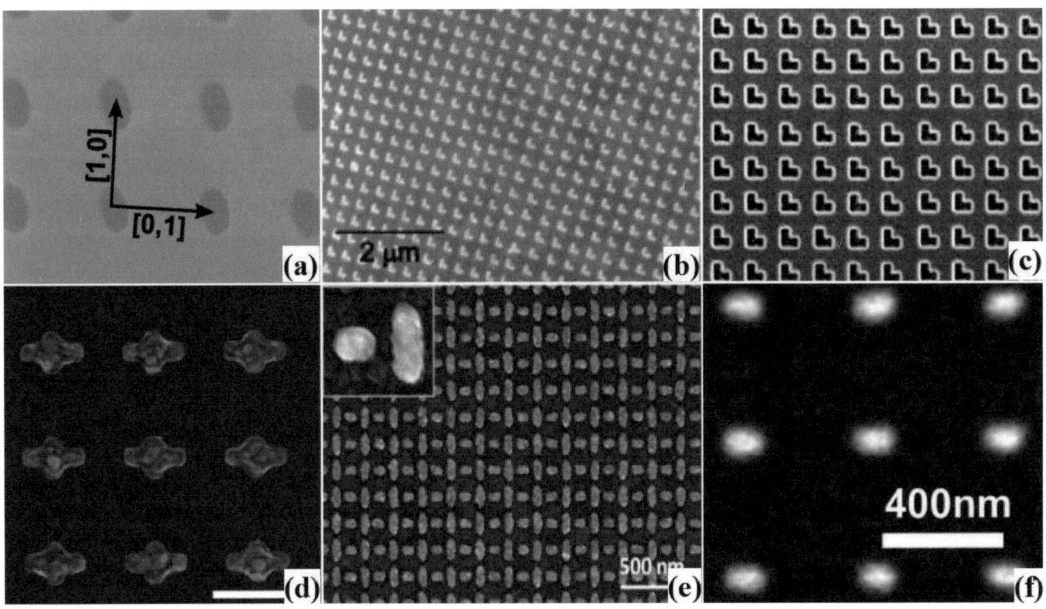

Figure 1. (a) Experimental realization of strong polarization controlling in the optical transmission through elliptical nanoholes. Reprinted by permission from [19] copyright 2004 American Physical Society. (b) Experimental realization of birefringence in two-dimensional L-shaped silver nanoparticles. Reprinted by permission from [20] copyright 2008 American Chemical Society. (c) Experimental realization of manipulating the polarization state of the electromagnetic waves and achieving giant optical rotation in near infrared wavelength by L-shaped nanoholes array in silver film. Reprinted by permission from [21] copyright 2008, AIP Publishing LLC. (d) Experimental realization of chromatic plasmonic polarizers by optical nanorods. Reprinted by permission from [22] copyright 2012 American Chemical Society. (e) Experimental realization of broadband optical meta-waveplates by crossed nanodipoles. Reprinted by permission from [23] copyright 2013 American Chemical Society. (f) Experimental realization of reflecting wave plates constructed with nanodipoles. Reprinted by permission from [24] copyright 2014 Optical Society of America.

http://dx.doi.org/10.5772/66036

effective polarization state manipulation of electromagnetic waves in reflection mode. On the other hand, applications in modern nanophotonics always require nanoscale polarization state manipulation in transmission mode. Thus, sandwich-like metamaterials is out of work in this situation. Recently, the proposition of few-layer metamaterials makes it possible to realize effective polarization state manipulation in transmission mode as shown in **Figure 2c** and **d** [29, 30]. The interference and near-field coupling between layers in few-layer metamaterials ensure that the energy of electromagnetic waves can be strongly redistributed and can effectively interact with the structures, resulting in polarization state manipulation with high efficiency and broadband in transmission mode. However, the above-mentioned metamaterials are all based on metallic structures, and thus the electromagnetic waves absorption and subsequent heat conversion in sandwich-like and few-layer metamaterials inevitably increase which impeded the applications of metallic structure-based metamaterials for polarization state manipulation

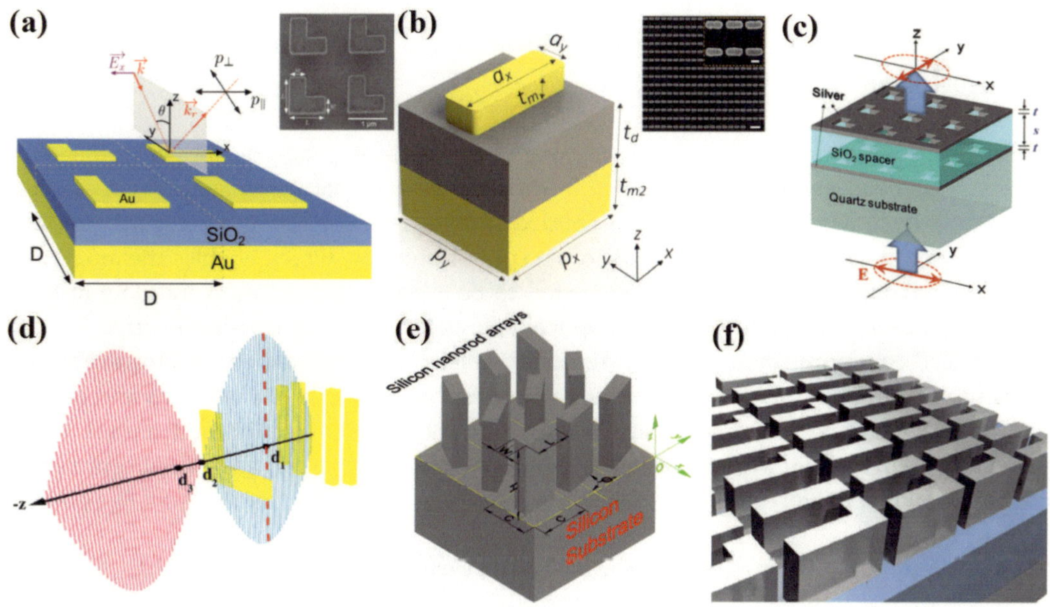

Figure 2. (a)Experimental realization of wide band and efficient linear polarization conversion with plasmonic planar antenna. Reprinted by permission from [27] copyright 2014 AIP Publishing LLC. (b) Experimental realization of wide-angle reflective plasmonic metasurface-based half-wave and quarter-wave plates that have a high polarization conversion efficiency and reflection magnitude over a broad bandwidth in the visible-to-near IR wavelength range. Reprinted by permission from Macmillan Publishers Ltd [28], Scientific reports copyright 2016. (c) Experimental realization of a plasmonic assisted Fabry-Perot cavity in a metal/insulator/metal trilayer structure with L-shaped hole arrays inside, which significantly contribute to the mechanism to realize a nearly complete polarization conversion in optical transmissions at near-infrared wavelength. Reprinted by permission from [29] copyright 2010, AIP Publishing LLC. (d) Simulated realization of a high performance broadband asymmetric polarization conversion composed of an L-shaped gold particle and a gold nanoantenna array for the near-infrared regime. Reprinted by permission from [30] copyright 2015 Springer International Publishing AG. (e) Simulated realization of all-silicon nanorod-based Dammann gratings with accurate phase controlling, and strong polarization conversion. Reprinted by permission from [31] copyright 2015 Optical Society of America. (f) Experimental realization of silicon-based infrared metasurfaces as high performance linear-to-circular polarization converter. Reprinted with permission from Macmillan Publishers Ltd, [32] Nature Communications copyright 2014.

of electromagnetic waves. Recently, a different approach has emerged. The Mie resonance in high-index dielectric structures provides a novel way to realize anisotropic optical resonance [31, 32]. Metamaterials based on all-dielectric nanoparticles (as shown in **Figure 2e** and **f**) overcome the energy loss of electromagnetic waves in metallic structure-based metamaterials and become an ideal selection for polarization state manipulation of electromagnetic waves.

In summary, metamaterials with the development of it can realize the effective controlling of amplitude and phase of electromagnetic waves in two orthogonal directions, which provide infinite possibilities for arbitrary manipulation of polarization state of electromagnetic waves in nanoscale. Thus, the role of metamaterials for polarization manipulation devices in nanophotonics is no substitute. With the improvement of the relative research, metamaterial-based polarization manipulation devices have been widely proposed in recent years which will be discussed in detail in the next section.

3. Fundamental applications of metamaterials for polarization state manipulation in nanophotonics

As mentioned above, polarization state is one of the intrinsic properties of electromagnetic waves, which can always be resolved into different orthogonal basis, such as x-polarized and y-polarized, left-handed and right-handed waves. It is worth mentioning that the left-right-handed polarization corresponds to the spin of photons, which enables metamaterials with great abilities of polarization conversion to be a possible quantum candidate in large scale. In modern optics and photonic applications, polarization conversion is often utilized in advanced communication, sensing, displayer, noise reduction, and so on. However, most of these applications come down to three categories: conversion of polarization and generation of vector beams, wave plates, and asymmetric transmission. In this section, we aim at providing an overall view on the applications and evaluating the possibilities of commercial utilization of metamaterials.

3.1. Conversion of polarization and generation of vector beams

Metamaterials have shown to the world their unique design flexibility, compactness, and highly novel characteristics at its very first birth. Unlike the traditional method, in which polarization is often controlled with polarized molecules (such as liquid crystals) [33], birefringent crystals [18], or magneto-optic phenomenon (Faraday Effect) [34], metamaterials can directly manipulate strength and phase of electromagnetic waves at subwavelength scale. In the eyes of a metamaterials' researcher, polarization of electromagnetic waves can be modified pixel by pixel (unit cell of the design), thus arbitrarily polarized outputs on Poincare Sphere can be easily achieved [35].

Throughout the recent works, there are two kinds of metamaterials polarization converters that have shown their powerful ability to handle polarization. One is as shown in **Figure 3a**, which is an enhanced optical rotator of the zero-order transmitted electromagnetic waves through a silver film with an array of perforated S-shaped holes [36]. The fundamental mechanisms of these kinds of polarization rotators are as follows: the incident wave interacts with the sub-wavelength structure and creates orthogonal polarized component due to the surface plasmon polaritons (SPPs), or localized surface plasmons (LSPs), or both. With elaborate design, the original polarized electric component can be cancelled due to near-field interference, and the orthogonal one is left over, thus a cross-polarization converter is achieved. Other states of polarization can also be accomplished with similar method because the complete orthogonal basis has been acquired. The transmitted polarization depends on the thickness of the device, as shown in **Figure 3b**, which exhibits a tiny ellipticity, indicating the transmission can be roughly treated as a linearly polarized one [21, 31, 37].

Recently another design almost dominates the area of polarization conversion with metamaterials, as shown in **Figure 3(c)** to **3(f)**. This kind of designs simultaneously manipulate the amplitude and phase of electric-magnetic components [35], [38]-[41], resulting in much higher degree of freedom for polarization conversion. In **Figure 3(c)**, the incident linearly polarized (LP) wave can be converted into left-circularly polarized (LCP) and right-circularly polarized (RCP) beams at sub-wavelength scale [38]. The difference of phase retardation between the LCP and RCP beams can be easily modified by varying the geometrical parameters of the nano-apertures, leading to continuously controllable optical activity. It's worth mentioning that most designs generate orthogonal electric components by the chiral-

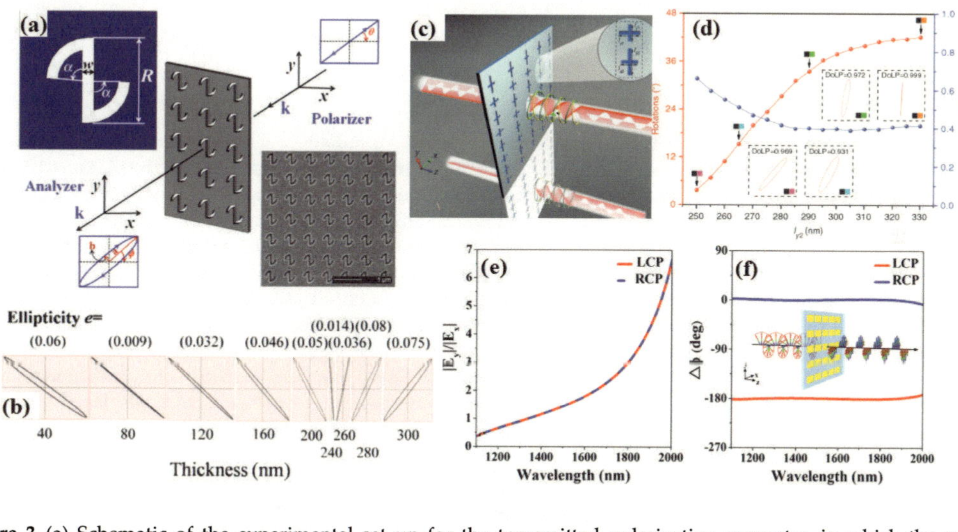

Figure 3. (a) Schematic of the experimental set up for the transmitted polarization converter, in which the sample made from silver is typically built with focused ion beam (FIB) method. In the experiment, the incident wave remains linearly polarized along the x axis. (b) The transmitted elliptical polarization states and the corresponding ellipticities for different samples with different thicknesses. Reprinted by permission from [36] copyright 2013 American Physical Society. (c) Sketch of the optical activity process with non-chiral plasmonic metasurfaces, in which the transmitted linearly polarized wave is synthesized with two subwavelength located circularly polarized wave. (d) Calculated optical rotations (red lines) and amplitude transmissions (blue lines) at a wavelength of 990 nm. The insets show the polarization states and the corresponding degree of linear polarization (DoLP) for four metasurfaces. Reprinted by permission from Macmillan Publishers Ltd [38], Light: Science and Applications, copyright 2016. (e) Simulated amplitude ratio and (f) phase difference of the electric components of the transmitted waves for right- and left-circularly polarized incidence, respectively. Inset: artistic rendering of the circular-to-linear (CTL) polarization converter. Reprinted by permission from Macmillan Publishers Ltd [39], Scientific reports, copyright 2015.

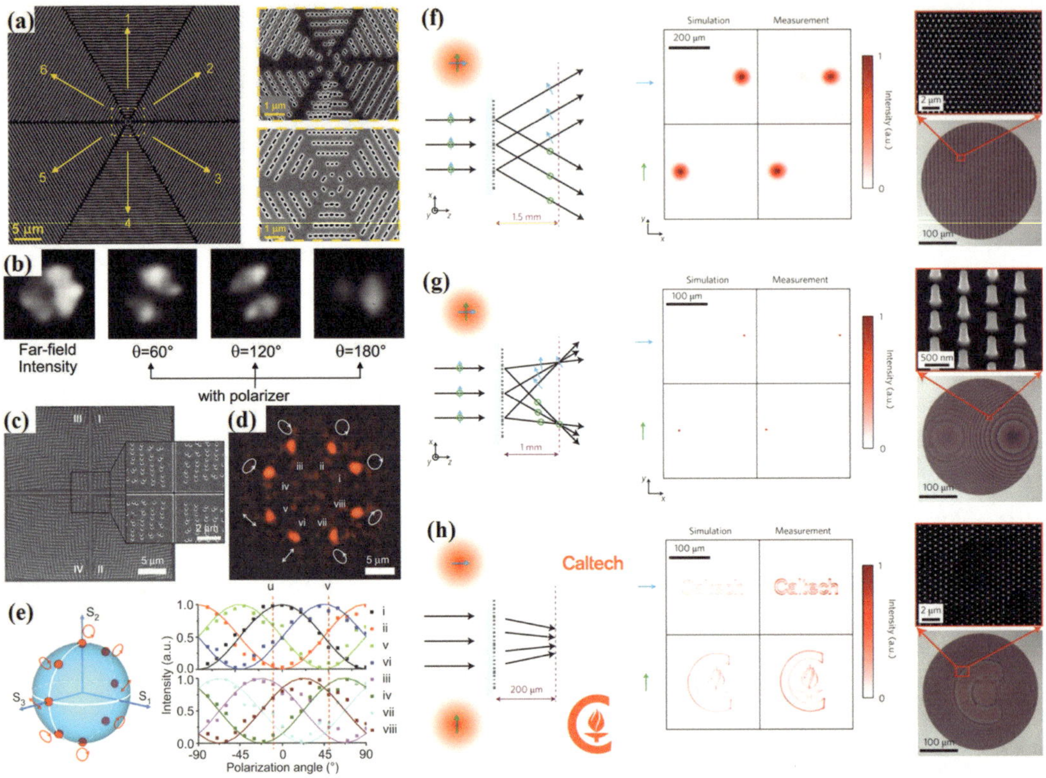

Figure 4. (a) SEM image of the plasmonic metasurface to generate a radially polarized beam. Each region of the plasmonic metasurface is filled with one type of nano-aperture pairs marked by the number. (b) Measured far-field intensity profiles of the radially polarized beam, with an analyzer oriented at various angles in front of the CCD camera. Reprinted by permission from [14] copyright 2015 by John Wiley and Sons. (c) SEM image of the polarization generator for the eight-foci process. (d) Focus image without polarization analysis, each hot spot of which represents one of the polarization states of the foci. (e) The polarization states of the foci on the Poincare sphere, and Si stands for the *i*th Stoke's parameter. Reprinted by permission from Macmillan Publishers Ltd [42], Light: Science and Applications, copyright 2015. (f) A polarization beam splitter that separate the *x*- and *y*-polarized waves and deflects them by different angles, and the angles can be chosen at will. (g) A device that separates *x*- and *y*-polarized waves and focuses them to two different points. The two different points can be chosen at will. (h) A polarization-switchable phase hologram that generates two arbitrary patterns for *x*- and *y*-polarized waves. In this case the word "Caltech" is displayed for input *x* polarization and an icon is displayed for input *y* polarization. Reprinted by permission from Macmillan Publishers Ltd [35], Nature nanotechnology copyright 2015.

ity of the materials or nano-structures, however the design in **Figure 3(c)** experimentally and numerically investigated an alternative approach to realize and control the optical activity with non-chiral plasmonic metasurfaces. The optical rotation was calculated in **Figure 3(d)**, and rotation with arbitrary values from 3° to 42° is obtained, with relatively high-amplitude transmission beyond 40%. Circularly to linearly polarized wave can also be achieved in [39], as shown in **Figure 3(e)** and **3(f)**. The inset in **Figure 3(f)** depicts the artistic rendering of the design. As amplitude ratio varies throughout the working waveband, indicating different orientation of transmitted wave when differing the incident wavelength, this device can hardly called a broadband quarter-wave plate. However, it is still a novel polarization converter, and may be utilized in a metasurface displayer to distinguish different colors. Since this design is a time reversal symmetric system (taking no account of the thermal loss), it can also be applied to convert linearly polarized wave to circularly polarized one.

As mentioned above, the subwavelength characteristics of metamaterials enable people to modify polarization of electromagnetic waves pixel by pixel (unit cell of the design), and to generate wave with different polarization in each cross-section of the outputs. As shown in **Figure 4a**, a radially polarized beam is generated by appropriately arranging the nano-apertures [14]. Although the six regions are discontinuous with each other, the achieved vector beam is continuous due to the subwavelength characteristics of the structure, which is always a principle of the metamaterials' design. A far-field intensity profile of the radially polarized beam is shown in **Figure 4b**. Furthermore, with the in-plane field of SPPs combined, all types of polarization states can be achieved simultaneously [42], as shown in **Figure 4c–e**. This remarkable consequence results from that the in-plane field of SPPs with proper polarization states and phases can be selectively scattered out to the desired electromagnetic wave beams. This design offers a novel route to achieve the full control of optical polarizations. However, all these devices are plasmonic based and the efficiency is limited by the intrinsic absorption loss. Although many researchers struggled to overcome this limitation by introducing multilayered structures or performing at reflectance mode [24, 27], and so on, it can hardly be solved thoroughly due to the high conductivity of metals. In recent years, high-contrast dielectric designs have been developed to enlarge the working efficiency and practicability of metamaterials [26]. As shown in **Figure 4f–h**, full control of polarization for beam splitter, lens, and phase hologram is achieved with a simple layer of silicon elliptical nanoposts [35], with an experimentally measured efficiency ranging from 72 to 97%. The complete and simultaneous control over the polarization and phase profiles of electromagnetic waves offered by the proposed platform and the design technique enables the realization of integrated nano-optic devices, which is one of the greatest steps in modern optics and photonics.

3.2. Wave plates accomplished by metamaterials

Although full control of polarization is the primary issue with respect to polarization, wave plates are still worth discussing due to their wide applications in applied optics. Normally wave plates are manufactured from birefringent crystals, which are really successful due to their high efficiency and accuracy. However, simultaneously achieving broadband and wide-angle properties are almost impossible because of the limitation of the crystals' dispersive properties. In contrast, metamaterials once again provide a promising pathway toward the perfect wave plates with thickness less than a micrometer. As shown in **Figure 5a**, a broadband half-wave plate is accomplished by a nanorod layer and a metallic reflective ground plane sandwiched by a dielectric layer [28]. An *s*-polarized wave incident from an angle of θ_i ($\varphi_i = 135°$) is converted into a *p*-polarized wave in reflectance mode, with reflection greater than 92% from 640 to 1290 nm, and with near unitary polarization conversion ratio (PCR) even when increasing the incident angle to 40°. This phenomenon results from the asymmetric length of the long and short sides of the nanorod, which responds to the incident beams with different phase delay and radiative intensity. Similarly, a quarter-wave plate can also be obtained with careful adjustment of dimensions of the structure. As shown in **Figure 5c**, the calculated results are in good agreement with the experimental ones. Although the designed

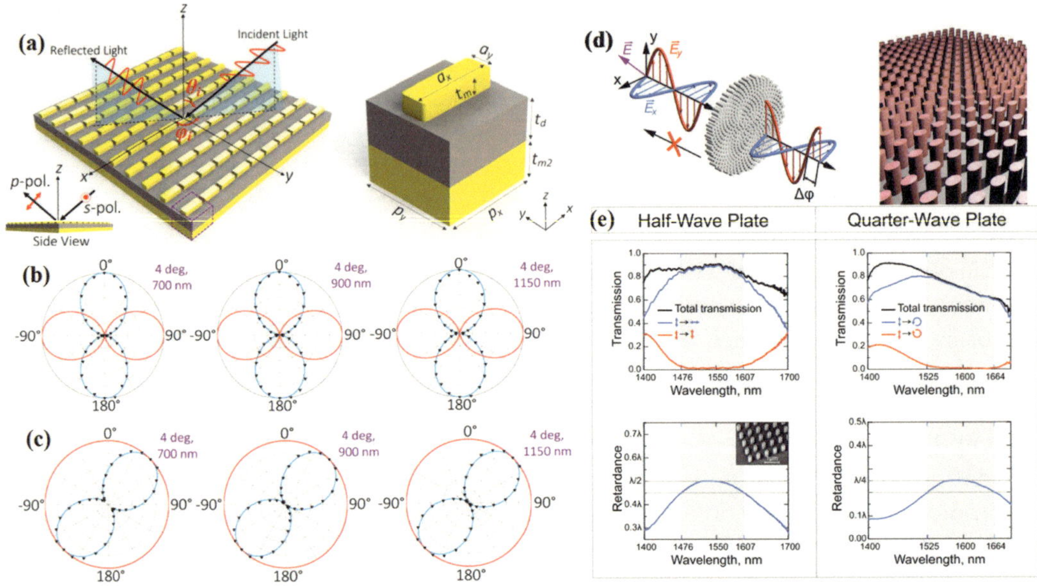

Figure 5. (a) Schematic of the metasurface-based half-wave plate. An *s*-polarized wave incident from an angle of θ_i (φ_i = 135°) is converted into a *p*-polarized wave in reflectance mode. The unit cell shows the sandwiched metallic structure. (b) Theoretically predicted (lines) and experimentally measured (triangles) polarization states in the plane perpendicular to the wave vector at 700, 900, and 1150 nm incidence, respectively. (c) Theoretically predicted (lines) and experimentally measured (triangles) polarization states in the plane perpendicular to the wave vector at 700, 900, and 1150 nm incidence, respectively, for the design of quarter-wave plate. Reprinted by permission from Macmillan Publishers Ltd [28], Scientific reports copyright 2016. (d) An artistic view of a meta-device for spatially variant polarization control operating in a broad spectral range. The device is composed by a single layer of elliptical dielectric nanoposts. (e) High efficient half-wave plate and quarter-wave plate achieved by the dielectric nanoposts. Reprinted by permission from [43].

wave plates are broadband and wide-angled, especially with high efficiency, a critical drawback still exists that they work in reflectance mode, compared to which transmission mode is often preferable because of its intrinsic convenience for experimental and commercial utilization. One way to solve this problem is to use high-contrast dielectric metasurfaces [43], which possess the ability to interact with electromagnetic waves at extremely confined spots without any heat dissipation, as depicted in **Figure 5d** and **e**. The designed half-wave plate and quarter-wave plate maintain a high transmission in the working waveband, and the undesired component of polarization is suppressed at a negligible level.

3.3. Asymmetric transmission

Another application on polarization is asymmetric transmission, which are often utilized in integrated photonic systems for communications and information processing. This unique phenomenon is often achieved by reducing the structural symmetry and converting to different polarization states. The corresponding process can be completely characterized by the 2 × 2 ray-transfer matrix in the paraxial ray-optics approximation. We can write

http://dx.doi.org/10.5772/66036

down the T matrix connecting the generally complex amplitudes of the incident and the transmitted field:

$$\begin{pmatrix} E_x^o \\ E_y^o \end{pmatrix} = \begin{pmatrix} T_{xx} & T_{xy} \\ T_{yx} & T_{yy} \end{pmatrix} \begin{pmatrix} E_x^i \\ E_y^i \end{pmatrix} = \begin{pmatrix} A & B \\ C & D \end{pmatrix} \begin{pmatrix} E_x^i \\ E_y^i \end{pmatrix} \tag{5}$$

where 'o' and 'i' denote the incident and output field, respectively. If the structure is rotated by 180° with respect to the x axis (assuming the wave vector is along z axis), it can be demonstrated that the **T** matrix (backward) can be written as [44]:

$$\mathbf{T}^b = \begin{pmatrix} A & -C \\ -B & D \end{pmatrix} \tag{6}$$

Similarly for circularly polarized incidence the **T** matrices are:

$$\mathbf{T}_{circ}^f = \begin{pmatrix} T_{++} & T_{+-} \\ T_{-+} & T_{--} \end{pmatrix} \tag{7}$$

and

$$\mathbf{T}_{circ}^b = \begin{pmatrix} T_{++} & -T_{-+} \\ -T_{+-} & T_{--} \end{pmatrix} \tag{8}$$

where the linearly and circularly polarized matrices are related by

$$\begin{pmatrix} T_{++} & T_{-+} \\ T_{+-} & T_{--} \end{pmatrix} = \frac{1}{2} \begin{pmatrix} A + D + i(B - C) & A - D - i(B + C) \\ A - D + i(B + C) & A + D - i(B - C) \end{pmatrix} \tag{9}$$

The difference between the **T** matrices for opposite propagating direction is the reason for asymmetric transmission, which can be defined as $\Delta = |T_{11}^f|^2 + |T_{21}^f|^2 - |T_{11}^b|^2 - |T_{21}^b|^2$. For linear polarization$\Delta = |C|^2 - |B|^2$, and for circular polarization$\Delta = |T_{-+}|^2 - |T_{+-}|^2$.

One of the designed metasurfaces with asymmetric transmission is shown in **Figure 6a** and **b**, which demonstrates a highly efficient and broadband asymmetric transmission of linearly polarized millimeter waves [45]. The remarkable consequence results from the tri-layer configuration, which can be promoted to other designs of metasurfaces and this will be discussed in detail in Section 5. At optical waveband, asymmetric transmission can also be achieved, as shown in **Figure 6c**. The fabricated devices designed for operation at central wavelength of 532 and 633 nm, exhibit broadband, efficient asymmetric optical transmission with contrast ratios exceeding 14 dB [46]. The FDTD-simulated amplitude of the magnetic field at an arbitrary time is shown in **Figure 6d**. It is clear to see that with forward incidence (from Side A to B), the transmitted field displays a profile of diffraction, and with backward incidence (from Side B to A), the transmitted field is blocked. As stated previously, the primary way of achieving asymmetric transmission is by reducing the structural symmetry and converting

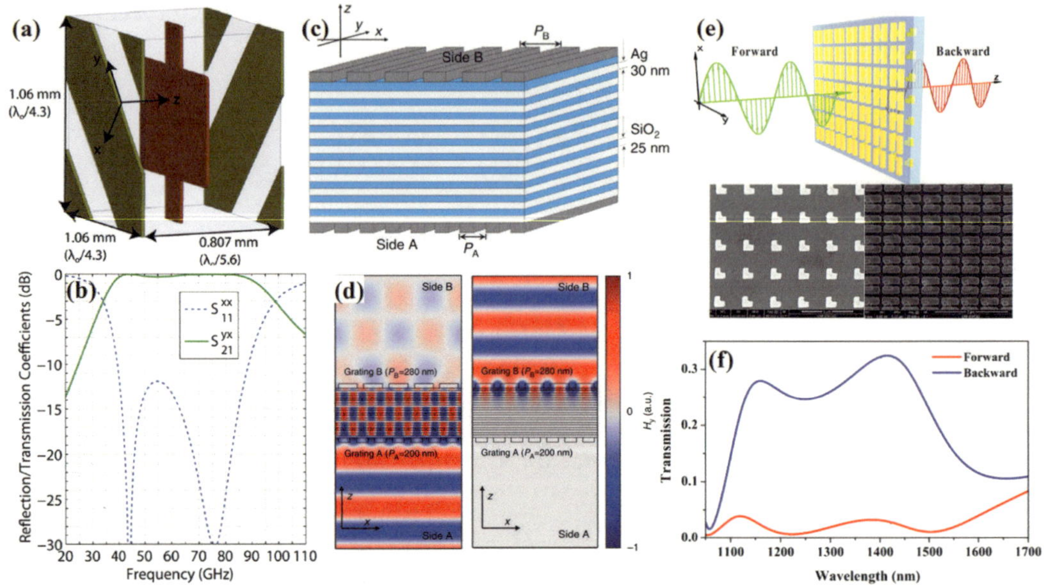

Figure 6. (a) Schematic of the tri-layer metasurface which enables (b) broadband and highly asymmetric transmission of linearly polarized waves. Reprinted by permission from [45] copyright 2014 American Physical Society. (c) Schematic diagram of the Ag/SiO_2 hyperbolic metamaterial device, which exhibits asymmetric transmission shown in (d). Electromagnetic waves illuminate from side A to side B on the left graph in (d) and side B to A on the right one. Reprinted by permission from Macmillan Publishers Ltd [46], Nature communications copyright 2014. (e) Asymmetric transmission achieved by a bi-layer metasurface consisting of an L-shaped layer and a nanorod layer. (f) Transmission of the x-polarized waves from forward and backward, respectively. Reprinted by permission from [47] copyright 2014 AIP Publishing LLC.

between polarization states, thus the metasurfaces can actually be more concise, as shown in **Figure 6e**. The hybrid metasurface consists of an L-shaped metallic layer and a nanorod layer. With x-polarized incidence, the forward and backward transmission reveals a great difference between 1200 and 1500 nm [47], the simulated results of which are illustrated in **Figure 6f**. It should be noticed that asymmetric transmission is different from optical isolator, which can only be attained with nonreciprocal-active devices.

In summary, full control of polarization can now be obtained with metallic or high-contrast dielectric-based metamaterials, the remarkable abilities of which can be attributed to the in-depth subwavelength design of metamaterials. Practical applications, such as vector beams, wave plates, and asymmetric transmission devices, can be accomplished with high efficiency, and other particular designs such as broadband or wide-angle one can also be acquired at will. Other novel applications of polarization integrated devices will be discussed in the next section.

4. Polarization integrated metamaterials devices

As discussed in the previous section, metamaterials has made remarkable development in nanophotonics devices for polarization state manipulation of electromagnetic waves. With the design of individual unit cell of metamaterials instead of the entire structure, metamateri-

http://dx.doi.org/10.5772/66036

als can not only manipulate the amplitude and phase of the electromagnetic waves but also realize the controlling of the wavefront of electromagnetic waves, thus it provides an ideal way to get the optical devices on demand and shows great impact on nanophotonics. These days, with the ever-increasing demand for integrated optical devices, the research direction is shifting toward achieving tunable, integrated, and novel functionalities.

4.1. Tunable metamaterials devices

Electromagnetic waves controlling in nanophotonics integration always require the devices with the tunability of waveband and functionalities. The polarization state manipulation of electromagnetic waves in earlier metamaterials for different working wavebands is always realized by accurately fabricating different nanostructures, which is an inherent drawback for integration [48, 49]. One way to overcome this drawback is to use tunable metamaterials, which relies on integrating metamaterials with optically active materials such as liquid crystals, semiconductors, phase-change material, and nonlinear media [50–54]. The optical response of these metamaterials can be actively controlled by external stimulus, such as electric field, magnetic field, voltage, or temperature. Among all these techniques, voltage control is one of the simplest ways in practical operations. Graphene is a monolayer of hexagonally arranged carbon atoms that can support the excitation of surface plasmons and its optical response shows a strong dependence on the Fermi energy, which can be dynamically controlled by a gate voltage [55–57]. Therefore, graphene is a promising electrically tunable plasmonic material. The investigation of tunable plasmons in graphene nanostructures has led to the proposition and demonstration of a variety of devices for polarization state manipulation of electromagnetic waves. **Figure 3a** shows a mid-IR highly tunable optical polarization converter composed of asymmetric graphene nano-crosses [58]. It can convert linearly polarized wave to circularly and elliptically polarized wave or exhibit a giant optical activity at different wavelengths. The transmitted wavelength and polarization states can also be dynamically tuned by varying the Fermi energy of graphene (as shown in **Figure 7b** and **c**), without reoptimizing and refabricating the nanostructures. This devices is potentially useful in applications, such as vibrational circular dichroism spectroscopy, ellipsometry, and integration of other optical devices for polarization manipulation, detection, and sensing at the nanoscale. **Figure 7d** shows a mid-IR highly wavelength-tunable broadband cross-polarization converter based on L-shaped graphene nanostructures [59]. It can convert linearly polarized wave to its cross-polarization in the reflection mode. The polarization conversion can be dynamically tuned and realize a broadband effect by varying the Fermi energy (as shown in **Figure 7e** and **f**). This tunable polarizers (or polarization switchers) provide an alternate way for the waveband controlling of polarization state manipulation. A step further, not only the waveband and functionalities of polarization state manipulation can be tuned with graphene-based metamaterials, **Figure 7g** shows tunable wavefront controlling of cross-polarized electromagnetic waves based on periodically patterned graphene nano-crosses in the infrared regime [60]. With this device, the wavefront of cross-polarized circular refraction waves can be effectively controlled with the polarization conversion induced geometric phase and the working waveband can be dynamically tuned (as shown in **Figure 7h** and **i**). This active wavefront controlling device can be treated as polarization and spectral beam splitters at nanoscale.

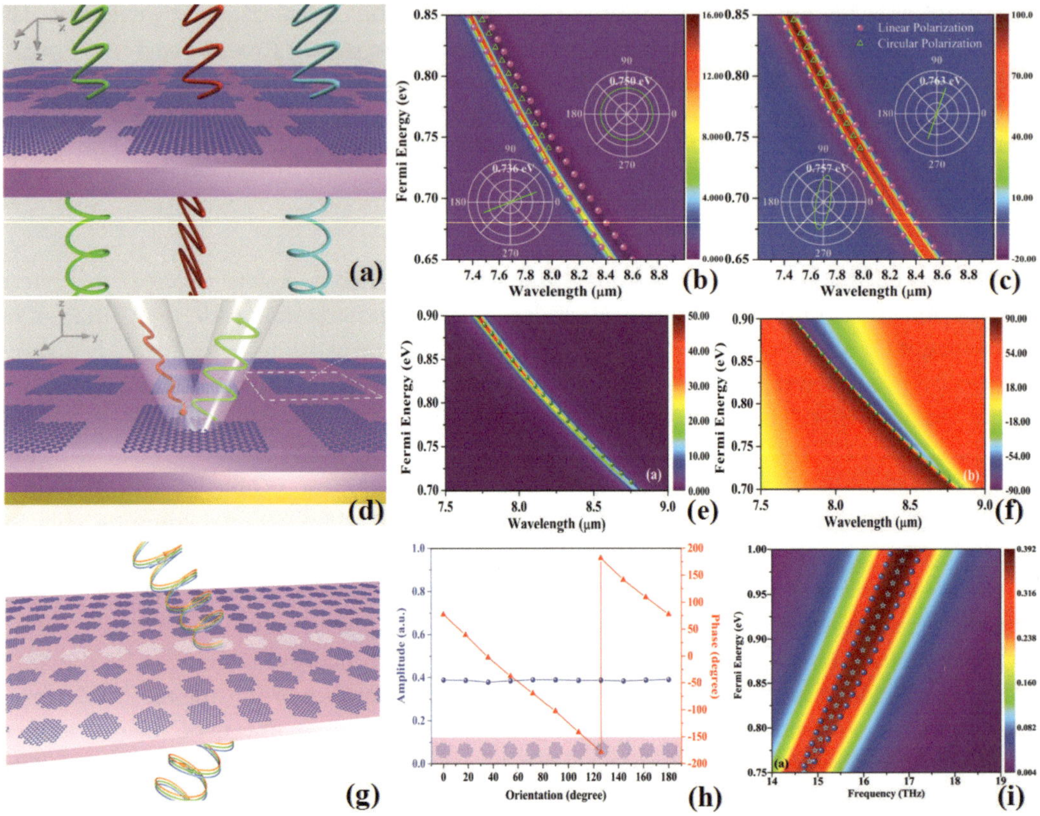

Figure 7. (a) Simulated realization of highly tunable optical polarization converter. (b) Calculated amplitude ratio and (c) phase difference as a function of the Fermi energy and wavelength. Inset images show the transmitted polarization state for different Fermi energies at a wavelength of 7.92 μm. The incident wave is linearly polarized with a polarization angle of 45°. Reprinted by permission from [58] copyright 2013 Optical Society of America. (d) Simulated realization of highly wavelength-tunable broadband cross-polarization converter. (e) Calculated amplitude ratio and (f) phase difference between two orthogonal polarization states as a function of Fermi energy and wavelength. The olive balls indicate that the reflection phase difference is 0°. Reprinted by permission from [59] copyright 2013 AIP Publishing LLC. (g) Simulated realization of highly tunable broadband wavefront controlling. (h) Calculated transmission amplitude and phase profile of the refraction wave along the graphene nano-cross array. (i) Amplitude of anomalous refraction of LCP wave as a function of Fermi energy and frequency, at normal incidence of RCP wave. The pentagrams and balls indicate the maximum amplitude and the extracted equal amplitude of anomalous refraction, respectively. Reprinted by permission from [60] copyright 2012 by John Wiley and Sons.

In general, tunable metamaterial devices (especially graphene-based one) provide an effect way to realize the polarization state manipulation of electromagnetic waves for nanophotonics integration because of the tunability of waveband and functionalities they have.

4.2. Novel applications of polarization integrated metamaterials

As mentioned above, metamaterial-based polarization converter can not only realize the transform of polarization state of electromagnetic waves, but also control the wavefront of the cross-polarized wave by inducing the geometric phase gradient via individual unit cell design. This character makes metamaterial devices become an ideal unit for integrated manipulation of electromagnetic waves, thus providing endless possibilities for novel applications in nanophotonics.

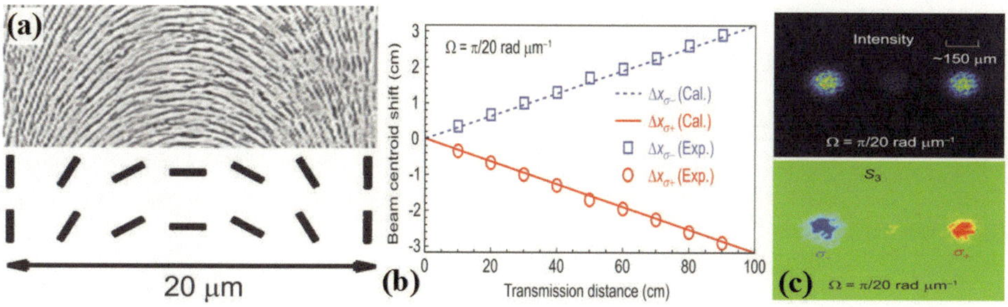

Figure 8. (a) The detailed geometry of the metamaterial over one period and a schematic illustration of the local optical axis (slow axis). (b) The calculated (Cal.) and experimental (Exp.) results of spin-dependent real-space shift induced by the momentum shift during beam with a geometric phase gradient of$\Omega = \pi/20$ rad μm -1. (c) Intensities recorded by the CCD and the corresponding S_3 parameters after a linearly polarized beam passes through metamaterial. The observation plane is located 10 cm away from the metamaterial. Reprinted by permission from Macmillan Publishers Ltd [65], Light: Science and Applications, copyright 2015.

4.2.1. Gain spin-orbit interaction of electromagnetic waves with metamaterial devices

Recently, spin-orbit interaction of electromagnetic waves attracted lots of attention [61]. The spin-orbit interaction of electromagnetic waves are analogous to the spin-orbit interaction of relativistic quantum particles and electrons in solids with a spatial scale of the order of the wavelength of electromagnetic waves, the RCP and LCP polarization states correspond to two spin states of photons. Thus, traditional geometrical optics always neglects the wavelength-scale spin-orbit interaction phenomena. With the development of nanophotonics and plasmonics, spin-orbit interaction phenomena play an important role at the subwavelength scales and bring novel functionalities to optical nanodevices. Photonic spin Hall effect, which manifests itself as the mutual interplay between the photon spin (polarization) and the trajectory (orbital angular momentum) of electromagnetic waves, is one of the basic classes of numerous spin-orbit interactions phenomena at the subwavelength scales. Traditional approaches to realize the photonic spin Hall effect are always associated with the evolution of the propagation direction of electromagnetic waves. However, the photonic spin Hall effect in these approaches are generally very weak, and the induced spin-dependent subwavelength shifts are also exceedingly tiny which prevent them to real applications in nanophotonics. Spin-dependent geometric phase gradient in metamaterials provide an alternative method to realize a gain on spin-orbit interaction at nanoscale [40, 62–64]. **Figure 8a–c** shows an experimental demonstration of a giant photonic spin Hall effect at a visible wavelength in a dielectric-based metamaterial device with spin-dependent geometric phase gradient [65]. The spin-dependent shift induced by geometric phase gradient is sufficiently large to be observed directly compared with traditional approaches. These kinds of devices bridge the gap between spin-based photonics and nanophotonics and thus provide an opportunity for manipulating the spin and orbital angular momentum of electromagnetic waves.

4.2.2. Information coding and wave coding with metamaterial devices

Modern optical communication always involves the effective device for information coding. Polarization integrated metamaterial devices provide an alternative way to realize the information coding via manipulating and detecting the polarization state of electromagnetic waves. Recently, metamaterial devices with the ability to dynamically tune the polarization state of

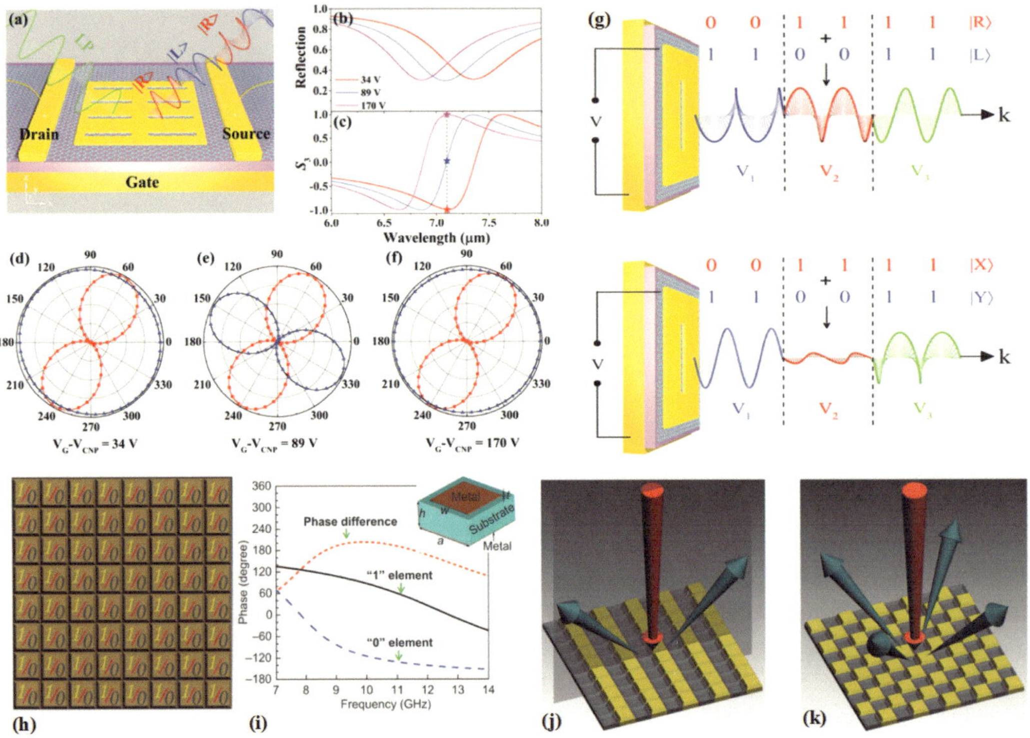

Figure 9. (a) Schematic illustration of the optical polarization encoding process based on the circular orthogonal polarization basis. (b) and (c) Simulated amplitude and normalized Stokes parameter S_3 spectra of the reflected electromagnetic wave from the nano-aperture array at different gate voltages under incident wave with a polarization direction of 61° relative to the *x*-axis. The stars indicate the values of S_3 under different gate voltages at 7.1 μm. (d)–(f) Simulated polarization state in the plane perpendicular to the wave vector at 7.1 μm for different gate voltages. Red and blue curves correspond to the incident and reflected wave, respectively. (g) Schematic illustration of realizing the PDM technique by the proposed metamaterial device with the circular and linear orthogonal polarization basis according to the superposition principle of polarization states. Polarization states of electromagnetic waves correspond to three difference gate voltages. Reprinted by permission from [66] copyright 2016 by John Wiley and Sons. (h) The 1-bit digital metamaterial device composed of only two types of elements: "0" and "1." (i) A square metallic patch unit structure (inset) to realize the "0" and "1" elements and the corresponding phase responses in a range of frequencies. Two 1-bit periodic coding metasurfaces to control the scattering of beams by designing the coding sequences of "0" and "1" elements: (j) the 010101.../010101... code and (k) 010101.../101010... code. Reprinted by permission from Macmillan Publishers Ltd [67], Light: Science and Applications, copyright 2015.

electromagnetic waves have attracted enormous interest because such devices can be employed for realizing not only the polarization encoding but also polarization-division multiplexing, which is a crucial technique that can significantly increase the transmission capacity of a single physical channel. The traditional technique for realizing polarization-division multiplexing requires a complex optical system and cumbersome volume. Therefore, a metamaterial-based polarization modulator offers a new approach for simplifying the optical process and miniaturizing the required volume. **Figure 9a** shows a metamaterial device by integrating a single layer of graphene with an anisotropic metamaterial, which can dynamically modulate the polarization state of electromagnetic waves with a wide tunable range in mid-infrared wavelengths

[66]. By switching gate voltage applied on the graphene among three different values, the incident LP wave can be dynamically converted into LCP, RCP, or linearly cross-polarized one in the reflection direction, as shown in **Figure 9b–f**. Based on these polarization-control characteristics, the proposed device can realize polarization encoding and the polarization-division multiplexing technique, as shown in **Figure 9g**. This design profoundly affects a wide range of modern optical communication devices, fulfills the demand of faster information transfer and processing, and opens a route to on-chip integration of metasurfaces with electronics.

Another approach named "coding metamaterial" has also been paid great attention by the scientific community, which is composed of several types of unit cells, with different constant phase responses, respectively [67–70]. **Figure 9h** shows a 1-bit coding metamaterial, composed of two types of unit cells, with 0 and π phase responses (as shown in **Figure 9i**), which is named "0" and "1" elements, respectively [67]. This coding metamaterial device can simply

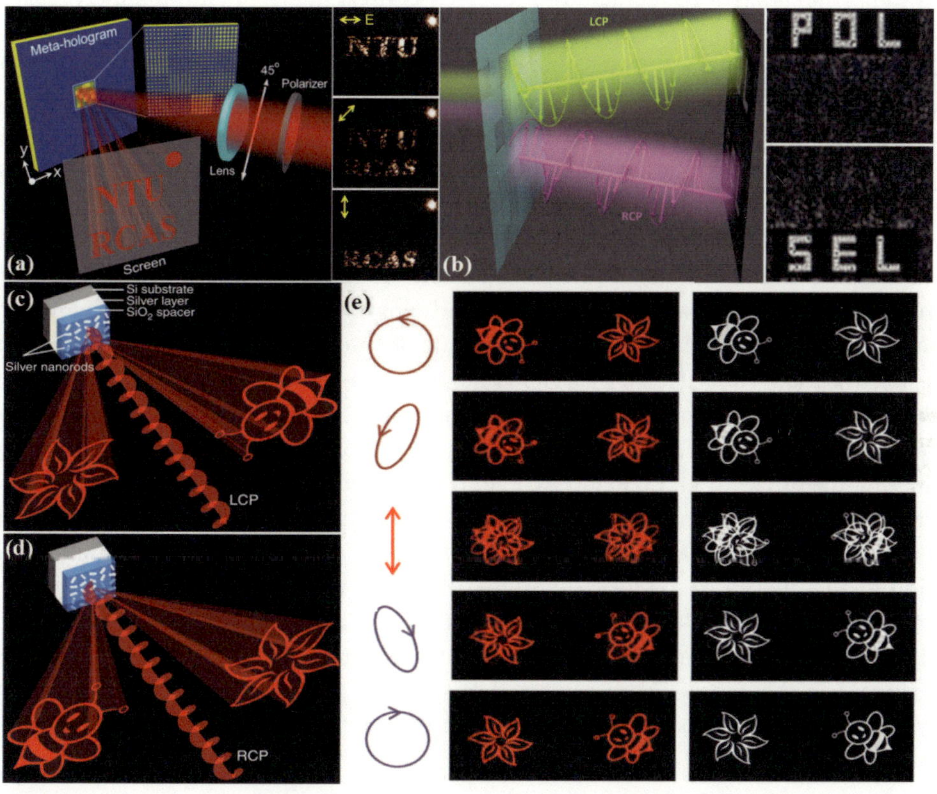

Figure 10. (a) An experimental demonstration of a polarization-switchable phase hologram and its experimental results. Reprinted by permission from [75] copyright 2013 American Chemical Society. (b) A schematic representation of polarization selective beam shaping in dielectric metamaterial device and its simulated results. Reprinted by permission from [76] copyright 2015 Optical Society of American. (c) and (d) Schematics of the helicity multiplexed hologram in nanorods based metamaterial device. (e) Reconstructed images versus incident polarization states in nanorod-based metamaterial device. The figures in the middle column and right column represent the experimental results and the corresponding simulation results, respectively. Reprinted by permission from Macmillan Publishers Ltd [77], Nature Communications copyright 2015.

manipulate electromagnetic waves through different coding sequences of "0" and "1" elements. For example, under the periodic coding sequence of 010101.../010101..., the normally incident beam will mainly be reflected to two symmetrically oriented directions by the metasurface, whereas under the periodic coding sequence of 010101.../101010.../010101.../101010 ..., the normally incident beam will mainly be reflected to four symmetrically oriented directions, as illustrated in **Figure 9j** and **k**. Based on the concept of coding metamaterials, we are able to not only control electromagnetic waves by changing the coding sequences of "0" and "1" (or "00," "01," "'10," and "11") elements but also create actual digital metamaterial device and programmable metamaterial device by dynamically manipulating the coding sequence. This design provides an effective way to realize the controlling of the radiation beams of antennas and reduce the scattering features of targets.

4.2.3. Polarization-switchable phase holograms with metamaterial devices

One of the prime goals for metamaterial research is realizing the manipulation of refractive index via using the concept of effective index and designing structures at the subwavelength scale. Metamaterials composed of subwavelength structures allow for the use of effective medium approach to describe their electromagnetic waves response in phase and amplitude sensitive to both shape and orientation of the structures. With this character, holograms which manifest itself as reconstructing predesigned images have been advanced dramatically by using metamaterial devices [71–77]. These days, polarization-switchable holograms that can separate the readout electromagnetic waves by its polarization to reconstruct different holographic images, lead to various applications such as image processing and multilevel optical switching. **Figure 10a** shows a plasmonic meta-hologram using metamaterial devices [75]. The meta-hologram sample consists of pixels made of 6 × 6 cross-nanoantennas of 16 different shapes for the phase modulation that yields polarization-controlled dual images. **Figure 10b** shows the demonstration of arbitrary polarization selective beam shaping with a dielectric metamaterial device [76]. Taking advantage of the conjugate phase modulation obtained by illuminating the device with LCP and RCP, two independent images, for the two orthogonal polarization states has been demonstrated. And **Figure 10c–e** show an experimental demonstration of helicity multiplexed metasurface hologram capable of achieving high efficiency and high image quality in the visible and near-infrared range [77]. Unlike previously demonstrated polarization multiplexed holograms that are sensitive to linear polarization, two off-axis images are interchangeable in one identical hologram by controlling the helicity of the incident wave in this work. These devices show endless possibilities for the development of holograms and provide an effective way for data storage and information processing.

5. Challenge in future

In previous sections, we overlooked the whole area of metamaterial-based polarization converter, and undoubtedly the achievements made by the researchers are remarkable: full control of polarization now can be accomplished with a single subwavelength film. However, there are still some challenges that stand in front of our path to the full control of electromagnetic waves. For example, how can we in depth manipulate polarization, phase, and amplitude of the incidence independently? How can we integrate different functional designs together, which are compatible with existing photonic devices on chip, such as silicon-based complementary metal-oxide-semiconductor (CMOS) photonics platform? How can we get

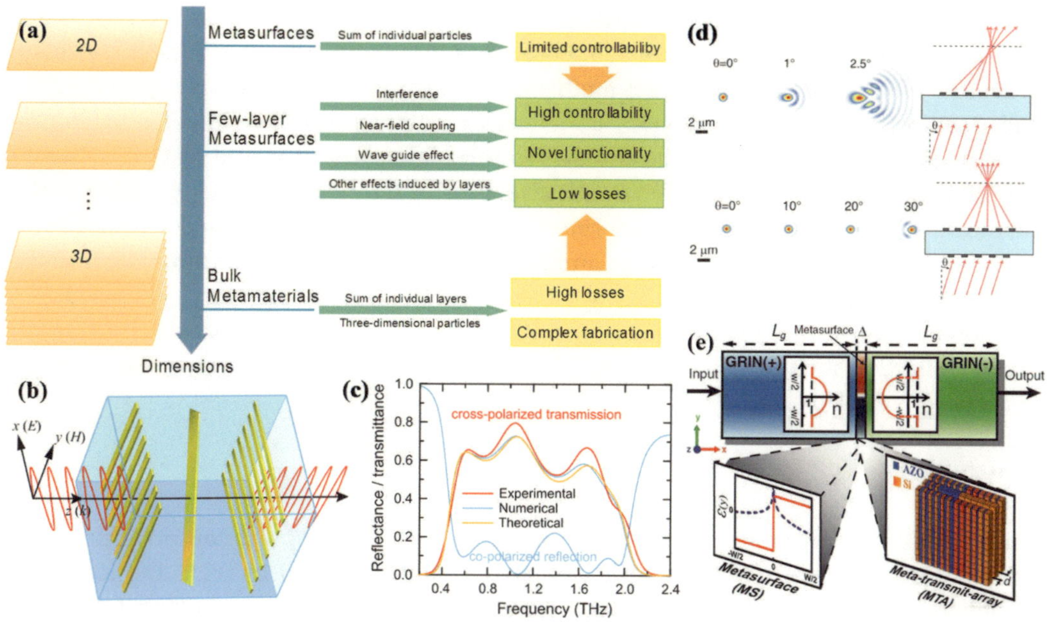

Figure 11. (a) Schematic representation of different metamaterials and their properties induced by various effects. Reprinted by permission from [25] copyright 2015 by John Wiley and Sons, Weinheim. (b) Schematic representation of the unit cell of a tri-layer metasurface linear polarization converter and (c) cross-polarized transmittance obtain through experimental measurements, numerical simulations, and theoretical calculations, together with the numerical copolarized reflection. Reprinted by permission from [81] copyright 2013 AAAS. (d) Simulated focal plane intensity for different incident angle with or without the correcting metasurface. Reprinted by permission from [82]. (e) Illustration of the system that can perform mathematical operations. A properly designed metasurface is sandwiched between two GRIN structures with positive and negative parameters. Reprinted by permission from [83] copyright 2014 AAAS.

tremendously active or tunable devices without refabricating the metamaterials? How can we achieve ultrashort response with metamaterials to obtain ultrafast detect?

One way to solving these problems may be utilizing few-layer metasurfaces [25], the principle of which is illustrated in **Figure 11a**. The properties of metasurfaces are sum of the individual particles (unit cells), which also means their functionalities are limited by each of the particles. For example, electric-magnetic dipoles supported by metallic nanoantennas and Mie-scattering by dielectric particles are often insufficient for independently manipulating each degree of freedom in electromagnetic waves, since each degree tends to mix up at the resonant frequency. However, other effects occur when adding another layer or layers upon the metasurfaces, such as interference [78], near-field coupling [79], waveguide effects [80], and so on, which may take advantage of each layer and even far better properties can be achieved with suitable designs. Compared with bulk metamaterials, few-layer metasurfaces are much more sufficient with easier fabrication, which are highly crucial for practical utilization. **Figure 11b** shows a transmitted cross-polarization converter consisting of two orthogonally positioned gratings and a low-efficiency polarization converter in the middle [81]. A Fabry-Pérot resonance occurs within the ultrathin polarization rotator, which can enhance the polarization conversion efficiency tremendously. The experimental and numerical results are shown in **Figure 11c**, which can be further optimized according to the authors. Few-layer metasurfaces can also be utilized as correcting layers [82]. As shown in **Figure 11d**, single-layer metasurfaces can easily obtain foci with hyperbolic phase distribution. However, aberration happens with oblique incidence, which is harmful as a camera lens. This problem can be solved by adding another correcting metasurface, and the incident angle can be increased to 30°, as shown in **Figure 11d**. Another really promising consequence achieved by few-layer metasurfaces is mathematical operation. As shown in **Figure 11e**, Silva et al. demonstrated a method to carry out complex mathematical operations in the Fourier domain [83]. The operation can be divided to three parts: the first and last parts involve Fourier and inverse Fourier transforms, which are realized by a graded-index (GRIN) dielectric index. Actual operations are performed by carefully designing the metasurface in the second part. If the permeability and permittivity of the metasurfaces conform with a certain relationship, differentiation, integration, and convolution can be precisely calculated. This technique requires total manipulation of the wavefront of incident electromagnetic wave with no reflection and normal wave, and this can only be realized by few-layer metasurfaces.

6. Summary

In this chapter, we have reviewed the recent progress in the field of polarization state manipulation of electromagnetic waves with metamaterials and discussed its wide applications in nanophotonics. Polarization state is one of the most important characteristics for electromagnetic waves, therefore arbitrary manipulation of polarization state with meta-materials provides endless possibilities for direct manipulation of electromagnetic waves and offers a powerful way for design of novel devices. With in-depth subwavelength designs, full control of polarization can now be obtained with metallic or high-contrast dielectric-based meta-materials. Furthermore, few-layer metasurfaces may provide more degree of freedom to manipulate electromagnetic waves without decrease of efficiency or increase of fabricating complexity compared to bulk metamaterials or traditional metasurfaces. With the development of metamaterial fabrication technology, meta-material devices for polarization state manipulation of electromagnetic waves are expected to have deep impact on real applications in our life.

Author details

Shuqi Chen*, Wenwei Liu, Zhancheng Li, Hua Cheng and Jianguo Tian

*Address all correspondence to: schen@nankai.edu.cn

The Key Laboratory of Weak Light Nonlinear Photonics, Ministry of Education, School of Physics and TEDA Institute of Applied Physics, Nankai University, Tianjin, China

References

[1] Zheludev N. I. The road ahead for metamaterials. Science. 2010;328(5978):582–583. DOI: 10.1126/science.1186756

[2] Zheludev N. I. Obtaining optical properties on demand. Science. 2015;348(6238):973–974. DOI: 10.1126/science.aac4360

[3] Liu Y., Zhang X. Metamaterials: a new frontier of science and technology. Chemical Society Reviews. 2011;40(5):2497–2507. DOI: 10.1039/C0CS00184H

[4] Meinzer N., Barnes W. L., Hooper I. R. Plasmonic meta-atoms and metasurfaces. Nature Photonics. 2014;8(12):889–898. DOI: 10.1038/nphoton.2014.247

[5] Estakhri N. M., Alù A. Recent progress in gradient metasurfaces. JOSA B. 2016;33(2):A21–A30. DOI: 10.1364/JOSAB.33.000A21

[6] Schurig D., Mock J. J., Justice B. J., Cummer S. A., Pendry J. B., Starr A. F., et al. Metamaterial electromagnetic cloak at microwave frequencies. Science. 2006;314(5801):977–980. DOI: 10.1126/science.1133628

[7] Pendry J. B., Schurig D., Smith D. R. Controlling electromagnetic fields. Science. 2006;312(5781):1780–1782. DOI: 10.1126/science.1125907

[8] Ni X., Wong Z. J., Mrejen M., Wang Y., Zhang X. An ultrathin invisibility skin cloak for visible light. Science. 2015;349(6254):1310–1314. DOI: 10.1126/science.aac9411

[9] Kildishev A. V., Boltasseva A., Shalaev V. M. Planar photonics with metasurfaces. Science. 2013;339(6125):1232009. DOI: 10.1126/science.1232009

[10] Yu P., Chen S., Li J., Cheng H., Li Z., Liu W., et al. Dynamically tunable plasmonic lens between the near and far fields based on composite nanorings illuminated with radially polarized light. Plasmonics. 2015;10(3):625–631. DOI: 10.1007/s11468-014-9848-3

[11] Shelby R. A., Smith D. R., Schultz, S. Experimental verification of a negative index of refraction. Science. 2001;292(5514):77–79. DOI: 10.1126/science.1058847

[12] Khorasaninejad M., Chen W. T., Devlin R. C., Oh J., Zhu A. Y., Capasso, F. Metalenses at visible wavelengths: diffraction-limited focusing and subwavelength resolution imaging. Science. 2016;352(6290):1190–1194. DOI: 10.1126/science.aaf6644

[13] Liu Z., Li Z., Liu Z., Li J., Cheng H., Yu, P., et al. High-performance broadband circularly polarized beam deflector by mirror effect of multinanorod metasurfaces. Advanced Functional Materials. 2015;25(34):5428-5434. DOI: 10.1002/adfm.201502046

[14] Li J., Chen S., Yang H., Li J., Yu P., Cheng H., et al. Simultaneous control of light polarization and phase distributions using plasmonic metasurfaces. Advanced Functional Materials. 2015;25(5):704–710. DOI: 10.1002/adfm.201403669

[15] Litchinitser, N. M. Photonic multitasking enabled with geometric phase. Science. 2016;352(6290):1177–1178. DOI: 10.1126/science.aaf8391

[16] Zhu A. Y., Kuznetsov A. I., Luk'yanchuk B., Engheta N., Genevet P. Traditional and emerging materials for optical metasurfaces. Nanophotonics. 2016; DOI: 10.1515/nanoph-2016–0032

[17] Yu N., Capasso F. Flat optics with designer metasurfaces. Nature materials. 2014;13(2):139–150. DOI: 10.1038/nmat3839

[18] Saleh B. E., Teich M. C., Masters B. R. Fundamentals of photonics, Second Edition. Journal of Biomedical Optics. 2008;13(4):9901. DOI: 10.1117/1.2976006

[19] Gordon R., Brolo A. G., McKinnon A., Rajora A., Leathem B., Kavanagh K. L. Strong polarization in the optical transmission through elliptical nanohole arrays. Physical Review Letters. 2004;92(3):037401. DOI: 10.1103/PhysRevLett.92.037401

[20] Sung J., Sukharev M., Hicks E. M., Van Duyne R. P., Seideman T., Spears K. G. Nanoparticle spectroscopy: birefringence in two-dimensional arrays of L-shaped silver nanoparticles. The Journal of Physical Chemistry C. 2008;112(9):3252–3260. DOI: 10.1021/jp077389y

[21] Li T., Liu H., Wang S. M., Yin X. G., Wang F. M., Zhu S. N., et al. Manipulating optical rotation in extraordinary transmission by hybrid plasmonic excitations. Applied Physics Letters. 2008;93(2):021110. DOI: 10.1063/1.2958214

[22] Ellenbogen T., Seo K., Crozier K. B. Chromatic plasmonic polarizers for active visible color filtering and polarimetry. Nano Letters. 2012;12(2):1026–1031. DOI: 10·1021/nl204257g

[23] Zhao Y., Alu` A. Tailoring the dispersion of plasmonic nanorods to realize broadband optical meta-waveplates. Nano Letters. 2013;13(3):1086–1091. DOI: 10.1021/nl304392b

[24] Pors A., Nielsen M. G., Della Valle G., Willatzen M., Albrektsen O., Bozhevolnyi S. I. Plasmonic metamaterial wave retarders in reflection by orthogonally oriented detuned electrical dipoles. Optics Letters. 2011;36(9):1626–1628. DOI: 10.1364/OL.36.001626

[25] Cheng H., Liu Z., Chen S., Tian J. Emergent functionality and controllability in few-layer metasurfaces. Advanced Materials. 2015;27(36):5410–5421. DOI: 10.1002/adma.201501506

[26] Jahani S., Jacob Z. All-dielectric metamaterials. Nature Nanotechnology. 2016;11(1):23–26. DOI: 10.1038/nnano.2015.304

[27] Lévesque Q., Makhsiyan M., Bouchon P., Pardo F., Jaeck J., Bardou N. Plasmonic planar antenna for wideband and efficient linear polarization conversion. Applied Physics Letters. 2014;104(11):111105. DOI: 10.1063/1.4869127

[28] Jiang Z. H., Lin L., Ma D., Yun S., Werner D. H., Liu Z., et al. Broadband and wide field-of-view plasmonic metasurface-enabled waveplates. Scientific Reports. 2014;4:7511. DOI: 10.1038/srep07511

[29] Li T., Wang S. M., Cao J. X., Liu H., Zhu S. N. Cavity-involved plasmonic metamaterial for optical polarization conversion. Applied Physics Letters. 2010;97(26):261113. DOI: 10.1063/1.3533912

[30] Li Z., Chen S., Liu W., Cheng H., Liu Z., Li J, et al. High performance broadband asymmetric polarization conversion due to polarization-dependent reflection. Plasmonics. 2015;10(6):1703–1711. DOI: 10.1007/s11468-015-9986-2

[31] Li Z., Zheng G., He P. A., Li S., Deng Q., Zhao J., et al. All-silicon nanorod-based Dammann gratings. Optics Letters. 2015;40(18):4285–4288. DOI: 10.1364/OL.40.004285

[32] Wu C., Arju N., Kelp G., Fan J. A., Dominguez J., Gonzales E. Spectrally selective chiral silicon metasurfaces based on infrared Fano resonances. Nature Communications. 2014;5:3892. DOI: 10.1038/ncomms4892

[33] Stalder M., Schadt M. Linearly polarized light with axial symmetry generated by liquid-crystal. Optics Letters. 1996;21(23): 1948–1950. DOI: 10.1364/OL.21.001948

[34] Prati E. Propagation in gyroelectromagnetic guiding systems. Journal of Electromagnetic Waves and Applications. 2003;17(8): 1177–1196. DOI: 10.1163/156939303322519810

[35] Arbabi A., Horie Y., Bagheri M., Faraon A. Dielectric metasurfaces for complete control of phase and polarization with subwavelength spatial resolution and high transmission. Nature Nanotechnology. 2015;10:937–943. DOI: 10.1038/NNANO.2015.186

[36] Wu S., Zhang Z., Zhang Y., Zhang K., Zhou L., Zhang X., et al. Enhanced rotation of the polarization of a light beam transmitted through a silver film with an array of perforated S-shaped holes. Physical Review Letters. 2013;110(20): 207401. DOI: 10.1103/PhysRevLett.110.207401

[37] Liu W., Chen S., Li Z., Cheng H., Yu P., Li J., et al. Realization of broadband cross-polarization conversion in transmission mode in the terahertz region using a single-layer metasurface. Optics Letters. 2015;40(13):3185–3188. DOI: 10.1364/OL.40.003185

[38] Yu P., Li J., Tang C., Cheng H., Liu Z., Li Z., et al. Controllable optical activity with non-chiral plasmonic metasurfaces. Light: Science and Applications. 2016;5(7):e16096. DOI: 10.1038/lsa.2016.96

[39] Li Z., Liu W., Cheng H., Chen S., Tian J. Realizing broadband and invertible linear-to-circular polarization converter with ultrathin single-layer metasurface. Scientific Reports. 2015;5:18106. DOI: 10.1038/srep18106

[40] Yin X., Ye Z., Rho J., Wang Y., Zhang X. Photonic spin Hall effect at metasurfaces. Science. 2013;339(6126): 1405–1407. DOI: 10.1126/science.1231758

[41] Yang Y., Wang W., Moitra P., Kravchenko I. I., Briggs D. P., Valentine, J. Dielectric meta-reflect array for broadband linear polarization conversion and optical vortex generation. Nano Letters. 2014;14(3): 1394–1399. DOI: 10.1021/nl4044482

[42] Li L., Li T., Tang X. M., Wang S. M., Wang Q. J., Zhu S. N. Plasmonic polarization generator in well-routed beaming. Light: Science and Applications. 2015;4(9):e330. DOI: 10.1038/lsa.2015.103

[43] Kruk S., Hopkins B., Kravchenko I., Miroshnichenko A., Neshev D. N., Kivshar Y. S. Broadband highly-efficient dielectric metadevices for polarization control. APL Photonics. 2016; 1(3). DOI: 10.1063/1.4949007

[44] Menzel C., Helgert C., Rockstuhl C., Kley E. B., Tünnermann A., Pertsch T., et al. Asymmetric transmission of linearly polarized light at optical metamaterials. Physical Review Letters. 2010;104(25):253902. DOI: 10.1103/PhysRevLett.104.253902

[45] Pfeiffer C., Grbic A. Bianisotropic metasurfaces for optimal polarization control: analysis and synthesis. Physical Review Applied. 2014;2(4):044011. DOI: 10.1103/PhysRevApplied.2.044011

[46] Xu T., Lezec H. J. Visible-frequency asymmetric transmission devices incorporating a hyperbolic metamaterial. Nature Communications. 2014;5:4141. DOI: 10.1038/ncomms5141

[47] Li Z., Chen S., Tang C., Liu W., Cheng H., Liu Z., et al. Broadband diodelike asymmetric transmission of linearly polarized light in ultrathin hybrid metamaterial. Applied Physics Letters. 2014;105(20):201103. DOI: 10.1063/1.4902162

[48] Ye Y., He S. 90° polarization rotator using a bilayered chiral metamaterial with giant optical activity. Applied Physics Letters. 2010;96(20):203501. DOI: 10.1063/1.3429683

[49] Han J., Li H., Fan Y., Wei Z., Wu C., Cao Y., et al. An ultrathin twist-structure polarization transformer based on fish-scale metallic wires. Applied Physics Letters. 2011;98(15):151908. DOI: 10.1063/1.3580608

[50] Khatua S., Chang W. S., Swanglap P., Olson J., Link, S. Active modulation of nanorod plasmons. Nano Letters. 2011;11(9):3797–3802. DOI: 10.1021/nl201876r

[51] Shen N. H., Massaouti M., Gokkavas M., Manceau J. M., Ozbay E., Kafesaki M., et al. Optically implemented broadband blueshift switch in the terahertz regime. Physical Review Letters. 2011;106(3):037403. DOI: 10.1103/PhysRevLett.106.037403

[52] Zharov A. A., Shadrivov I. V., Kivshar Y. S. Nonlinear properties of left-handed metamaterials. Physical Review Letters. 2003;91(3):037401. DOI: 10.1103/PhysRevLett.91.037401

[53] Shaltout A. M., Kildishev A. V., Shalaev V. M. Evolution of photonic metasurfaces: from static to dynamic. JOSA B. 2016;33(3):501–510. DOI: 10.1364/JOSAB.33.000501

[54] Zheludev N. I., Plum E. Reconfigurable nanomechanical photonic metamaterials. Nature Nanotechnology. 2016;11(1):16–22. DOI: 10.1038/nnano.2015.302

[55] Novoselov K. S., Geim A. K., Morozov S. V., Jiang D., Zhang Y., Dubonos S. V., et al. Electric field effect in atomically thin carbon films. Science. 2004;306(5696):666–669. DOI: 10.1126/science.1102896

[56] Emani N. K., Chung T. F., Ni X., Kildishev A. V., Chen Y. P., Boltasseva A. Electrically tunable damping of plasmonic resonances with graphene. Nano Letters. 2012;12(10): 5202–5206. DOI: 10.1021/nl302322t

[57] Valmorra F., Scalari G., Maissen C., Fu W., Schönenberger C., Choi J. W., et al. Low-bias active control of terahertz waves by coupling large-area CVD graphene to a terahertz metamaterial. Nano Letters. 2013;13(7):3193–3198. DOI: 10.1021/nl4012547

[58] Cheng H., Chen S., Yu P., Li J., Deng L., Tian J. Mid-infrared tunable optical polarization converter composed of asymmetric graphene nanocrosses. Optics Letters. 2013;38(9):1567–1569. DOI: 10.1364/OL.38.001567

[59] Cheng H., Chen S., Yu P., Li J., Xie B., Li Z., et al. Dynamically tunable broadband mid-infrared cross polarization converter based on graphene metamaterial. Applied Physics Letters. 2013;103(22):223102. DOI: 10.1063/1.4833757

[60] Cheng H., Chen S., Yu P., Liu W., Li Z., Li J., et al. Dynamically tunable broadband infrared anomalous refraction based on graphene metasurfaces. Advanced Optical Materials. 015;3(12):1744–1749. DOI: 10.1002/adom.201500285

[61] Bliokh K. Y., Rodríguez-Fortuño F. J., Nori F., Zayats A. V. Spin-orbit interactions of light. Nature Photonics. 2015;9(12):796–808. DOI: 10.1038/NPHOTON.2010.201

[62] Ling X., Zhou X., Shu W., Luo H., Wen S. Realization of tunable photonic spin Hall effect by tailoring the Pancharatnam-Berry phase. Scientific Reports. 2014;4:5557. DOI: 10.1038/srep05557

[63] Shu W., Ke Y., Liu Y., Ling X., Luo H., Yin X. Radial spin Hall effect of light. Physical Review A. 2016;93(1):013839. DOI: 10.1103/PhysRevA.93.013839

[64] Liu Y., Ling X., Yi X., Zhou X., Chen S., Ke Y., et al. Photonic spin Hall effect in dielectric metasurfaces with rotational symmetry breaking. Optics Letters. 2015;40(5):756–759. DOI: 10.1364/OL.40.000756

[65] Ling X., Zhou X., Yi X., Shu W., Liu Y., Chen S., et al. Giant photonic spin Hall effect in momentum space in a structured metamaterial with spatially varying birefringence. Light: Science and Applications. 2015;4(5):e290. DOI: 10.1038/lsa.2015.63

[66] Li J., Yu P., Cheng H., Liu W., Li Z., Xie B., et al. Optical polarization encoding using graphene-loaded plasmonic metasurfaces. Advanced Optical Materials. 2016;4(1):91–98. DOI: 10.1002/adom.201500398

[67] Cui T. J., Qi M. Q., Wan X., Zhao J., Cheng Q. Coding metamaterials, digital metamaterials and programmable metamaterials. Light: Science and Applications. 2014;3(10):e218. DOI: 10.1038/lsa.2014.99

[68] Gao L. H., Cheng Q., Yang J., Ma S. J., Zhao J., Liu S., et al. Broadband diffusion of terahertz waves by multi-bit coding metasurfaces. Light: Science and Applications. 2015;4(9):e324. DOI: doi:10.1038/lsa.2015.97

[69] Liu S., Cui T. J., Xu Q., Bao D., Du L., Wan X., et al. Anisotropic coding metamaterials and their powerful manipulation of differently polarized terahertz waves. Light: Science and Applications. 2016;5(5):e16076. DOI: 10.1038/lsa.2016.76

[70] Wang K., Zhao J., Cheng Q., Cui T. J. Broadband and broad-angle low-scattering metasurface based on hybrid optimization algorithm. Scientific Reports. 2014;4:5935. DOI: 10.1038/srep05935

[71] Zheng G., Mühlenbernd H., Kenney M., Li G., Zentgraf T., Zhang, S. Metasurface holograms reaching 80% efficiency. Nature Nanotechnology. 2015;10(4):308–312. DOI: 10.1038/NNANO.2015.2

[72] Yifat Y., Eitan M., Iluz Z., Hanein Y., Boag A., Scheuer J. Highly efficient and broadband wide-angle holography using patch-dipole nanoantenna reflectarrays. Nano Letters. 2014;14(5):2485–2490. DOI: 10.1021/nl5001696

[73] Ni X., Kildishev A. V., Shalaev V. M. Metasurface holograms for visible light. Nature Communications. 2013;4:2807. DOI: 10.1038/ncomms3807

[74] Khorasaninejad M., Ambrosio A., Kanhaiya P., Capasso, F. Broadband and chiral binary dielectric meta-holograms. Science Advances. 2016;2(5):e1501258. DOI: 10.1126/sciadv.1501258

[75] Chen W. T., Yang K. Y., Wang C. M., Huang Y. W., Sun G., Chiang I. D., et al. High-efficiency broadband meta-hologram with polarization-controlled dual images. Nano Letters. 2013;14(1):225–230. DOI: 10.1021/nl403811d

[76] Desiatov B., Mazurski N., Fainman Y., Levy U. Polarization selective beam shaping using nanoscale dielectric metasurfaces. Optics Express. 2015;23(17):22611–22618. DOI: 10.1364/OE.23.022611

[77] Wen D., Yue F., Li G., Zheng G., Chan K., Chen S., et al. Helicity multiplexed broadband metasurface holograms. Nature Communications. 2015;6:8241. DOI: 10.1038/ncomms9241

[78] Chen H. T. Interference theory of metamaterial perfect absorbers. Optics Express. 2012;20(7):7165–7172. DOI: 10.1364/OE.20.007165

[79] Lee J., Tymchenko M., Argyropoulos C., Chen P. Y., Lu F., Demmerle F., et al . Giant nonlinear response from plasmonic metasurfaces coupled to intersubband transitions. Nature. 2014;511(7507):65–69. DOI: 10.1038/nature13455

[80] Holloway C. L., Kuester E. F., Gordon J. A., O'Hara J., Booth J., Smith, D. R. An overview of the theory and applications of metasurfaces: the two-dimensional equivalents of metamaterials. IEEE Antennas and Propagation Magazine. 2012;54(2):10–35. DOI: 10.1109/MAP.2012.6230714

[81] Grady N. K., Heyes J. E., Chowdhury D. R., Zeng Y., Reiten M. T., Azad A. K., et al. Terahertz metamaterials for linear polarization conversion and anomalous refraction. Science. 2013;340(6138):1304–1307. DOI: 10.1126/science.1235399

[82] Arbabi A., Arbabi E., Kamali S. M., Horie Y., Han S., and Faraon, A. Miniature optical planar camera based on a wide-angle metasurface doublet corrected for monochromatic aberrations. arXiv. Forthcoming. DOI: arXiv:1604.06160

[83] Silva A., Monticone F., Castaldi G., Galdi V., Alù A., Engheta N. Performing mathematical operations with metamaterials. Science. 2014;343(6167):160–163. DOI: 10.1126/cience.1242818

Chapter 10

Recent Progress in Far-Field Optical Metalenses

Mahin Naserpour, Mahdieh Hashemi and
Carlos J. Zapata-Rodríguez

Additional information is available at the end of the chapter

http://dx.doi.org/10.5772/66467

Abstract

In this chapter, a review of the recent advances in optical metalenses is presented, with special emphasis in their experimental implementation. First, the Huygens' principle applied to ultrathin engineered metamaterials is introduced for the purpose of giving curvature to the wavefront of free-space wave fields. Primary designs based on metallic nanoslits and holey screens occasionally with variant width are first examined. Holographic plasmonic lenses are also explored offering a promising route to realize nanophotonic components. More recent metasurfaces based on nano-antenna resonators, either plasmonic or high-index dielectric, are analyzed in detail. Furthermore, 2D material lenses in the scale of a few nanometers enabling the thinnest lenses to date are here considered. Finally, dynamically reconfigurable focusing devices are reported for creating a scenario with new functionalities.

Keywords: metamaterials, optical lenses, subwavelength nanostructures

1. Introduction

The advent of artificially structured media, coined as electromagnetic metamaterials, enables the observation of tailored properties not occurring in natural materials, opening up a new scenario in physics and technology [1]. For instance, negative-index metamaterials with simultaneously negative electric permittivity and magnetic permeability exhibit intriguing reversed electromagnetic properties like phase and energy velocity of opposite directions, leading to a reversed Snell's law. As a consequence, a thick planar slab made of this sort of novel materials, immersed in air, provides real images of an object set behind, which under some circumstances contains details with a resolution beyond the diffraction limit [2]. Furthermore, a concave negative-index lens would evidence a positive focal length and it might also bring an incident plane wave into a focus [3]. However, optimized structures with reduced losses have been mostly developed at microwave frequencies, dramatically reducing the above-mentioned applied possibilities in nanophotonics and optics [4].

For such spectral window, the well-known concept of lensing with Fresnel zone plates was efficiently materialized over ultrathin metallic films including nanoslits and holes with varying width (and occasionally shape) to achieve high transmission meta-screens with controlled phase distribution induced by elementary plasmonic modes with engineered effective index. As a stimulating alternative, metasurfaces with reduced dimensionality in the direction of light propagation confine the light-matter interaction into a surface thus drastically reducing the inherent losses of the constituting meta-atoms. Possibly the most revolutionary approach to metasurfaces-based focusing devices came to us soon consisting in the inclusion of optical nano-antennas arranged over transparent substrates and flat mirrors, thus achieving an abrupt phase change of the scattered field with controlled polarization pattern. By directly comparing with three-dimensional architectures, simpler fabrication processes through for instance electron beam lithography and its accessible implementation into optoelectronic devices establish the relevant features of these metasurfaces.

Although several reviews concerning plasmonic Fresnel-type lenses, hyperlenses and metalenses are present in the literature [5, 6], including an open-access book chapter published by InTechOpen [7], a continuous effort is necessary in order to provide the current state of the art of such fascinating and fast-evolving topic. Metalenses with applications in photonics (IR and visible wavelengths) are of our interest here, leaving to supplementary reviews the analysis of reflectarrays and array lenses for uses in low-frequency regimes [8]. Our purpose is then introducing the above-mentioned preliminary results but mainly focus on recent advances in the field, not thoroughly discussed yet. For instance, a revolutionary all-dielectric approach to plasmonic nano-antenna metasurfaces is presently being developed for applications in focusing and others where high-index nanoresonators are the constituents of the gradient nanostructure to provide a full control over the local amplitude, phase and polarization of the scattered field under minimal dissipation effects. On the other hand, two-dimensional materials such as graphene sheets and transition-metal dichalcogenide semiconductors have emerged as promising candidates for miniaturized optoelectronic devices and a few novel proposals enabling the thinnest optical lenses in the scale of a few nanometers. Furthermore, versatile platforms have experimentally been demonstrated for creating dynamically reconfigurable focusing devices. All these examples make the scene of new concepts and materials for the design and fabrication of the next generation metalenses.

Here, we present a review including the recent advances in the field of gradient-metasurfaces lenses. For that purpose, we first introduce the fundamental analysis enabling the control of the wavefront with metasurface structures. We then look into primary designs based on holey metallic screens and nanoslits with variant widths and more recent proposals based on nano-antenna resonators. Next we analyze in detail innovative assemblies including all-dielectric high-index nanoresonators as the constituents of the metasurface. A special emphasis will be set in atomically thin lenses enabling the thinnest lenses to date. Holographic lenses are also examined and we present the possibilities that they could offer in numerous useful applications. Finally, we consider reconfigurable lenses allowing the design of powerful new functionalities.

2. Fresnel zone plate metalenses

Conventional refractive lenses rely on gradual phase accumulation via light propagation through a bulk material polished to a specific surface topology to shape the incident beam. In general terms, metamaterial lenses will mold the phase front curvature of the light passing through the constituents of the photonic device. For the waves of individual building blocks to be in phase at the focal distance f from the lens, the phase delay as a function of distance x_t from the center of the lens, as predicted from the Huygens' principle, has to be

$$\phi(x_t) = \phi(0) + \frac{2\pi n}{\lambda}\left(\sqrt{f^2 + x_t^2} - f\right), \tag{1}$$

where n is the index of refraction of the medium propagating in either transmission or reflection and λ is the vacuum wavelength of the incident light. Note that for one-dimensional architectures, $x_t = x$ represents the corresponding spatial coordinate, whereas for two-dimensional arrangements $x_t = \sqrt{x^2 + y^2}$ stands for a radial coordinate. In this section we will analyze novel proposals for focusing wave fields based on the concept of diffraction. That is, modification of the complex amplitude of a given electromagnetic field by interacting with single-apertured or multiple-apertured opaque screens. We are mostly concerned on holey metallic films, though nanostructured metamaterials can be used to stop the contribution of light in the prescribed Fresnel zone plates.

Fresnel zone plate lenses are practical energy collectors, which are currently in use in the electromagnetic spectrum spanning from the microwave to X-rays. The principle of operation of Fresnel zone plate lenses is based on the wave nature of light. The wavefront arriving at the lens is divided into sections, or zones. The specifications employed to characterize these zones are well established in the literature. For simplicity, let us first consider a central nanoslit and a set of off-axis nanoslits suitably placed in the metallic film. In the transmissive mode, the amplitude of the diffracted field is in practical terms zero in the plane immediately behind the zones where the metal is deposited. The phase difference between the light radiated from the mth surrounding slit and the light transmitted from the central nanoslit at the focusing spot position is about $\Delta\phi(x_m) = \phi(x_m) - \phi(0)$, where $m = \pm 1, \pm 2, \pm 3 \ldots$ and x_m is the center position of the mth slit. The constructive interference should occur when $\Delta\phi$ is equal to a multiple of 2π radians. In this case, the focusing intensity in the structure will be further improved due to the multiple-beam interference. The Fresnel zone plates are composed of concentric circles with their radii proportional to the square root of integers times the wavelength. The diffracted wave fields in the focal region can be accurately estimated by using for instance the Debye diffraction integral [9], provided that the focusing geometry is characterized by a high Fresnel number. From the practical point of view, by using nanoimprint lithography one can use molds to pattern PMMA on silicon substrates and PMMA patterns later being transferred to metals by lift-off, reaching a 20 nm minimum feature size [10].

In recent years, metallic nanostructures based on surface plasmon polaritons (SPPs) and exhibiting extraordinary enhancement of transmission is appealing for researchers. The conversion from SPP waves, which can propagate inside slits and holes much smaller that the wavelength, to propagation waves in the quasi-far-field region takes place by diffraction from the subwavelength zones of a given zone plate metalens. In this way, it was experimentally demonstrated that a Ag film-based microzone plate can combine such SPP wave effect and Fresnel zone plate focusing together [11]. Further noble metals such as gold can be used instead. In the cases given above, plasmonic waves do not contribute to far-field focusing. Importantly, the lens performance is significantly altered for the incidence in different polarization directions.

Following the basic proposal given above, a nano-focusing structure was experimentally demonstrated with high focusing efficiencies and easy fabrication by using a T-shape microslit surrounded by multislits in Au films [12]. In practical terms, it should be taken into account the propagation loss of SPPs owing to the Ohmic loss of Au and the fabrication roughness by focused ion beam (FIB). To realize the focusing, the T-shape microslit (microslit width of w = 1200 nm and groove depth of d = 150 nm) was suitably designed to be surrounded by multislits with the same slit width.

As a natural progress in the field, numerous types of Fresnel 2D zone metalenses subsequently appeared including for instance circular and elliptical nanopinholes. In Ref. [13], an optical nanosieve with circular symmetry was proposed, which is composed of 7240 subwavelength holes located at 22 concentric rings, where the holes in each ring are uniformly allocated and equally sized. However, more sophisticated designs can be found. For instance, light can concentrate into multiple, discrete spots by exploiting an evolutionary algorithm to optimize a structured optical material based on the discretization of a surface into a two-dimensional subwavelength lattice [14]. Polarization can be controlled by using the slits as antennas, as shown in **Figure 1**, acting as local linear polarizers. In such a way one may modulate an optical field in amplitude, binary phase and polarization for the cross-polarized component of the scattered field, for the generation of vectorial optical fields [15].

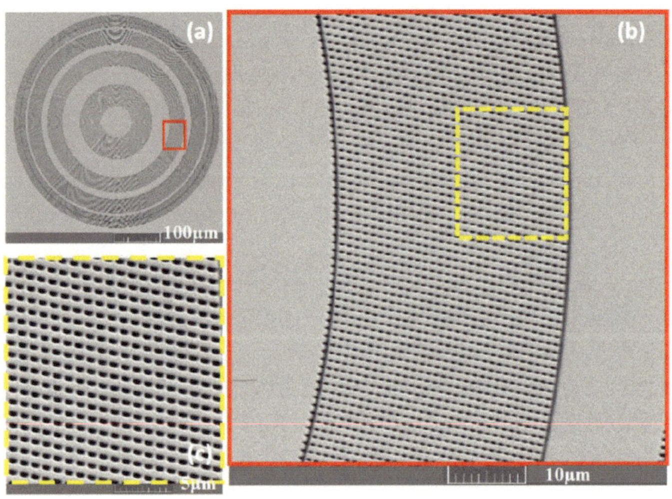

Figure 1. SEM image of a slot-antenna-based metalens with 450 μm diameter. (a) Overall structure from top view. (b) Part of ring for the area enclosed with red solid line in (a). (c) Zoomed-in picture for the area enclosed with yellow-dashed line in (b). Reprinted with permission from [15] of copyright ©2015 Optical Society of America.

http://dx.doi.org/10.5772/66467

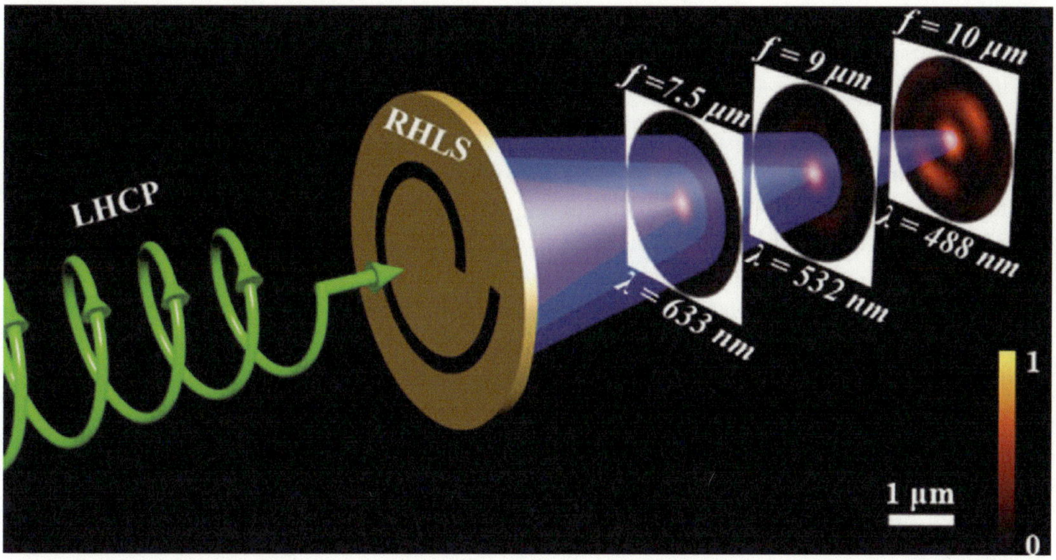

Figure 2. Right-hand logarithmic spiral exhibiting a broadband response under left-hand circularly polarized illumination. Reprinted with permission from [17] of copyright ©2015 WILEY-VCH Verlag GmbH & Co.

In some circumstances, the number of slits and any other kind of aperture mapping a metallic thin layer can be substantially reduced in order to generate a focused beam. One of the first experimental examples we found consists of a subwavelength annular aperture made on a silver film, where the transmitted light of a 442 nm incident laser was focused at several micrometers behind the silver structure at a tiny spot (354 nm), also exhibiting a remarkable depth of focus [16]. The resultant wave field is essentially a quasi-nondiffracting Bessel beam, whose depth of focus and transverse central spot can be tuned by changing the diameter of the subwavelength annular aperture. More recently, a logarithmic spiral nanoslit was proposed to converge an incident wave field of opposite handedness to that of the nanoslit into a subwavelength spot, a fact that is caused by a strong dependence on the incident photon spin [17]. By varying the nanoslit width, different incident wavelengths interfere constructively at different positions, thus configuring a sort of switchable and focus-tunable structure, as depicted in **Figure 2**. Anyhow, such inherent dispersive behavior attributed to diffraction can be corrected [18], enabling ultra-broadband achromatic focusing that is necessary for some applications. These kinds of lenses were recently fabricated using single-step grayscale lithography, where linear grooves with a designed height are set on a photoresist [19].

2.1. Super-oscillatory metalenses

A new idea in optical imaging has emerged which exploits the recently predicted and observed effect of optical super-oscillations. The key element of this new super-resolution technique is a super-oscillatory lens, which is a nanostructured mask that creates a focus beyond the diffraction limit, in principle of any prescribed size, without the contribution of evanescent waves [20]. However, the subwavelength hotspot is created in a low-intensity super-oscillatory region, thus only receiving a minimal fraction of the total power, with a simultaneous larger high-intensity halo, the latter limiting the field of view of this focusing device in imaging.

One of the first experimental demonstrations of super-oscillatory metalenses consisted of a quasicrystal array of nanoholes in a metal screen producing bright foci sparsely distributed in the focal plane [21]. Although nearly any array of small holes will create a diffraction pattern of foci when illuminated by a point source, a suitable choice of nanohole pattern is required for lensing applications [22]. Particularly, a Penrose array of 200 nm holes enables to generate an isolated hotspot of electromagnetic radiation at some distances [23]. Furthermore, a displacement of the point source leads to a linear shift of the image point, thus mimicking a function of the lens.

To design the binary mask, the radial coordinate can be divided into a number of concentric annuli, each of which had either unit or zero transmittance. In Ref. [24] 25 transparent regions of varying size was finally reached. Specifically, the ring pattern of the super-oscillatory lens with outer diameter of 40 μm was manufactured by FIB milling of a 100-nm-thick aluminum film supported on a round glass substrate and mounted as a microscope illuminating lens. When illuminated with a laser at $\lambda = 640$ nm, it generates a focal hotspot of 185 nm in diameter, ~25 times the intensity of the incident light and located at a distance of 10.3 μm from the film. The super-oscillation concept can be extended into the vectorial-field regime to work with circularly polarized light [25].

3. SPP-driven holographic metalenses

Surface plasmon polaritons cannot be used directly to focus light by free-space propagation since they propagate at the interface between a metal and a dielectric. SPPs and free-space beams are often coupled through periodic gratings. A metal grating with a period Λ on top of the metal layer provides an additional wave vector, in such a way that an incoming electromagnetic wave at an angle θ will satisfy the condition of momentum conservation:

$$k_d \sin(\theta) + 2\pi m/\Lambda = \pm\mathrm{Re}(k_{\mathrm{SPP}}) \tag{2}$$

where k_d and k_{SPP} are the dielectric and SPP wave numbers, respectively and m stands for an integer.

By employing holographic-based techniques for modulating the grating, one can systematically control the amplitude and phase of the coupled free-space beam. Thus, a planar beam transformer can be obtained by means of gratings with different periods for the input and output coupling. Following such procedure, in Ref. [26] it was experimentally demonstrated the coupling of SPPs into focused free-space beams, as shown in **Figure 3**, as well as into accelerating airy beams and vortex beams.

Much earlier, it was theoretically shown how, by patterning the exit plane of an aperture in a metal film, the angular distribution of the transmitted light in the far-field region can be shaped, producing highly focused beams at resonant wavelengths [27]. Since the diffracted beams are modulated by the nanometric grooves, through adjusting the parameters of the grooves such as their width, depth, period and number, the diffracted beams can be fully manipulated resulting in a tailored ultracompact lens [28]. The architecture can exhibit radial symmetry, so the plasmonic lens would consist of an annular slit and concentric grooves which, under radially polarized illumination, produces a focus spot. With a proper design, the inherent chromatic dispersion can be compensated and thus focusing dual-wavelength SPPs to the same focal plane [29]. Furthermore, the manipulation of plasmonic beaming fields can be based on the variation of dielectric surface gratings instead of using grooves. Plasmonic off-axis beaming and focusing of light by the use of asymmetric or nonperiodic dielectric gratings around a metallic slit has been experimentally demonstrated in Ref. [30].

Note that the present procedure based on plasmonics is suitable for terahertz beamforming. In this case, metallic grooves with a subwavelength spacing can couple to free-space propagating wave fields by means of spoof surface plasmon polaritons. Bullseye structures made of concentric scatterers periodically incorporated at a wavelength scale experimentally have demonstrated the launching of surface waves into free space to define a focal beam [31].

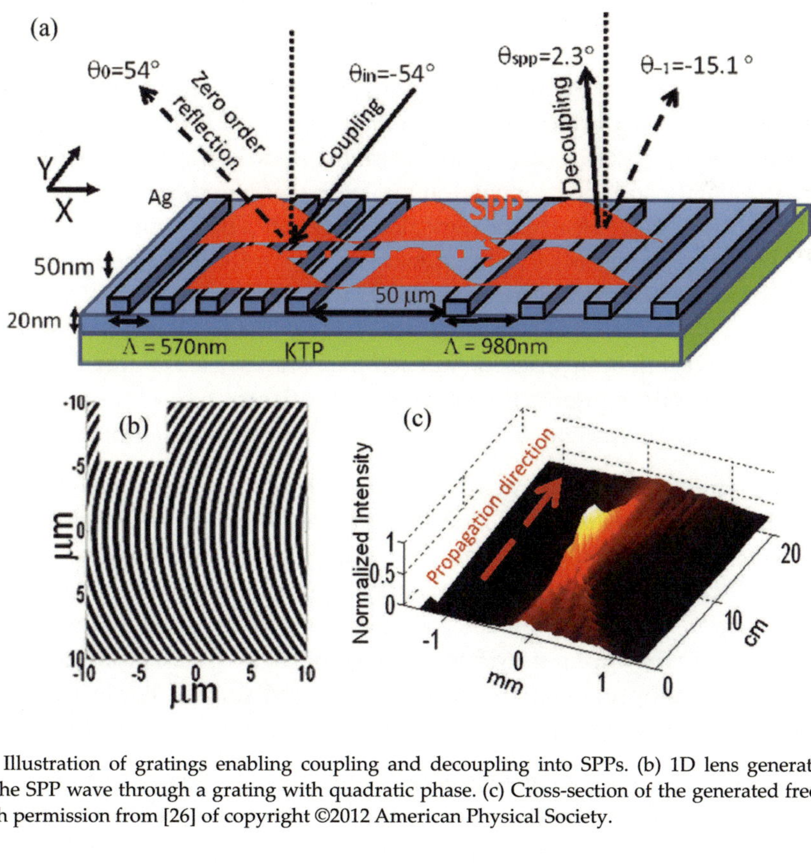

Figure 3. (a) Illustration of gratings enabling coupling and decoupling into SPPs. (b) 1D lens generated by output coupling of the SPP wave through a grating with quadratic phase. (c) Cross-section of the generated free-space beam. Adapted with permission from [26] of copyright ©2012 American Physical Society.

4. Gradient-index metalenses

In this section we consider different proposals to locally modulate the effective index of refraction undergone by wave fields traversing the apertures that constitute a metalens. The basic degrees of freedom include changing the width of the slits, assuming a cylindrical holey lens, their relative positions and the shape of the entrance and exit surfaces. An early idea inspired in thick left-handed metamaterial lenses is based on creating metallic lenses with curved surfaces, resembling glass lenses in their shape, such that each nanoslit transmits light with a phase retardation which is controlled by the metal thickness [32]. However, efficient fabrication processes may benefit from keeping the film thickness fixed but tailoring alternate geometrical parameters of the holey metallic device. Shi et al. first proposed a lens design based on an array of nanoslits which have constant depth but variant widths [33]. The variation of the slit width is associated with a change in the effective index of the transmitted field mode. The latter principle of operation was massively followed by others.

The basic building block of the patterned metallic lens is a narrow slit surrounded by metallic walls. By illuminating with a plane wave whose electric field is polarized perpendicularly to the orientation of the slits, the fundamental transverse magnetic (TM) mode is excited in the metal-insulator-metal structure following a dispersion relationship as

$$\tanh\left(\frac{w}{2}\sqrt{\beta^2 - k_0^2 \epsilon_d}\right) = -\frac{\epsilon_d \sqrt{\beta^2 - k_0^2 \epsilon_m}}{\epsilon_m \sqrt{\beta^2 - k_0^2 \epsilon_d}} \quad (3)$$

which relates the modal propagation constant β to the free space propagation constant $k_0 = 2\pi/\lambda$, the permittivity of the metal ϵ_m and the dielectric ϵ_d inside the slit and the slit width w. The real and imaginary parts of the effective refractive index n_{eff}, calculated as β/k_0, determine the phase velocity and the propagation loss of the SPP modes, respectively. The phase delay introduced by a single pass in the nanoslit is estimated by βd and strongly depends on the slit width, where d represents the film thickness. The narrower slit introduces the larger phase delay. Thus, a structure that consists of slits with increasing width from the center to the side creates a curved wavefront. Simultaneously, the field at the exit surface of the film is also modulated since the transmission of each nanoslit varies as a function of slit width. Finally, the influence of the interaction between two adjacent nanoslits on the phase delay has been systematically investigated using the finite-difference time-domain method [34].

Verslegers et al. reported the first experimental demonstration of far-field lensing using this type of gradient-index nanoslit array [35, 36]. The fabricated planar lens consists of an array of nanoscale slits, with range in width from 80 nm at the center of the array to 150 nm on the side, in an optically thick gold film, as illustrated in **Figure 4**. The gold film is evaporated onto a fused silica substrate using electron beam evaporation, which is later patterned by milling through the film. After this exploratory finding, the interest for gradient-index metalens grew tremendously followed by a large number of experimental studies. For instance, Chen et al. [37] fabricated plasmonic lenses in the visible range using nanoslits in an aluminum film.

Flat metalenses offer some advantageous accessibility to manufacture and integrate in complex devices; however, they suffer from severe geometrical restrictions to attain extremely high numerical apertures. An alternative design relies on sculpturing a concave surface. In

http://dx.doi.org/10.5772/66467

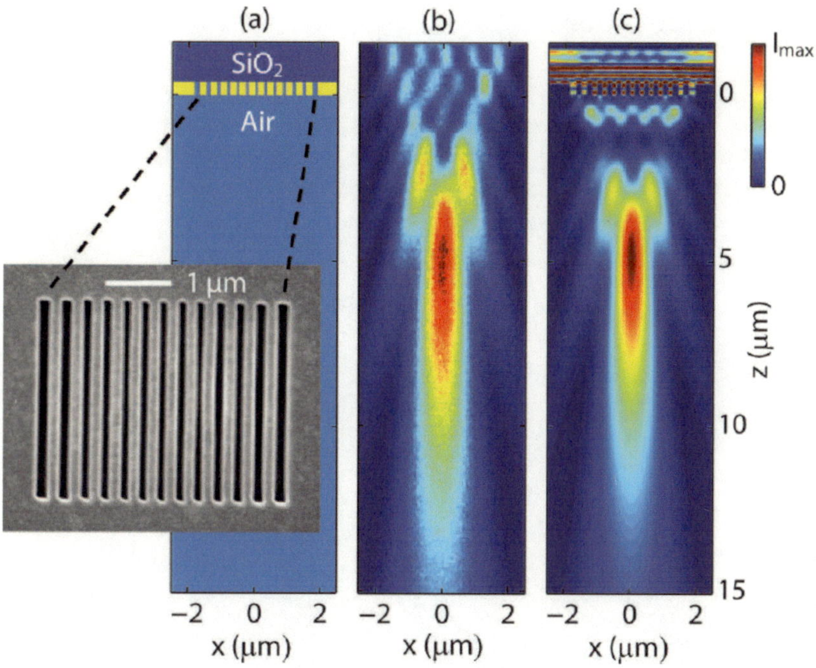

Figure 4. Planar lens based on nanoscale slit array in metallic film. (a) Geometry of the gradient-index lens. The inset shows a scanning electron micrograph of the structure as viewed from the air-side. (b) Focusing pattern measured by confocal scanning optical microscopy. (c) Finite-difference frequency-domain simulated focusing pattern of the field intensity through the center of the slits. Reprinted with permission from [35] of copyright ©2009 American Chemical Society.

such manner an aplanatic metasurface patterned on a spherical substrate has been proposed to focus light without coma and spherical aberrations [38]. Note that nonplanar metacoatings composed of subwavelength metal-dielectric arrays can not only focus an incoming light [39] but also let it accelerate in the near field [40]. Such basic concept can be utilized to engineer gradient-index ultrathin coatings, in a parallelizable assembly, as focusing elements with high efficiency [41, 42].

Differently from works conducted previously by other groups, it was investigated by numerical simulations and experiments the manipulation of an incident electric field that is parallel to the slits (TE-polarized plane wave) passing through arrays of nanometric spatially varying near-resonant slits perforated in a silver film [43]. Slits illuminated with the TE polarization exhibit a cut-off wavelength together with a resonance for transmission, enabling a fast phase modulation of the transmitted field, which is absent in the case of the TM polarization. Similarly, the dispersion of metal-dielectric-metal plasmonic waveguides is exploited to artificially mimic an epsilon-near-zero medium at optical wavelengths by working near the cut-off of the TE_1 mode [44].

An important drawback of the nanoslit lens design is its polarization dependence, which can be eliminated by extending the concepts of gradient-index metalenses into two dimensions. For example, a planar, holey metal lens was experimentally made as a set of concentric

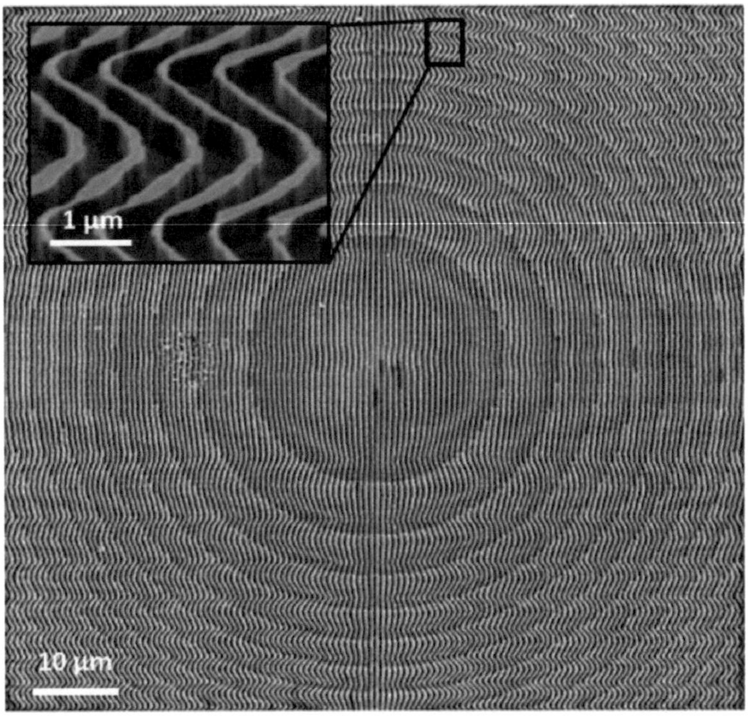

Figure 5. Scanning micrograph of the entire fabricated grating lens for f =100 µm (top view). The inset emphasizes the region with two phase jumps. Reprinted with permission from [48] of copyright ©2013 Optical Society of America.

circular arrays of nanoscale holes milled in a subwavelength-thick metal film [45]. In this case, each nanohole used as a phase-shifting element acts as a finite-length, circular, single-mode waveguide with a radius-dependent mode index.

Note that laterally propagating SPPs excited by the periodic nanohole array play an important role in light transmission through a metal-dielectric hole array. Therefore, the novel idea is not exploiting propagating waves in a plasmonic waveguide but using resonances. An early experimental demonstration of two-dimensional focusing was based on planar plasmonic lenses formed by an array of subwavelength cross-shaped apertures in a thin metal film, where the apertures have dimensions that vary spatially across the device. In this case, the lenses utilize localized surface plasmon resonances occurring within the apertures that are accompanied by controlled phase shifts in the transmitted field by means of a modulation of the arm length of the crosses [46].

4.1. Metalenses based on dielectric gratings

As an important category of the gradient-index metalenses, focusing elements based on semiconductor high-contrast gratings has been widely investigated. The basic geometry of the high-contrast grating consists of an array of nanoscale ridges completely surrounded by low refractive index materials (e.g., air). By using a nonperiodic design of the subwavelength grating pattern allows a dramatic control of the phase front of the reflected beam, without affecting the high reflectivity of the mirror [47]. Note that wavefront-engineered diffraction gratings also enable high aperture three-dimensional focusing in free space. As an example, **Figure 5** shows a grating structure, made of amorphous silicon on a glass substrate, which is designed by locally modulating the period and duty cycle of the grating for a wavelength of 980 nm and optimized for operation in reflection, normal incidence and TM polarization [48]. In particular, a path-finding algorithm allows to find a path in the parameter space of the grating that continuously connects two points of the desired phase evolution.

In a complementary scheme, one may design a metalens by writing a space-variant nanograting which results in a space-variant effective birefringence. Then, by suitably engineering the local orientation of the nanograting, a Pancharatnam-Berry phase lens can be achieved [49]. In Ref. [50] it was reported a dielectric gradient metasurface optical element with a 120-nm-wide and 100-nm-high Si nanobeam as a basic building block. When the metalens is uniformly illuminated from the substrate side with right circularly polarized light at a wavelength of 550 nm, it concentrates light into a left circularly polarized focal spot.

5. Focusing with nano-antennas

Metasurfaces have their conceptual roots in early works on subwavelength gratings. The concept of optical phase discontinuities provides a different path for designing flat lenses and has been used in the demonstration of metasurfaces capable of beaming light in directions characterized by generalized laws of reflection and refraction [51]. Under this approach, the control of the wavefront no longer relies on the phase shift accumulated during the propagation of light but is achieved by radiation as it scatters off the optically thin resonators comprising the metasurface.

5.1. Plasmonic metasurfaces

Light focusing was experimentally demonstrated in free space at telecom wavelength using a 60 nm thick gold metasurface [52] as shown in **Figure 6**. In the fabricated metalens with 3 cm focal distance, eight different plasmonic V-shaped antennas were designed to scatter light in cross-polarization with relatively constant amplitudes and an incremental phase of $\pi/4$ between neighbors. However, the variety in design of nanoresonators varying their shape and spatial distribution is certainly broad. For instance, a metal-insulator-metal configuration in which the top metal layer consists of a periodic arrangement of differently sized bricks, thus functioning as a metasurface in close vicinity of a metal film, was designed to work as a parabolic reflector [53].

Instead of molding linearly polarized wave fields, suitable designs of the elementary resonant nano-antennas and their spatial distribution can be used to manipulate the wavefronts and polarization of azimuthally polarized and radially polarized light. By means of elliptical optical antennas, which instead of converting one linear polarization to the other as in previous works, enable incident right-handed circularly polarized light to be almost converted into left-handed circularly polarized light [54]. In the latter case, the abrupt phase change occurs

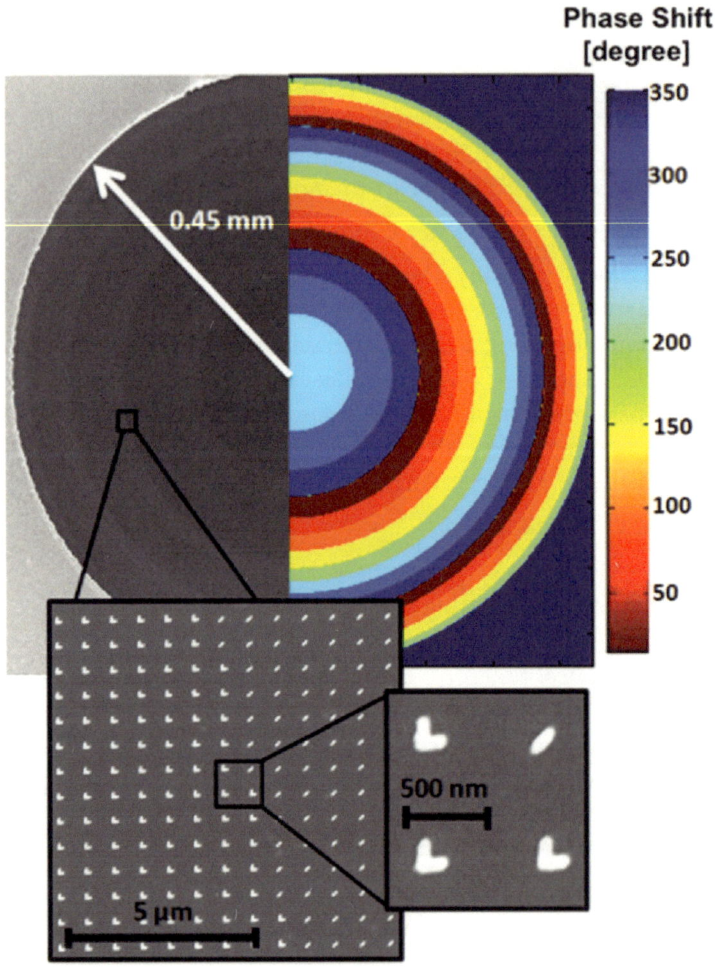

Figure 6. SEM image of the fabricated lens (left). The corresponding phase shift profile is displayed on the right. Insets: close up of patterned antennas. Reprinted with permission from [52] of copyright ©2012 American Chemical Society.

for circularly polarized light that is converted to its opposite helicity, in such a way that the created focused light carries orbital angular momentum.

One can apply the same working principle by exploiting the shift of wave vector that results from the geometric Pancharatnam-Berry phase. For instance, a plasmonic bipolar lens consisting of dipole nano-antennas with controlled directional orientation was designed to exhibit helicity-controllable real and virtual focal planes, as well as magnified and demagnified imaging at visible and near-infrared wavelengths [55]. We point out that, also, gradient-rotation split-ring antenna metasurfaces were designed and experimentally demonstrated as spin-to-orbital angular momentum beam converters to simultaneously generate and separate pure optical vortices [56].

In contrast to conventional nano-antennas, it is possible to use an inverted design built on Babinet's principle. In this case, similarly shaped nano-voids (Babinet-inverted, or complementary nano-antennas) have been milled in a thin metallic film, which provides a significantly higher signal-to-noise ratio [57]. More recently, helical devices composed of more than 121 thousands of spatially rotated nanosieves have been fabricated to achieve full manipulations of optical vortices by controlling the geometric phase of spin light, enabling for instance to create multifoci vortex lenses [58]. This approach has been also utilized in the development of innovative terahertz imaging systems.

The metasurfaces demonstrated so far are conveniently designed to manipulate only the phase profile of the outgoing electromagnetic waves. A complete manipulation of the propagation of light, on the other hand, requires simultaneous control of amplitude and phase. To implement this approach for governing both phase and amplitude, a metasurface was designed that consists of C-shaped antennas. The scattered fields from both the symmetric and antisymmetric modes contribute to an orthogonally polarized output wave, whose phase and amplitude can be engineered by adjusting the shape parameters of the antenna [59]. By employing the designed C-shape split-ring resonators, a metasurface with a hyperboloidal phase profile is arranged to engineer the transmitted wavefront in the same way as a conventional lens. This architecture enabling scattered fields with controlled phase and amplitude distribution has also inspired the design of holographic-based multifocal metalenses [60].

In connection with Fresnel zone plates discussed in Section 2, it is worth noting that metasurfaces consisting of cross-shaped and rod-shaped optical nano-antennas can be used to construct diffractive lenses. Since the transmission function of the optical nano-antenna depends on its resonance, by creating metasurfaces out of wavelength and polarization selective optical nano-antennas, the total transmission function can be modulated and construct a metalens that operates as a binary-amplitude Fresnel zone plate.

5.2. Dielectric nanoresonators

A vast majority of previous designs based on plasmonic metasurfaces and gratings cannot provide a performance comparable to conventional curved lenses. The implementation of Huygens' sources at optical frequencies utilizing the strong electric and magnetic resonances of high-permittivity all-dielectric nanoparticles in the near IR spectral range is currently a very active route. Using dielectric nanoresonators as phase shift elements, metasurfaces enable wavefront molding with experimental demonstrations of beam bending and lensing [61]. For instance, cylindrical converging wave fields can be produced by means of a lens consisting of an aperiodic arrangement of coupled rectangular dielectric resonators whose scattering phases are in addition engineered to achieve dispersion-free focusing at telecommunication wavelengths. The achromatic metalens can be fabricated by depositing 400 nm amorphous silicon on a fused silica substrate, where the rectangular dielectric resonators were defined by electron-beam lithography [62]. Such metalenses address an increased need for low-cost, lightweight and compact optical elements that can easily be assembled to electronic and mechanical components.

Many more proposals based on all-dielectric metasurfaces can be found in the literature to produce spherical wavefronts. Full-phase coverage combined with high efficiency in transmission are experimentally confirmed using for instance silicon nanodisks and square silicon

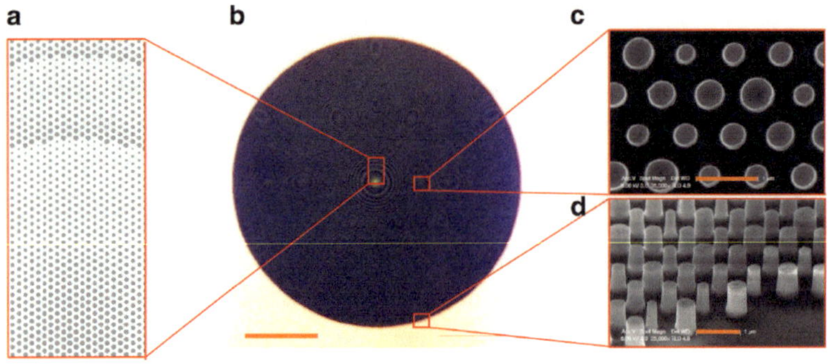

Figure 7. (a) Schematic illustration of the aperiodic high-contrast transmitarray used to realize a high-NA dielectric microlens. (b) Optical microscope image of the fabricated metalens with large NA. Scale bar, 100 μm. (c and d) SEM images of the silicon posts forming the microlens. Scale bars, 1 μm. Reprinted with permission from [64] of copyright ©2015 Macmillan Publishers Limited.

ridges [63]. Utilizing lithographically patterned arrays of elliptical silicon nanowires enables in addition to control the birefringence of the material and thus to create dynamically reconfigurable lenses in response to the polarization of the illumination.

Particularly interesting is the reported high-NA lens illustrated in **Figure 7**, which is composed of dielectric nanoposts that are fabricated in an amorphous silicon film [64]. Such polarization-insensitive, micron-thick, high-contrast transmitarray microlens produces a focal spot as small as 0.57 λ with a measured focusing efficiency of 82%. Furthermore, complete control of polarization and phase can be attained using high-contrast dielectric elliptical nanoposts [65]. This novel metasurface platform achieves a measured efficiency of up to 97%, enabling the realization of free-space transmissive optical lenses for the simultaneous generation and focusing of radially and azimuthally polarized light from linearly polarized beams, even separating x- and y-polarized light and focusing them to two different points. Note that these lenses provide the high NA required for collimation of mid-IR quantum cascade lasers.

Silicon can be replaced by new high-index dielectric materials for the assembly of efficient focusing metasurfaces. For example, a metalens that consists of TiO_2 nanofins on a glass substrate has been designed and fabricated, with a NA = 0.8 and an efficiency as high as 86% in the visible range [66]. In this case, the required phase is conferred by rotation of the nanofin according to the geometric Pancharatnam-Berry phase, showing that they are able to provide diffraction-limited focal spots at arbitrary design wavelengths. Note that these sort of planar metalenses exhibit super-dispersive characteristics, enabling for instance to simultaneously form two images with opposite helicity of an object within the same field-of-view.

6. Reconfigurable metalenses

The inclusion of active tunability in static plasmonic devices greatly enhances their functionality. Index-variable materials are often incorporated in plasmonic devices and optical metasurfaces, including liquid crystals, vanadium dioxide, silicon and other materials. Thus, one may manipulate the optical phase of the guided modes excited in graded-index metalenses in order to achieve a certain degree of tunability in the focusing behavior of the photonic device. Perhaps one of the first proposals consisted of embedding nonlinear media in the slit region of metallic nano-optic lens in such a way that the focal length can be easily controlled by means of the intensity of incident light in the structures [67].

As mentioned above, the slits of a holey metalens can be filled with liquid crystals to take advantage of their index changing property. In this case, the focal length can be easily controlled by exposing the plasmonic metalens to a constant external electric field, where the change in refractive index is directly converted into a locally modulated phase shift for the incident light. Ishii et al. experimentally showed that the irradiance as well as the transmission profile are altered provided that the liquid crystals change their phase from the nematic state to the isotropic state [68]. Such approach also allows electric and magnetic resonances to be spectrally tuned in all-dielectric metasurfaces as well as switching of the anisotropy (temperature-dependent) of the optical response of the device [69].

A novel planar metalens shown in **Figure 8**, which consists of an array of slits that are filled with phase-change material $Ge_2SB_2Te_5$ (GST), has also been proposed to engineer the far-field focusing patterns [70]. By varying the crystallization level of GST from 0 to 90%, the transmitted electromagnetic phase modulation supported inside each slit can be as large as $0.56\,\pi$ at the working wavelength 1.55 μm. In fact, this geometrically fixed platform can be applied to a variety of devices such as visible-range reconfigurable bichromatic lenses, multifocus Fresnel zone plates and super-oscillatory lenses [71].

In the teraherz regime one may fill the slit array with InSb to create a plasmonic lens with tunable focal length by simply controlling the temperature. When an external magnetic field is applied, the cyclotron frequency of high electron mobility semiconductors such as InSb locates also in THz frequencies, enabling to tune the dielectric properties of the magneto-optical material. Based on the magneto-optical effect, a tunable metal/magneto-optic plasmonic lens for terahertz isolator was introduced which, in addition, exhibits the nonreciprocal transmission property [72].

Planar diffractive microfluidic lenses with switchable properties have also been reported, in particular integrating controlled dielectrophoresis for trapping suspended silicon and tungsten oxide nanoparticles [73]. These nanomaterials can be trapped to produce alternating opaque and transparent rings using the dielectrophoresis forces, thus suitably forming diffractive Fresnel zone plates capable to focus the incident light. The Fresnel zone plate is tuned for the visible light region and the lens can be turned on (dielectrophoresis applied) or off (dielectrophoresis removed) in a controlled manner.

Graphene, indium tin oxide and vanadium dioxide have been frequently employed, owing to their large optical tunability as a function of either voltage or temperature. For instance, the Fermi level of graphene can be easily tuned by applying bias voltage, which, in turn, affects the resonance of nearby plasmonic structures. As a result, active metasurfaces capable of delivering efficient real-time control and complex-amplitude manipulation have been proposed, as we will also analyze in the next section devoted to atomically thin metalenses.

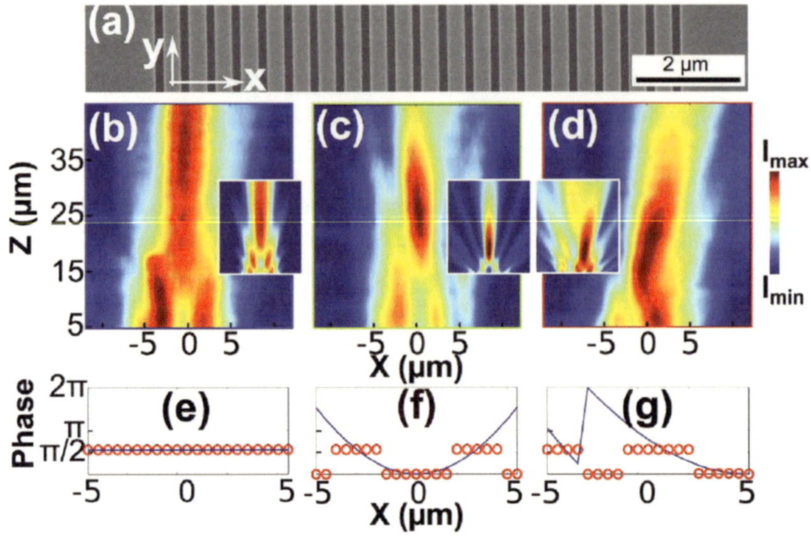

Figure 8. (a) SEM image of the fabricated planar lens before sputtering GST. (b, c and d) Focusing pattern by confocal scanning optical microscopy for amorphous GST in all slits without crystallization and GST being crystallized in the selected slits; insets show numerical simulations. (e, f and g) The calculated phase fronts (blue curves) and the binarized discrete phase distributions (red circles) which are anticipated from the samples in (b), (c) and (d), respectively. Reprinted with permission from [70].

Active tunability can also exploit the benefits of microelectromechanical systems (MEMS). In these techniques, the substrate supporting for the metallic structures itself moves and shifts the plasmonic resonances. As an example, the active plasmon lens shown in Ref. [74] combines a MEMS structure with a metallic width-variant nanoslit lens. The active lens has five movable nanoslits whose widths can be controlled by an external bias, allowing focusing or defocusing of the diffracted light via controlling the phase front. In the reported experiments, a 5-V bias can modify the slit widths initially fixed at $d_1 = 80$ nm, $d_2 = 120$ nm and $d_3 = 150$ nm, thereby changing the metalens transmission profile. Also, a mechanically tunable metasurface that includes a stretchable polydimethylsiloxane substrate has been proposed to change the relative position of arrayed plasmonic gold nanorods deposited above, thus controlling the local refraction angle continuously and thus the optical wavefront of the scattered field [75]. Following this method, a flat 1.7× zoom metalens was realized whose focal length can gradually be changed from 150 to 250 μm. Direct applications in all cases include optical trapping, optical routing and beam stirring at the micron scale.

7. Atomically thin lenses

The ultrathin metalens fabricated by Capasso et al. [52], which is based on plasmonic nanoresonators with a thickness of only 60 nm, was considered to be a milestone in the current photonic nanotechnology. As graphene is one of the thinnest and probably most promising materials nowadays, it was used for the design of atomically thin Fresnel zone plates, where in addition the lens performance can be tuned by adjusting the Fermi level of graphene and the number of layers. Early experimental demonstrations of graphene-based metalenses appeared soon after. Monolayers and multilayers of graphene were fabricated and giving form to Fresnel zones in order to produce diffractive lenses, allowing to operate in the visible and near-infrared frequency range [76]. In the reflection mode, the focusing efficiency increases with an increasing number of graphene layers. **Figure 9** shows a graphene Fresnel zone plate with a radius of about 50 μm and containing 24 zones, where the average surface roughness was measured as 3.47 nm, which corresponds to approximately 10 layers of graphene. In order to demonstrate the thinnest possible metalens, a single layer of graphene was also successfully patterned lithographically onto a thick glass. The graphene lenses were found to be thinner and easier to fabricate compared to the metasurface-based lenses, having the potential to revolutionize the design of compact optical systems, such as laser focusing for optical storage and fiber optic communication.

Graphene has been demonstrated to support surface plasmon polaritons and, thereby, represents an attractive alternative to metals for the design of plasmonic metasurfaces. Furthermore, graphene conductivity can be dynamically controlled by an external stimulus such as electric and/or magnetic field, voltage, or temperature and therefore the optical response of the metamaterial device potentially exhibits a versatile tunability. Metasurfaces based on 1D graphene nanoribbons, periodically patterned graphene nanocrosses, or alternatively nano-antenna plasmonic metasurfaces integrating single-layer graphene, have experimentally evidenced their capability to manipulate the wavefront of light, thereby used to create ultrathin lenses [77–79]. Moreover, a graphene cut-wire layer introducing a discontinuous Pancharatnam-Berry phase profile has been proposed as integrated in a metal-dielectric-graphene three-layer structure to improve the interaction of graphene nanostructures with incident waves [80]. By arranging two different graphene cut-wire resonators in one unit cell, one can excite two resonances and, as a result, 2π phase modulation with efficiency approaching 50% can be achieved in a wide range of frequencies.

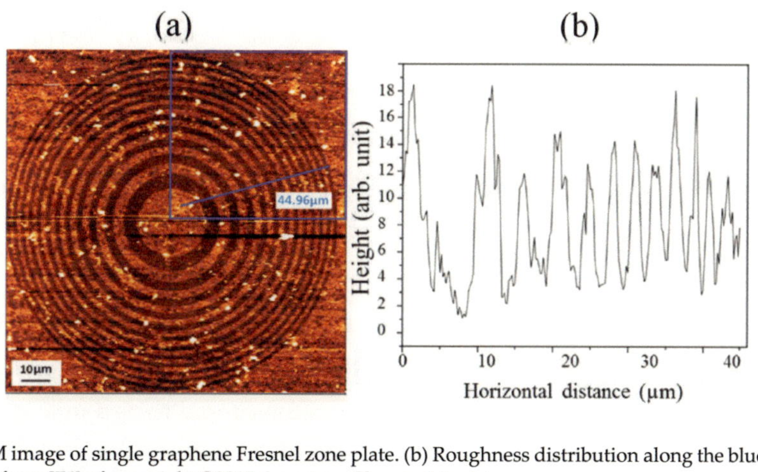

Figure 9. (a) AFM image of single graphene Fresnel zone plate. (b) Roughness distribution along the blue line. Reprinted with permission from [76] of copyright ©2015 American Chemical Society.

2D materials other than graphene can offer a superior performance for some specific applications and frequency ranges. For instance, a single-layer molybdenum disulfide provides an extremely high optical path length, which is roughly one order of magnitude larger in comparison with that experienced from a single layer of graphene. Such a giant optical path length has been exploited to design and fabricate probably the world's thinnest optical lens, enabling to curve the plane phase front of an optical beam within less than 6.3 nm thick that constitute a few layers of MoS_2 [81]. The bowl-shaped structure is 20 μm in diameter, where the MoS_2 thickness is gradually changed by using a FIB, thus serving as an atomically thin reflective concave lens. In addition, the refractive index of the layered MoS_2 can be largely tuned by applying an electric field, allowing the fabrication of microlenses with electrically tunable focal lengths.

8. Summary

In summary, we have reviewed and discussed the significant recent progress in the field of far-field metalenses in optical applications. The devices analyzed here exhibit important features directed toward solving the main objections related to conventional metamaterial lenses, such as transmission losses and fabrication. Some of the greatest achievements in this area are the staggering progress of current nanofabrication techniques, which have a higher capacity of resolution at the nanometer scale, the development of new materials enabling an increased performance in optical and IR metalenses and the design of novel subwavelength nanostructures with improved functionality. A number of proposals include tunability which allows the dynamic manipulation of complex wave fields, commonly on the diffraction limit and the versatile reconfiguration of metalenses. Owing to its countless advantages in terms of size, manufacturing simplicity and the wide range of applications, the metalenses may be considered indispensable elements conducting future optical devices such as integrated photonics, super-resolving imaging, optical trapping and quantum optics, to mention a few.

Author details

Mahin Naserpour[12], Mahdieh Hashemi[3] and Carlos J. Zapata-Rodríguez[1*]

*Address all correspondence to: carlos.zapata@uv.es

1 Departament of Optics and Optometry and Vision Science, University of Valencia, Burjassot, Spain

2 Department of Physics, College of Sciences, Shiraz University, Shiraz, Iran

3 Department of Physics, College of Science, Fasa University, Fasa, Iran

References

[1] Zheludev NI. A roadmap for metamaterials. Opt. Photonics News. 2011;**22**:30–35.

[2] Pendry JB. Negative refraction makes a perfect lens. Phys. Rev. Lett. 2000;**85**:3966–3969.

[3] Vodo P, Parimi PV, Lu WT, Sridhar S. Focusing by plano concave lens using negative refraction. Appl. Phys. Lett. 2005;**86**:201108.

[4] Valentine J, Zhang S, Zentgraf T, Ulin-Avila E, Genov DA , Bartal G, Zhang X. Three-dimensional optical metamaterial with a negative refractive index. Nature. 2008;**455**:376–379.

[5] Lu D, Liu Z. Hyperlenses and metalenses for far-field super-resolution imaging. Nat. Commun. 2012;**3**:1205–1214.

[6] Yu N, Capasso F. Flat optics with designer metasurfaces. Nat. Mater. 2014;**13**:139–150.

[7] Fu Y, Wang J, Zhang D. Plasmonic lenses. In: Kim KY, editor. Plasmonics - Principles and Applications. Rijeka, Croatia. InTech; 2012. pp. 183–228.

[8] Glybovski SB, Tretyakov SA, Belov PA, Kivshar YS, Simovski R. Metasurfaces: From microwaves to visible. Phys. Rep. 2016;**634**:1–72.

[9] Zapata-Rodríguez CJ. Debye representation of dispersive focused waves. J. Opt. Soc. Am. A. 2007;**24**:675–686.

[10] Li M, Wang J, Zhuang L, Chou SY. Fabrication of circular optical structures with a 20 nm minimum feature size using nanoimprint lithography. Appl. Phys. Lett. 2000;**76**:673–675.

[11] Fu Y, Zhou W, Lim L, Du C, Luo X. Plasmonic microzone plate: Super focusing at visible regime. Appl. Phys. Lett. 2007;**91**:061124.

[12] Chen J, Wang C, Lu G, Li W, Xiao J, Gong Q. Highly efficient nanofocusing based on a T-shape microslit surrounded with multi-slits. Opt. Express. 2012;**20**:17734–17740.

[13] Huang K, Liu H, Garcia-Vidal FJ, Hong M, Luk′yanchuk B, Teng J, Qiu C-W. Ultrahigh-capacity nonperiodic photon sieves operating in visible light. Nat. Commun. 2015;**6**:7059.

[14] Huntington MD, Lauhon LJ, Odom TW. Subwavelength lattice optics by evolutionary design. Nano Lett. 2014;**14**:7195–7200.

[15] Wang S, Abeysinghe DC, Zhan Q. Generation of vectorial optical fields with slot-antenna-based metasurface. Opt. Lett. 2015;**40**:4711–4714.

[16] Lin DZ, Chen CH, Chang CH, Cheng TD, Yeh CS, Lee CK. Subwavelength nondiffraction beam generated by a plasmonic lens. Appl. Phys. Lett. 2008;**92**:233106.

[17] Mehmood MQ, Liu H, Huang K, Mei S, Danner A, Luk'yanchuk B, Zhang S, Teng J, Maier SA, Qiu CW. Broadband spin-controlled focusing via logarithmic-spiral. Laser Photon. Rev. 2015;**9**:674–681.

[18] Zapata-Rodríguez CJ, Caballero MT. Isotropic compensation of diffraction-driven angular dispersion. Opt. Lett. 2007;**27**:2472–2474.

[19] Wang P, Mohammad N, Menon R. Chromatic aberration-corrected diffractive lenses for ultra-broadband focusing. Sci. Rerp. 2016;**6**:21545.

[20] Roy T, Rogers ET, Yuan G, Zheludev NI. Point spread function of the optical needle super-oscillatory lens. Appl. Phys. Lett. 2014;**104**:231109.

[21] Huang FM, Zheludev N, Chen Y, Garcia de Abajo FJ. Focusing of light by a nanohole array. Appl. Phys. Lett. 2007;**90**:091119.

[22] Zheludev NI. What diffraction limit? Nat. Mater. 2008;**7**:420–422.

[23] Huang FM, Kao TS, Fedotov VA, Chen Y, Zheludev NI. Nanohole array as a lens. Nano Lett. 2008;**8**:2469–2472.

[24] Rogers ET, Lindberg J, Roy T, Savo S, Chad JE, Dennis MR, Zheludev NI. A super-oscillatory lens optical microscope for subwavelength imaging. Nat. Mater. 2012;**11**:432–435.

[25] Yuan G, Rogers ET, Roy T, Adamo G, Shen Z, Zheludev NI. Planar super-oscillatory lens for subdiffraction optical needles at violet wavelengths. Sci. Rep. 2014;**4**:6333 .

[26] Dolev I, Epstein I, Arie A. Surface-plasmon holographic beam shaping. Phys. Rev. Lett. 2012;**109**:203903.

[27] García-Vidal FJ, Martín-Moreno L, Lezec HJ, Ebbesen TW. Focusing light with a single subwavelength aperture flanked by surface corrugations. Appl. Phys. Lett. 2003;**83**:4500–4502.

[28] Jia B, Shi H, Li J, Fu Y, Du C, Gu M. Near-field visualization of focal depth modulation by step corrugated plasmonic slits. Appl. Phys. Lett. 2009;**94**:151912.

[29] Venugopalan P, Zhang Q, Li X, Kuipers L, Gu M. Focusing dual-wavelength surface plasmons to the same focal plane by a far-field plasmonic lens. Opt. Lett. 2014;**39**:5744–5747.

[30] Lim Y, Hahn J, Kim S, Park J, Kim H, Lee B. Plasmonic light beaming manipulation and its detection using holographic microscopy. IEEE J. Quant. Electron. 2010;**46**:300–305.

[31] Monnai Y, Jahn D, Withayachumnankul W, Koch M, Shinoda H. Terahertz plasmonic Bessel beam former. Appl. Phys. Lett. 2015;**106**:021101.

[32] Sun Z, Kim HK. Refractive transmission of light and beam shaping with metallic nano-optic lenses. Appl. Phys. Lett. 2004;**85**:642–644.

[33] Shi H, Wang C, Du C, Luo X, Dong X, Gao H. Beam manipulating by metallic nano-slits with variant widths. Opt. Express. 2005;13:6815–6820.

[34] Zhu Y, Yuan W, Yu Y, Diao J. Metallic planar lens formed by coupled width-variable nanoslits for superfocusing. Opt. Express. 2015;**23**:20124–20131.

[35] Verslegers L, Catrysse PB, Yu Z, White JS, Barnard ES, Brongersma ML, Fan S. Planar lenses based on nanoscale slit arrays in a metallic film. Nano Lett. 2009;**9**:235–238.

[36] Verslegers L, Catrysse PB, Yu Z, Fan S. Planar metallic nanoscale slit lenses for angle compensation. Appl. Phys. Lett. 2009;**95**:071112.

[37] Chen Q, Cumming DRS. Visible light focusing demonstrated by plasmonic lenses based on nano-slits in an aluminum film. Opt. Express. 2010;**18**:14788–14793.

[38] Aieta F, Genevet P, Kats M, Capasso F. Aberrations of flat lenses and aplanatic metasurfaces. Opt. Express. 2013;**21**:31530–31539.

[39] Naserpour M, Zapata-Rodríguez CJ, Díaz-Aviñó C, Hashemi M, Miret JJ. Ultrathin high-index metasurfaces for shaping focused beams. Appl. Opt. 2015;**54**:7586–7591.

[40] Naserpour M, Zapata-Rodríguez CJ, Zakery A, Díaz-Aviñó C, Miret JJ. Accelerating wide-angle converging waves in the near field. J. Opt. 2015;**17**:015602.

[41] Pfeiffer C, Grbic A. Cascaded metasurfaces for complete phase and polarization control. Appl. Phys. Lett. 2013;**102**:231116.

[42] Naserpour M, Zapata-Rodríguez CJ, Díaz-Aviñó C, Hashemi M. Metacoatings for wavelength-scale, high-numerical-aperture plano–concave focusing lenses. J. Opt. Soc. Am. B. 2016;**33**:2120–2128.

[43] Goh XM, Lin L, Roberts A. Planar focusing elements using spatially varying near-resonant aperture arrays. Opt. Express. 2010;**18**:11683–11688.

[44] Pacheco-Peña V, Navarro-Cía M, Beruete M. Epsilon-near-zero metalenses operating in the visible. Opt. Laser Technol. 2016;**80**:162–168.

[45] Ishii S, Shalaev VM, Kildishev AV. Holey metal lenses: Sieving single modes with proper phases. Nano Lett. 2013;**13**:159–163.

[46] Lin L, Goh XM, McGuinness LP, Roberts A. Plasmonic lenses formed by two-dimensional nanometric cross-shaped aperture arrays for Fresnel-region focusing. Nano Lett. 2010;**10**:1936–1940.

[47] Fattal D, Li J, Peng Z, Fiorentino M, Beausoleil RG. Flat dielectric grating reflectors with focusing abilities. Nat. Photonics. 2010;**4**:466–470.

[48] Klemm AB, Stellinga D, Martins ER, Lewis L, Huyet G, O'Faolain L, Krauss TF. Experimental high numerical aperture focusing with high contrast gratings. Opt. Lett. 2013;**38**:3410–3413.

[49] Ke Y, Liu Y, Zhou J, Liu Y, Luo H, Wen S. Photonic spin filter with dielectric metasurfaces. Opt. Express. 2015;**23**:33079–33086.

[50] Lin D, Fan P, Hasman E, Brongersma ML. Dielectric gradient metasurface optical elements. Science. 2014;**345**:298–302.

[51] Yu N, Genevet P, Kats MA, Aieta F, Tetienne JP, Capasso F, Gaburro Z. Light propagation with phase discontinuities: generalized laws of reflection and refraction. Science. 2011;**334**:333–337.

[52] Aieta F, Genevet P, Kats MA, Yu N, Blanchard R, Gaburro Z, Capasso F. Aberration-free ultrathin flat lenses and axicons at telecom wavelengths based on plasmonic metasurfaces. Nano Lett. 2012;**12**:4932–4936.

[53] Pors A, Nielsen MG, Eriksen R, Bozhevolnyi SI. Broadband focusing flat mirrors based on plasmonic gradient metasurfaces. Nano Lett. 2013;**13**:829–834.

[54] Ma X, Pu M, Li X, Huang C, Wang Y, Pan W, Zhao B, Cui J, Wang C, Zhao Z, Luo X. A planar chiral meta-surface for optical vortex generation and focusing. Sci. Rep. 2015;**5**:10365.

[55] Chen X, Huang L, Mühlenbernd H, Li G, Bai B, Tan Q, Jin G, Qiu CW, Zhang S, Zentgraf T. Dual-polarity plasmonic metalens for visible light. Nat. Commun. 2012;**3**:1198.

[56] Zeng J, Li L, Yang X, Gao J. Generating and separating twisted light by gradient-rotation split-ring antenna metasurfaces. Nano Lett. 2016;**16**:3101–3108.

[57] Ni X, Ishii S, Kildishev AV, Shalaev VM. Ultra-thin, planar, Babinet-inverted plasmonic metalenses. Light Sci. Appl. 2013;**2**:e72.

[58] Mei S, Mehmood MQ, Hussain S, Huang K, Ling X, Siew SY, Liu H, Teng J, Danner A, Qiu CW. Flat helical nanosieves. Adv. Funct. Mater. 2016; 26. DOI: 10.1002/adfm.201601345.

[59] Liu L, Zhang X, Kenney M, Su X, Xu N, Ouyang C, Shi Y, Han J, Zhang W, Zhang S. Broadband metasurfaces with simultaneous control of phase and amplitude. Adv. Mater. 2014;**26**:5031–5036.

[60] Hashemi M, Moazami A, Naserpour M, Zapata-Rodríguez CJ. A broadband multifocal metalens in the terahertz frequency range. Opt. Commun. 2016;**370**:306–310.

[61] West PR, Stewart JL, Kildishev AV, Shalaev VM, Shkunov VV, Strohkendl F, Zakharenkov YA, Dodds RK, Byren R. All-dielectric subwavelength metasurface focusing lens. Opt. Express. 2014;**22**:26212–26221.

[62] Aieta F, Kats MA, Genevet P, Capasso F. Multiwavelength achromatic metasurfaces by dispersive phase compensation. Science. 2015;**347**:1342–1345.

[63] Matsui T, Yamashita S, Wado H, Fujikawa H, Iizuka H. Flat grating lens utilizing widely variable transmission phase via guided-modes. Opt. Lett. 2015;**40**:25–28.

[64] Arbabi A, Horie Y, Ball AJ, Bagheri M, Faraon A. Subwavelength-thick lenses with high numerical apertures and large efficiency based on high-contrast transmitarrays. Nat. Commun. 2015;**6**:7069.

[65] Arbabi A, Horie Y, Bagheri M, Faraon A. Dielectric metasurfaces for complete control of phase and polarization with subwavelength spatial resolution and high transmission. Nat. Nanotechnol. 2015;**10**:937–943.

[66] Khorasaninejad M, Chen WT, Devlin RC, Oh J, Zhu AY, Capasso F. Metalenses at visible wavelengths: Diffraction-limited focusing and subwavelength resolution imaging. Science. 2016;**352**:1190–1194.

[67] Min C, Wang P, Jiao X, Deng Y, Ming H. Beam manipulating by metallic nano-optic lens containing nonlinear media. Opt. Express. 2007;**15**:9541–9546.

[68] Ishii S, Kildishev AV, Shalaev VM, Drachev VP. Controlling the wave focal structure of metallic nanoslit lenses with liquid crystals. Laser Phys. Lett. 2011;**8**:828–832.

[69] Sautter J, Staude I, Decker M, Rusak E, Neshev DN, Brener I, Kivshar YS. Active tuning of all-dielectric metasurfaces. ACS Nano. 2015;**9**:4308–4315.

[70] Chen Y, Li X, Sonnefraud Y, Fernández-Domínguez AI, Luo X, Hong M, Maier SA. Engineering the phase front of light with phase-change material based planar lenses. Sci. Rep. 2015;**5**:8660.

[71] Wang Q, Rogers ETF, Gholipour B, Wang CM, Yuan G, Teng J, Zheludev NI. Optically reconfigurable metasurfaces and photonic devices based on phase change materials. Nat. Photonics. 2016;**10**:60–65.

[72] Fan F, Chen S, Wang XH, Chang SJ. Tunable nonreciprocal terahertz transmission and enhancement based on metal/magneto-optic plasmonic lens. Opt. Express. 2013;**21**:8614–8621.

[73] Chrimes AF, Khodasevych I, Mitchell A, Rosengarten G, Kalantar-zadeh K. Dielectrophoretically controlled Fresnel zone plate. Lab Chip. 2015;**15**:1092–1100.

[74] Yamaguchi K, Yamanaka H, Ohtsu T, Ishii S. Electrically driven plasmon chip: Active plasmon lens in the visible range. Appl. Phys. Lett. 2016;**108**:111903.

[75] Ee HS, Agarwal R. Tunable metasurface and flat optical zoom lens on a stretchable substrate. Nano Lett. 2016;**16**:2818–2823.

[76] Kong XT, Khan AA, Kidambi PR, Deng S, Yetisen AK, Dlubak B, Hiralal P, Montelongo Y, Bowen J, Xavier S, Jiang K, Amaratunga GAJ, Hofmann S, Wilkinson TD , Qing Dai Q, Butt H. Graphene-based ultrathin flat lenses. ACS Photonics. 2015;**2**:200–207.

[77] Li Z, Yao K, Xia F, Shen S, Tian J, Liu Y. Graphene plasmonic metasurfaces to steer infrared light. Sci. Rep. 2015;**5**:12423.

[78] Cheng H, Chen S, Yu P, Liu W, Li Z, Li J, Xie B, Tian J. Dynamically tunable broadband infrared anomalous refraction based on graphene metasurfaces. Adv. Opt. Mater. 2015;**3**:1744–1749.

[79] Dabidian N, Kholmanov I, Khanikaev AB, Tatar K, Trendafilov S, Mousavi SH, Magnuson C, Ruoff RS, Shvets G. Electrical switching of infrared light using graphene integration with plasmonic Fano resonant metasurfaces. ACS Photonics. 2015;**2**:216–227.

[80] Hu X, Wen L, Song S, Chen Q. Tunable graphene metasurfaces by discontinuous Pancharatnam–Berry phase shift. Nanotechnology. 2015;**26**:505203.

[81] Yang J, Wang Z, Wang F, Xu R, Tao J, Zhang S, Qin Q, Luther-Davies B, Jagadish C, Yu Z, Lu Y. Atomically thin optical lenses and gratings. Light Sci. Appl. 2016;**5**:e16046.

Chapter 11

Planar Antennas for Reliable Multiband RF Communications

Mohammad AlibakhshiKenari,
Mohammad Naser-Moghadasi,
Ramazan Ali Sadeghzadeh, Bal Singh Virdee and
Ernesto Limiti

Additional information is available at the end of the chapter

http://dx.doi.org/10.5772/66675

Abstract

Multiband functionality in antennas has become a fundamental requirement to equip wireless devices with multiple communication standards so that they can utilize the electromagnetic spectrum more efficiently and effectively. This is necessary to ensure global portability and enhance system capacity. To meet these requirements, microstrip technology is increasingly being used in communication systems because it offers considerable size reduction, cost-effectiveness as they can be easily manufactured in mass production, are durable and can conform to planar or cylindrical surfaces. Unfortunately, such antennas suffer from intrinsically narrow bandwidth. To overcome this deficiency, various techniques have been investigated in the past. In this chapter, a novel approach is presented to design antennas for applications that cover radio frequency identification (RFID) and WiMAX systems.

Keywords: planar antenna, meandered strip-line feed, wideband antenna, multiband antenna

1. Introduction

Innovative design concepts that have been facilitated by cutting-edge technology are contributing toward boosting wireless communications. Antennas are an essential component in such systems; however the large dimensions and narrow operating range of conventional antennas preclude them from application in the next generation of wireless communications systems. This is because future wireless systems impose strict requirements from antennas, such as large impedance bandwidth to support multiple systems, small physical size, low cost designs, high efficiency and reliability.

To utilize the electromagnetic spectrum more efficiently and effectively it has become necessary to equip portable wireless devices with multiple communication standards. This necessitates antennas to operate over a wideband [1, 2]. Nowadays microstrip integrated technology (MIT) is being increasingly used in the design of antennas for application in wireless communication systems because it offers considerable size reduction, cost effectiveness as it allows easy manufacture in mass production, durability and enables antennas to be configured for mounting on various irregular surfaces [3, 4]. Unfortunately, such antennas have intrinsically narrow bandwidth. To circumvent this deficiency, numerous techniques have been investigated recently, which include embedding slit-lines in the patch antenna [5–9]; employing unconventional feeding structures [10, 11]; insertion of parasitic elements in the vicinity of the patch antenna [12]; employing thick substrates and/or higher dielectric constant substrates [13, 14]; loading the antenna with an arrangement of electromagnetic band-gap (EBG) structures [15]; using meta-surfaces [16] and employing metamaterial (MTM) unit cells [17–19]. These techniques certainly improve the impedance bandwidth of antennas; however, it is not sufficient to support multiple wireless communications systems. Another interesting technique to enhance the bandwidth of patch antennas uses meandered strip-line feed which has the advantage of being less complex to implement in practice [20].

In this chapter, a planar microstrip antenna is proposed for multiband wireless communications systems. Embedded in the antenna's radiation patch are an H-shaped slit and two inverted U-shaped capacitive slits. The antenna is fed through a meandered strip-line. The antennas described in this chapter are fabricated on RT/duroid® RO4003 substrate with permittivity of 3.38, thickness of 1.6 mm and $\tan\delta = 0.0022$. The final second antenna occupies an area of 13.5 mm × 12.7 mm or $0.036\lambda_0 \times 0.033\lambda_0$, where the free space wavelength (λ_0) is 0.8 GHz. Numerical parametric analyses was used to (1) determine the location of the slits in the rectangular patch and (2) the optimum meandered-line feed structure. Verification of the antenna performance was done through measurement. It will be shown later the proposed antenna exhibits properties of low radiation loss, low cross-polarization, ease of manufacture and integration in RF transceivers, and no use via holes. These characteristics make it suitable for multiband applications, such as Ultra High Frequency (UHF) Radio Frequency Identification (RFID), Global Positioning System (GPS), Personal Communication Systems (PCS), Digital Communication Systems (DCS), World-wide Interoperability for Microwave Access (WiMAX), Wireless Local Area Network (WLAN), WiFi, Bluetooth and other applications in the UHF, L, S, major part of C-bands.

2. MTM antenna design

The design technique described in this chapter offers expansion of the impedance bandwidth of the antenna without compromising its size and salient characteristics. This is achieved by inserting dielectric slits in the antenna patch and exciting it through a meandered strip-line. The antenna design employs an H-shape and inverted U-shape slit.

To minimize design complexity and reduce manufacturing cost the proposed antenna structure avoids the use of via-holes.

http://dx.doi.org/10.5772/66675

2.1. H-shaped slit patch antenna

The generic patch antenna configuration and its equivalent circuit model are shown in **Figure 1**. The slit essentially behave as series capacitor (C_L) and the meandered microstrip feed-line behaves as a shunt inductor (L_L). Other losses introduced by the antenna structure are represented by series right-handed resistance (R_R), shunt left-handed resistance (R_L), series conductance (G_L) and shunt conductance (G_R). Location and dimensions of the slit was optimized using 3D electromagnetic high frequency structural simulator (HFSS™) by ANSYS [21]. The electromagnetic fields and currents are concentrated in the vicinity of the patch by placing the slit at the center of patch. This prevents the fields from spreading on the ground plane of the antenna and thereby minimizes unwanted coupling. This technique contributes in the realization of a smaller patch antenna. The use of this antenna structure in an array would incur minimal mutual coupling which is important to de-correlate multipath channels in, for example, small cellular systems. The meandered microstrip feed-line ensures low radiation loss and low cross-polarization. It also helps to eliminate unwanted notch bands in the antenna's response due to impedance mismatch, thus providing bandwidth extension.

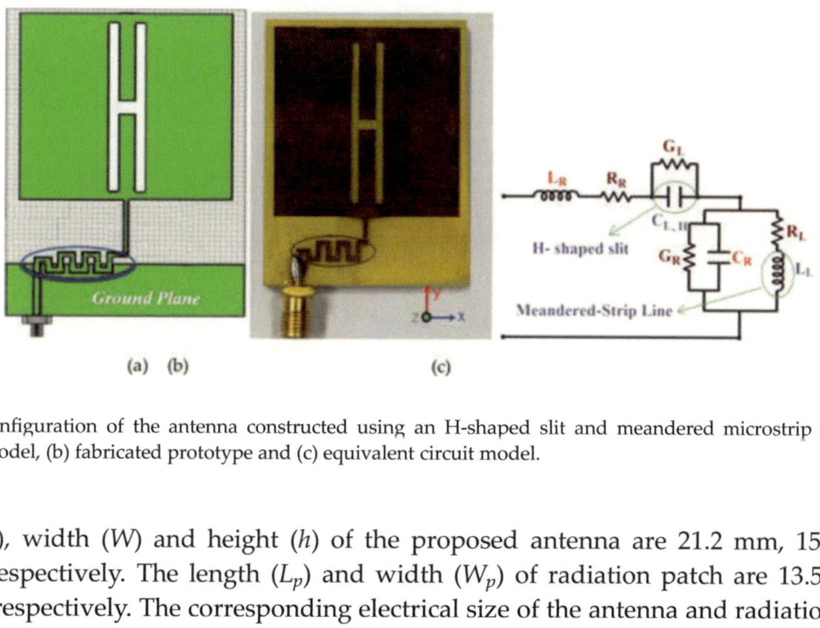

Figure 1. Configuration of the antenna constructed using an H-shaped slit and meandered microstrip feed-line, (a) simulation model, (b) fabricated prototype and (c) equivalent circuit model.

Length (L), width (W) and height (h) of the proposed antenna are 21.2 mm, 15 mm and 1.6 mm, respectively. The length (L_p) and width (W_p) of radiation patch are 13.5 mm and 12.7 mm, respectively. The corresponding electrical size of the antenna and radiation patch is $0.155\lambda_o \times 0.110\lambda_o \times 0.011\lambda_o$ and $0.099\lambda_o \times 0.093\lambda_o$, respectively, where λ_o is free space wavelength at 2.2 GHz. The antenna's simulated and measured reflection-coefficient (S_{11}) response is shown in **Figure 2**. The simulated impedance bandwidth of the antenna for $S_{11} < -10$ dB is 2.58 GHz (2.05–4.63 GHz), and a fractional bandwidth of 77.2%. The antenna has a measured bandwidth of 2.3 GHz (2.2–4.5 GHz), and a fractional bandwidth of 68.7%. The simulated result shows the antenna to resonate at $f_{r,sim}$ = 3.6 GHz3.6 GHz, which is very close to the measured result at $f_{r,measured}$ = 3.5 GHz3.5 GHz. The discrepancy between the simulated and measured bandwidth is attributed to manufacturing tolerance and imperfect soldering of the SMA connector.

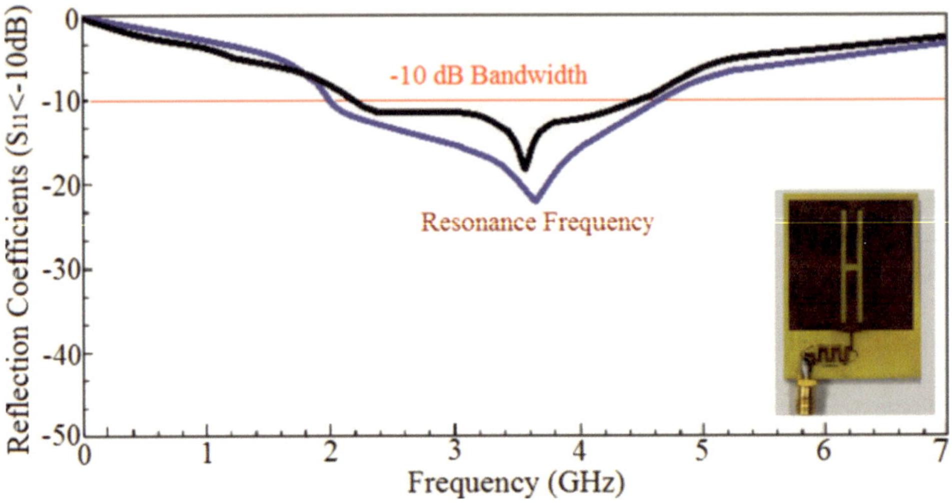

Figure 2. Simulated (blue line) and measured (black line) reflection-coefficient response of the antenna.

The measured gain and radiation efficiency of the antenna at spot frequencies of 2.2, 3.55 and 4.5 GHz are 0.65 dBi and 18.34%, 2.75 dBi and 47.15% and 1.90 dBi and 36.12%, respectively. The optimum measured gain and radiation efficiency of the antenna are 2.75 dBi and 45.15%, respectively, as shown in **Figure 3**, at f_r = 3.55 GHz3.55 GHz.

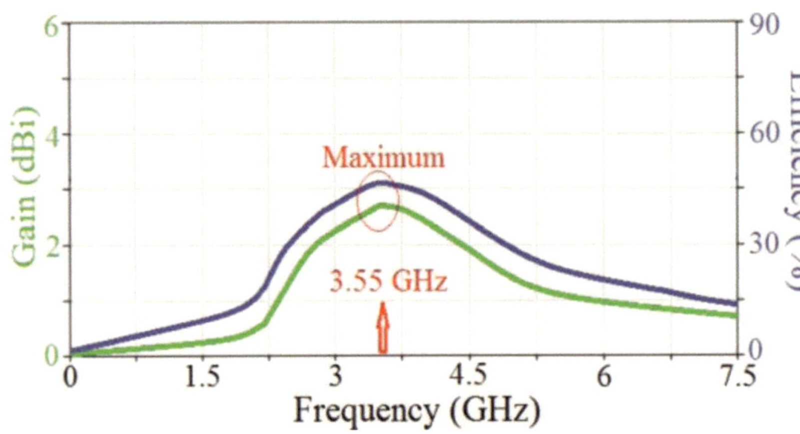

Figure 3. Measured gain and radiation efficiency response of the antenna.

Figure 4 shows the measured co-polarization and cross-polarization radiation patterns in the E(yz) and H(xz) planes at its resonance frequency of f_r = 3.55 GHz3.55 GHz. The measured radiation pattern shows the antenna radiates omni-directionally in the E-plane and bi-directionally in the H-plane. The copolarization and cross-polarization patterns are similar to a typical monopole antenna.

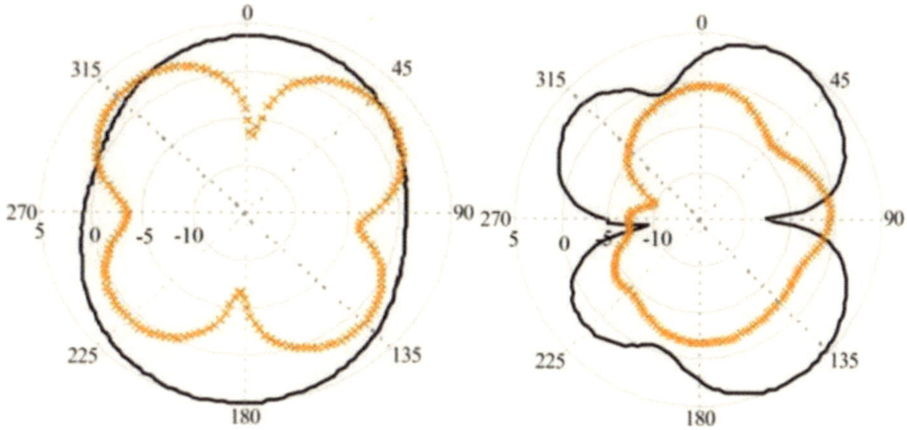

E(*yz*)-plane

H(*xz*)-plane

At f_r =3.55 GHz

Figure 4. Measured radiation patterns of the antenna at the resonance frequency of 3.55 GHz [Solid line: co-polarization, and crossed line: cross-polarization].

2.2. Patch antenna with inverted U-shape slits located on either of the H-shape slit

The above antenna was modified to improve its performance by inserting two inverted U-shaped slits on either side of the H-shape slit, as shown in **Figures 5** and **6**. The dimensions of the antenna structure are given in **Table 1**. The equivalent circuit model of the proposed symmetrical antenna structure inset in **Figure 7** consists of the composite right/left-handed transmission-line (CRLH-TL), where parasitic series reactance is represented by inductor L_R and shunt capacitor C_R. Series resonance is due to L_R and C_L, and shunt resonance due to C_R and L_L. At low frequency, C_L and L_L dominate and the transmission-line circuit exhibits left-handed characteristics; at high frequency, L_R and C_R dominate and the transmission-line circuit exhibits right-handed characteristics. The propagation constant of the resulting structure is given by Refs. [22–30]:

$$\gamma = \alpha + j\beta = \sqrt{ZY} \tag{1}$$

With

$$\beta(\omega) = s(\omega)\sqrt{\omega^2 L_R C_R + \frac{1}{\omega^2 L_L C_L} - \left(\frac{L_R}{L_L} + \frac{C_R}{C_L}\right)} \tag{2}$$

Where

$$s(\omega) = \begin{cases} -1\,\text{if}\omega < \omega_{se} = \min\left(\frac{1}{\sqrt{L_R C_L}}, \frac{1}{\sqrt{L_L C_R}}\right) \\ 0\,\text{if}\omega_{se} < \omega < \omega_{sh} \\ +1\,\text{if}\omega > \omega_{sh} = \max\left(\frac{1}{\sqrt{L_R C_L}}, \frac{1}{\sqrt{L_L C_R}}\right) \end{cases} \tag{3}$$

and

$$Z(\omega) = j(\omega L_R - 1/\omega C_L) \tag{4}$$

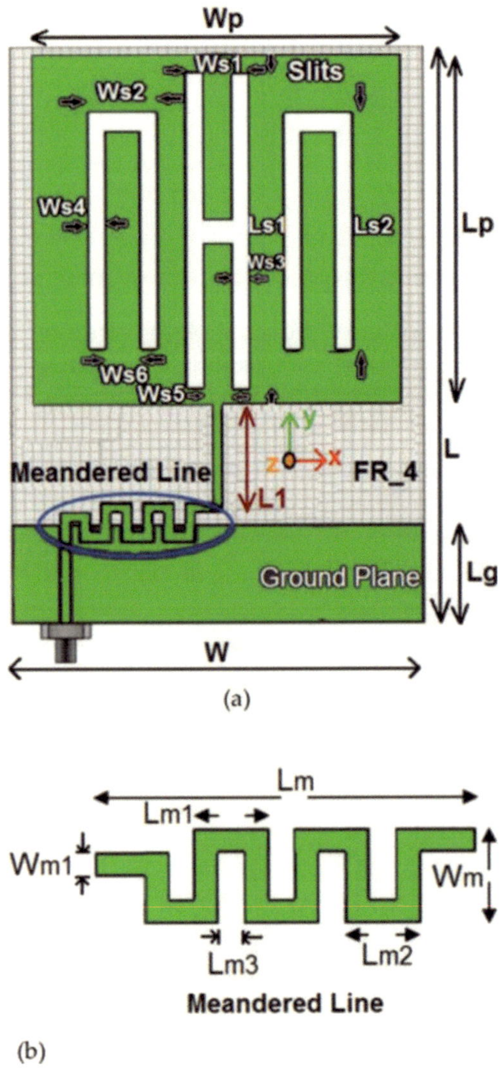

Figure 5. (a) Geometry of the MTM antenna and (b) meandered strip-line feed.

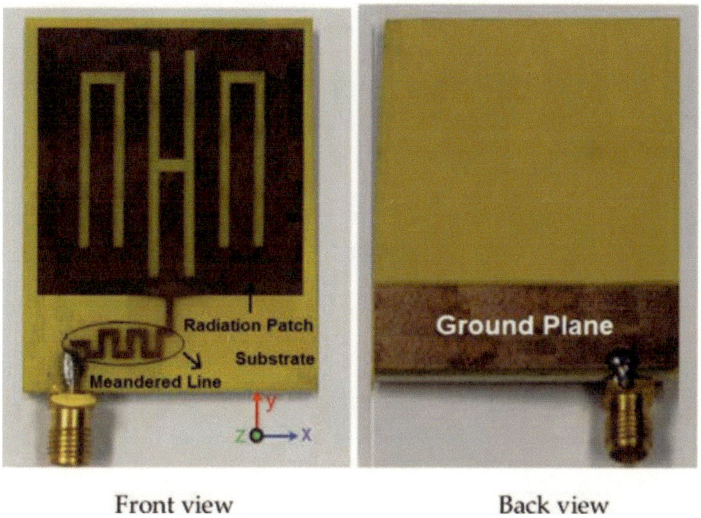

Figure 6. Fabricated prototype MTM antenna.

Dimensions (millimeters)									
L	W	W_P	L_p	L_g	L_{s1}	L_{s2}	W_{s1}	W_{s2}	W_{s3}
21.2	15	12.7	13.5	3.6	12	8.5	2.4	2.4	0.6
W_{s4}	W_{s5}	W_{s6}	L_1	L_m	L_{m1}	L_{m2}	L_{m3}	W_{m1}	W_m
0.6	1.2	1.2	3.5	4.5	0.9	0.9	0.3	0.3	1.2
Equivalent circuit Components (pF, nH, Ω, S)									
$C_{L,H}$	$C_{L,\pi Left}$	$C_{L,\pi Right}$	C_R	L_L	L_R	R_L	R_R	G_L	G_R
4.5	3.1	3.1	2	5.1	2.3	1.6	0.8	1.3	0.5

Table 1. Structural parameters of the proposed antenna and meandered strip-line feed.

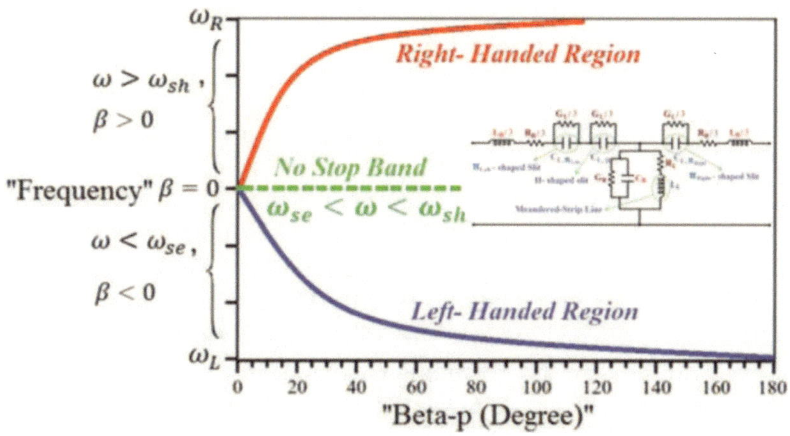

Figure 7. Dispersion diagram.

$$Y(\omega) = j(\omega C_R - 1/\omega L_L) \tag{5}$$

Parameters $\beta(\omega)$, $s(\omega)$, $Z(\omega)$ and $Y(\omega)$ are a function of frequency and represent dispersion, sign function, impedance and admittance of the antenna structure, respectively. The series and shunt resonance frequencies are given by:

$$\omega_{se} = \frac{1}{\sqrt{L_R C_L}} \tag{6}$$

$$\omega_{sh} = \frac{1}{\sqrt{L_L C_R}} \tag{7}$$

The phase and group velocities are given by:

$$v_p = \frac{\omega}{\beta} = \omega^2 \sqrt{L_L C_L} \tag{8}$$

$$v_g = \left(\frac{\partial \beta}{\partial \omega}\right)^{-1} = \omega^2 \sqrt{L_L C_L} \tag{9}$$

The antenna's dispersion diagram in **Figure 7** shows the bandwidth of structure changes from high-pass left-handed response with cut-off frequency ω_L to low-pass right-handed response with cut-off frequency ω_R with no obvious stop-band. The cut-off frequencies ω_L and ω_R are given by:

$$\omega_L = \frac{1}{\sqrt{L_L C_L}} \tag{10}$$

$$\omega_R = \frac{1}{\sqrt{L_R C_R}} \tag{11}$$

Individual slits embedded in the antenna resonate at specific frequencies, as shown in **Figure 8**. The resonance at 2.05 GHz is generated by the inverted U-slit on the left-hand side of the H-slit; the resonance at 3.7 GHz is generated by the H-slit and the resonance at 4.45 GHz results from the inverted U-slit on the right-hand side of the H-slit. Simulated and measured impedance bandwidth of the proposed antenna are 5.55 GHz (0.65–6.2 GHz) and 5.25 GHz (0.8–6.05 GHz), respectively; and the corresponding fractional bandwidths are 162.04 and 153.28%, respectively. These results confirm the antenna can operate over multiple wireless communications standards, in particular, UHF RFID, WLAN, WiMAX, WiFi, Bluetooth, GPS, PCS, and

DCS [31, 32]. Electrical size of the antenna at 800 MHz is $0.056\lambda_o \times 0.040\lambda_o \times 0.004\lambda_o$, which makes it eligible for application in various wireless systems.

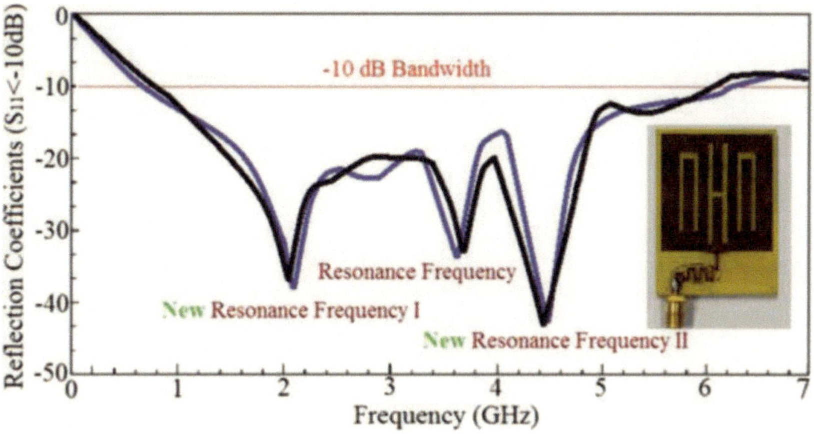

Figure 8. Simulated (blue line) and measured (black line) reflection-coefficient response of the MTM antenna.

The measured gain and radiation efficiency of the antenna in **Figure 9** at 0.8, 2.05, 3.7, 4.45 and 6.05 GHz are 0.95 dBi and 25.8%, 3.85 dBi and 63.1%, 4.73 dBi and 75.9%, 5.35 dBi and 84.1% and 3.05 dBi and 50.2%, respectively. The optimum gain and radiation efficiency of the antenna are 5.35 dBi and 84.1% at 4.45 GHz. **Figure 10** shows the antenna's measured radiation pattern at spot frequencies.

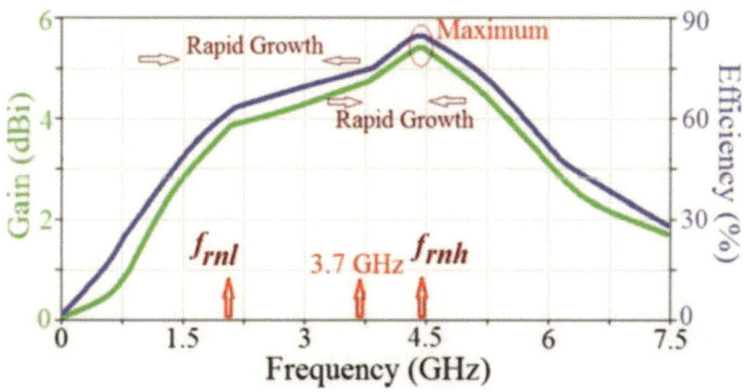

Figure 9. Measured gain and radiation efficiency of the MTM antenna.

The radiation field in the E-plane is omni-directional however this deteriorates at the first resonance frequency of 2.05 GHz. In the H-plane the antenna radiates bi-directionally across its operational bandwidth. This antenna provides the best cross-polarization compared to the above antennas. The current distribution over the antenna at its three resonance frequencies is shown in **Figure 11**.

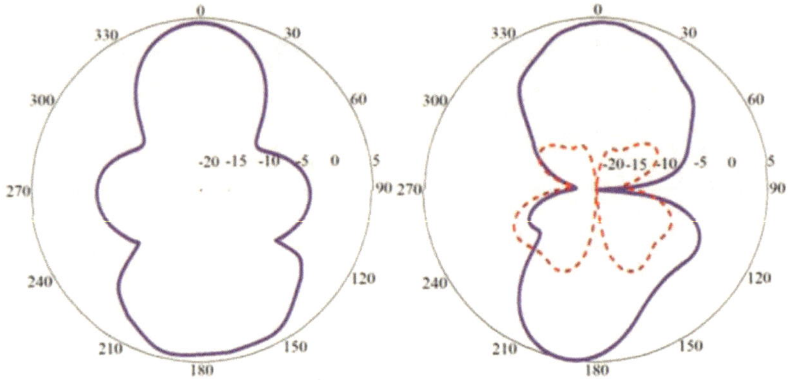

E(*yz*)-plane **H(*xz*)-plane**

At f_{r1} =2.05 GHz, due the left-hand side inverted U-slit

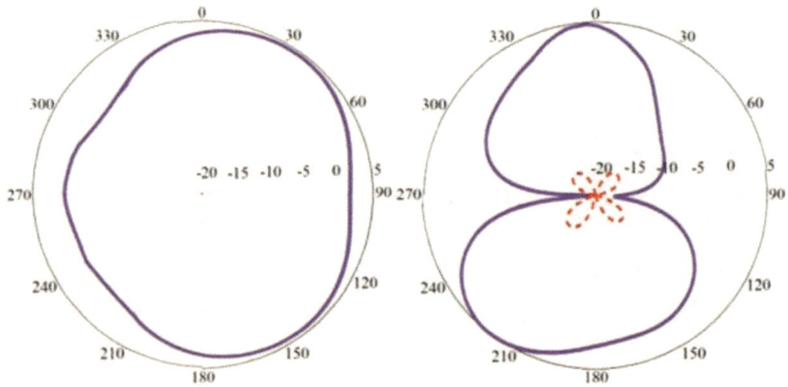

E(*yz*)-plane H(*xz*)-plane

At f_{r2} =3.7 GHz, due the middle H-slit

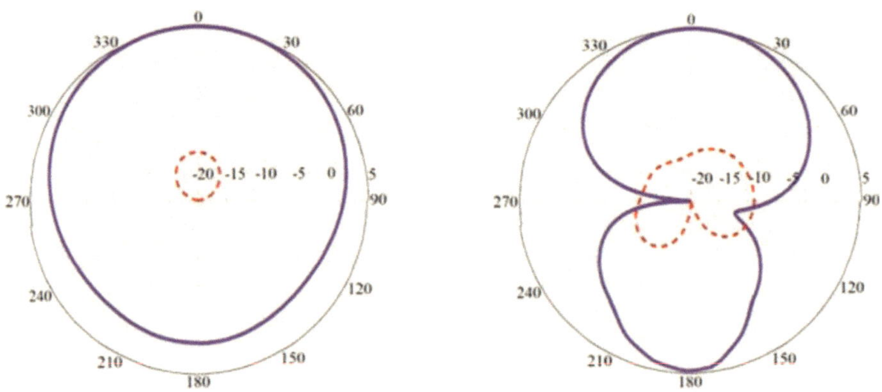

E(*yz*)-plane H(*xz*)-plane

At f_{r3} = 4.45 GHz, due the right-hand side inverted U-slit

Figure 10. Measured radiation patterns of the MTM antenna at 2.05, 3.7 and 4.45 GHz [Solid line: copolarizations, and dashed line: cross-polarizations]. Note: in some E-plane patterns the cross-polarization is not visible it is well below −30 dB.

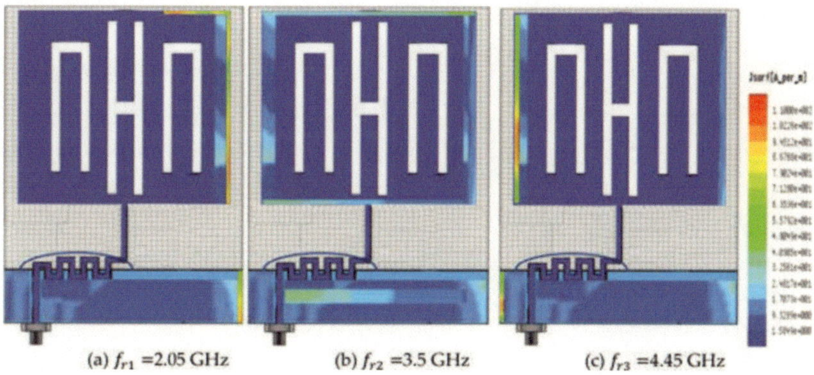

Figure 11. Current density distribution over the MTM antenna at various spot frequencies. (a) f_{r1} = 2.05 GHz; (b) f_{r2} = 3.5 GHz; (c) f_{r3} = 4.45 GHz.

3. Parametric study

A parametric study is necessary to understand the effect of the slits and the meandered stripline feed on the characteristics of the second antenna. The results in **Figure 12** shows that the length (L_{S1}) and width (W_{S3}) of H-shaped slit can substantially improve the impedance bandwidth of the antenna. Salient results are given in **Table 2** where all other structural parameters were kept fixed.

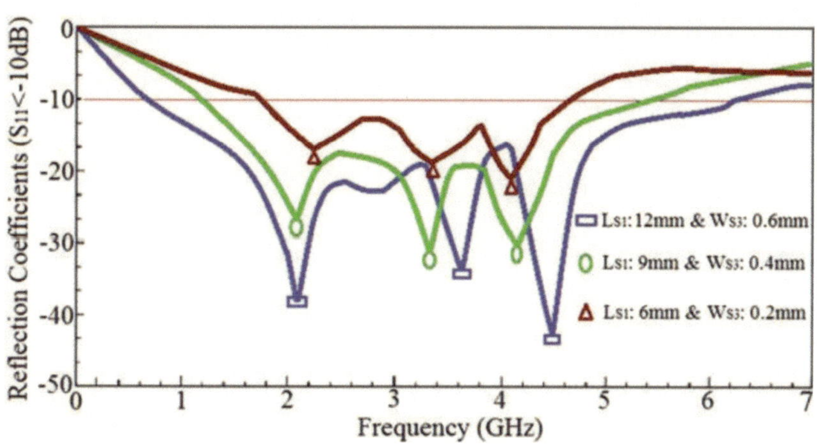

Figure 12. Effect on the antenna impedance bandwidth as a function of H-slit length (L_{S1}) and width (W_{S3}). All other structural parameters given in **Table 1** remain fixed.

The effect of inverted U-shaped slit's length (L_{S2}) and width (W_{S4}) on the antenna is shown in **Figure 13**. Figure 13 shows that by increasing L_{S2} and W_{S4} of the two inverted U-shaped slits the impedance bandwidth and impedance match of the antenna are improved significantly. The results of this analysis are given in **Table 3**.

Length and width	Frequency bandwidth
$L_{S1} = 6$ mm & $W_{S3} = 0.2$ mm	1.75–4.63 GHz ≈90.28%, S_{11} better than −20 dB
	$f_{r1} = 2.25$, $f_{r2} = 3.38$ and $f_{r3} = 4.1$ GHz
$L_{S1} = 9$ mm & $W_{S3} = 0.4$ mm	1.2–5.47 GHz ≈128.03%, S_{11} better than −30 dB
	$f_{r1} = 2.1$, $f_{r2} = 3.35$ and $f_{r3} = 4.15$ GHz
$L_{S1} = 12$ mm & $W_{S3} = 0.6$ mm	0.65–6.2 GHz ≈162.04%, S_{11} better than −40 dB
	$f_{r1} = 2.1$, $f_{r2} = 3.65$ and $f_{r3} = 4.5$ GHz

Table 2. Effect of length and width of H-shaped slit on the antenna bandwidth.

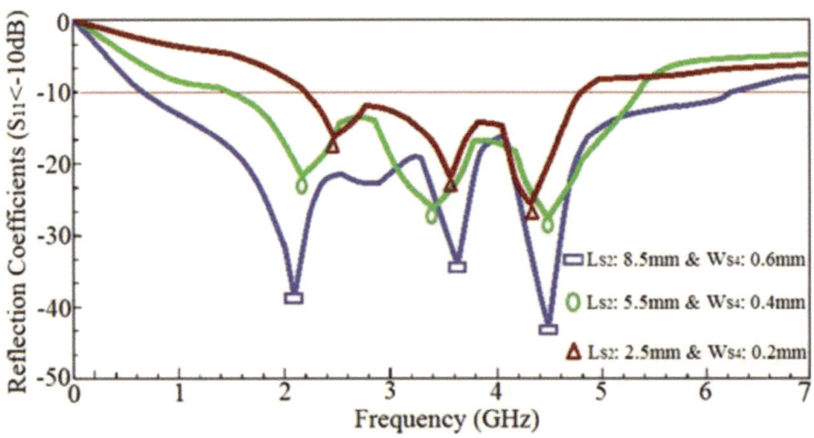

Figure 13. Parametric study on the antenna as a function of length (L_{S2}) and width (W_{S4}) of the inverted U-shaped slits. All other structural parameters in **Table 1** remain fixed.

Length and width	Frequency bandwidth
$L_{S2} = 2.5$ mm & $W_{S4} = 0.2$ mm	2.2–4.8 GHz ≈74.28%, S_{11} better than −20 dB
	$f_{r1} = 2.48$, $f_{r2} = 3.58$ and $f_{r3} = 4.34$ GHz
$L_{S2} = 5.5$ mm & $W_{S4} = 0.4$ mm	1.46–5.37 GHz ≈114.49%, S_{11}better than −30 dB
	$f_{r1} = 2.2$, $f_{r2} = 3.2$ and $f_{r3} = 4.48$ GHz
$L_{S2} = 8.5$ mm & $W_{S4} = 0.6$ mm	0.65–6.2 GHz ≈162.04%, S_{11} better than −40 dB
	$f_{r1} = 2.1$, $f_{r2} = 3.65$ and $f_{r3} = 4.5$ GHz

Table 3. Effect of length and width of the inverted U-shaped slits on the antenna bandwidth.

The effect on the antenna's performance by the meandered microstrip feed-line in **Figure 14** shows that it greatly contributes toward improving its impedance match. In fact by increasing the length of L_m, L_{m1} and L_{m2}, the width of W_m has virtually no effect on the impedance bandwidth of the antenna, however it significantly improves the antenna's impedance match. Details of these results are provided in **Table 4**.

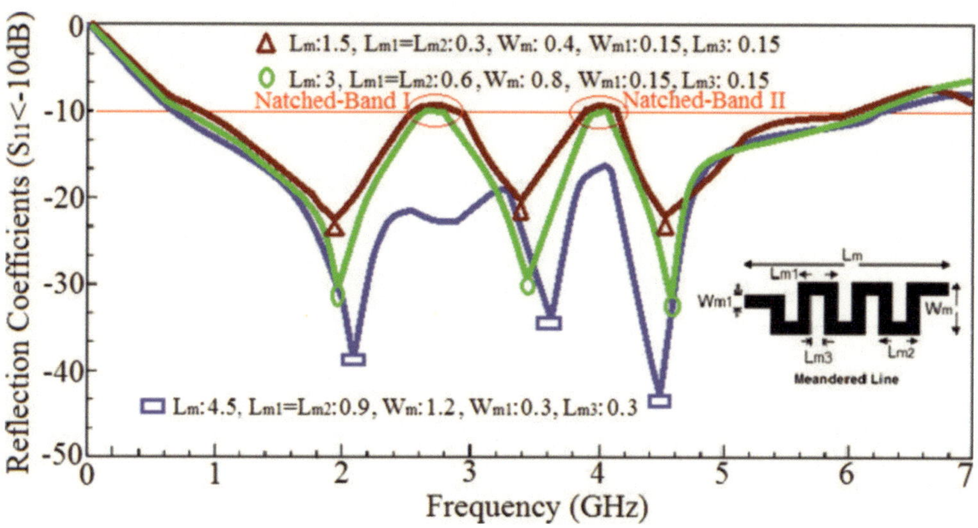

Figure 14. Antenna impedance bandwidth as a function of meandered strip-line size. All other structural parameters in **Table 1** remain fixed.

L_m	L_{m1}&L_{m2}	W_m	W_{m1}	L_{m3}	Notched-band
1.5	0.3	0.4	0.15	0.15	I: 2.55–2.95 GHz II: 3.85–4.15 GHzImpedance matching ≥ −20 dB
3	0.6	0.8	0.15	0.15	I: 2.6–2.8 GHz II: 3.95–4.05 GHzImpedance matching ≥ −30 dB
4.5	0.9	1.2	0.3	0.3	EliminatedImpedance matching ≥ −40 dB

Dimensions are in millimeters.

Table 4. Results of optimizing the meandered strip-line feed.

To summarize, a simple and effective technique has been demonstrated to extend the impedance bandwidth of patch antennas. This involves embedding three slits and exciting the antenna through a meandered strip-line. The radiating surface of the antenna is loaded with two inverted U-shape slits that are placed on either side of an H-shape slit. The antenna essentially behaves as a CRLH-TL structure. The antenna is shown to provide a fractional bandwidth of 223.27%, a maximum gain of 5.35 dBi and radiation efficiency of 84.12% at 4.45 GHz. Its radiation characteristics are similar to a monopole antenna. The proposed antenna should provide reliable wireless communication across UHF, L, S and C-bands.

Acknowledgements

The authors would like to give their special thanks to faculty of microelectronics for the financial supports.

Author details

Mohammad AlibakhshiKenari[1*], Mohammad Naser-Moghadasi[2], Ramazan Ali Sadeghzadeh[3], Bal Singh Virdee[4] and Ernesto Limiti[1]

*Address all correspondence to: Alibakhshikenari@ing.uniroma2.it

1 Department of Electronic Engineering, University of Rome Tor Vergata, Rome, Italy

2 Faculty of Engineering, Science and Research Branch, Islamic Azad University, Tehran, Iran

3 Faculty of Electrical Engineering, K. N. Toosi University of Technology, Tehran, Iran

4 London Metropolitan University, Center for Communications Technology, London, UK

References

[1] C.Y.D. Sim, F.R. Cai, and Y.P. Hsieh, "Multiband slot-ring antenna with single-and dual-capacitive coupled patch for wireless local area network/worldwide interoperability for microwave access operation", *IET Microw. Antenn. Propag. Lett.*, 2011; 5(15): 1830–1835. DOI: 10.1049/iet-map.2010.0462

[2] M. Alibakhshi-Kenari, M. Naser-Moghadasi, R.A. Sadeghzadeh, B.S. Virdee, and E. Limiti, "Dual-band RFID tag antenna based on the Hilbert-curve fractal for HF and UHF applications", *IET Circuits Dev. Sys.*, 2016; 10(2): 140–146. DOI: 10.1049/iet-cds.2015.0221

[3] J. William and R. Nakkeeran, "A new UWB slot antenna with rejection of WiMax and WLAN bands", *Appl. Comp. Electro. Soc. (ACES) J.*, 2010; 25(9): 787–793.

[4] S. Noghanian and M.K. Jung, "Ultra wide band planar slot antenna", *J. Electromagn. Waves Appl.*, 2008; 22(8–9): 1299–1308. DOI: 10.1163/156939308784158869

[5] M. Alibakhshi-Kenari, M. Naser-Moghadasi, and R.A. Sadeghzadah, "Bandwidth and radiation specifications enhancement of monopole antennas loaded with split ring resonators", *IET Microw. Antenn. P*, 2015; 9(14): 1487–1496. DOI: 10.1049/iet-map.2015.0172

[6] M. Alibakhshi-Kenari, M. Naser-Moghadasi, and R.A. Sadeghzadah, "The resonating MTM based miniaturized antennas for wide-band RF-microwave systems", *Microw. Opt. Technol. Lett.*, 2015; 57(10): 2339–2344. DOI: 10.1002/mop.29328

[7] M. Alibakhshi-Kenari, "Introducing the new wideband small plate antennas with engraved voids to form new geometries based on CRLH MTM-TLs for wireless applications", *Int. J. Microw. Wirel. Technol.*, 2014; 6(6): 629–637.

[8] M. Alibakhshi-Kenari, "A new compact UWB traveling-wave antenna based on CRLH-TLs for embedded electronic systems", *Int. J. Microw. Wirel. Technol.*, 2014: 1–4. DOI: 10.1017/S1759078714001020

[9] M. Alibakhshi-Kenari, "Printed planar patch antennas based on metamaterial", *Int. J. Elect. Lett.*, 2014; 2(1): 37–42. DOI: 10.1080/21681724.2013.874042

[10] H.-W. Liu, C.-H. Ku, and C.-F. Yang, "Novel CPW-fed planar monopole antenna for WiMAX/WLAN applications", *IEEE Antenn. Wirel. Propag. Lett.*, 2010; 9: 240–243. DOI: 10.1109/LAWP.2010.2044860

[11] Mohammad Alibakhshi-Kenari, Mohammad Naser-Moghadasi, R. A. Sadeghzadeh, Bal S. Virdee and Ernesto Limiti, "Traveling-Wave Antenna Based on Metamaterial Transmission Line Structure for Use in Multiple Wireless Communication Applications", AEUE Elsevier- International Journal of Electronics and Communications, Volume 70, Issue 12, December 2016, Pages 1645–1650.

[12] X.-L. Ma, W. Shao, and G.-Q. He, "A novel dual narrow band-notched CPW-Fed UWB slot antenna with parasitic strips", *Appl. Comp. Electro. Soc. (ACES) J.*, 2012; 27(7): 581–586.

[13] A. Sharma and G. Singh, "Design of single pin shorted three-dielectric-layered substrates rectangular patch microstrip antenna for communication systems", *Prog. Electromagn. Res. Lett.*, 2008; 2: 157–165.

[14] P.B.A. Fechine, A. Távora, L.C. Kretly, A.F.L. Almeida, M.R.P. Santos, F.N.A. Freire, and A.S. B. Sombra, "Microstrip antenna on a high dielectric constant substrate: $BaTiO_3$ (BTO)-$CaCu_3$ Ti_4O_{12} (CCTO) composite screen-printed thick films", *J. Elect. Mater.*, 2006; 35(10): 1848–1856.

[15] A. Pirhadi, H. Bahrami, and A. Mallahzadeh, "Electromagnetic band gap (EBG) superstrate resonator antenna design for monopulse radiation pattern", *Appl. Comp. Electro. Society (ACES) J.*, 2012; 27(11): 908–917.

[16] K.L. Chung and S. Chaimool, "Broadside gain and bandwidth enhancement of microstrip patch antenna using a MNZ-metasurface", *Microw. Opt. Technol. Lett.*, 2012; 54(2): 529–532.

[17] M. Alibakhshi-Kenari, M. Naser-Moghadasi, R.A. Sadeghzadeh, "Composite right-left handed based-antenna with wide applications in very-high frequency–ultra-high frequency bands for radio transceivers", *IET Microw. Antenn. P*, 2015; 9(15): 1713–1726. DOI: 10.1049/iet-map.2015.0308

[18] M. Alibakhshi-Kenari and M. Naser-Moghadasi, "Novel UWB miniaturized integrated antenna based on CRLH metamaterial transmission lines", *AEUE—Int. J. Elect. Commun.*, 2015; 69(8): 1143–1149.

[19] M. Alibakhshi-Kenari, M. Naser-Moghadasi, B.S. Virdee, A. Andújar, and J. Anguera, "Compact antenna based on a composite right/left handed transmission line", *Microw. Opt. Technol. Lett.*, 2015; 57(8): 1785–1788. DOI: 10.1002/mop.29191

[20] H.-W. Lai, K.-M. Luk, "Wideband patch antenna fed by printed meandering strip", *Microw. Opt. Technol. Lett.*, 2008; 50(1): 188–192.

[21] ANSYS®, High Frequency Structural Simulator (HFSS). ANSYS, Inc., USA.

[22] G.V. Eleftheriades, A. Grbic, and M. Antoniades, "Negative-refractive-index transmission-line metamaterials and enabling electromagnetic applications", *Proc. IEEE Antenn. Propag. Int. Symp.*, 2004: 1399–1402.

[23] A. Lai, C. Caloz, and T. Itoh, "Composite right/left-handed transmission line metamaterials", *IEEE Microw. Mag.*, 2004; 5(4): 34–50.

[24] C.-J. Lee, K.M.K.H. Leong, and T. Itoh, "Composite right/left-hand-handed transmission line based compact resonant antennas for RF module integration", *IEEE Trans. Antenn. Propag.*, 2006; 54(8): 2283–2291.

[25] C. Caloz and T. Itoh, "Application of the transmission line theory of left-handed (LH) materials to the realization of a microstrip 'LH line'", *Proc. IEEE Antenn. Propag. Soc. Int. Symp.*, 2002; 2: 412–415.

[26] Christophe Caloz and Tatsuo Itoh, "Electromagnetic Metamaterials: Transmission Line Theory and Microwave Applications", ISBN: 978-0-471-66985-2, 376 pages, December 2005, Wiley-IEEE Press.

[27] A. Sanada, C. Caloz, and T. Itoh, "Planar distributed structure with negative refractive index", *IEEE Trans. Microw. Theory Tech.*, 2004; 52(4): 1252–1263.

[28] C. Caloz, T. Itoh, and A. Rennings, "CRLH traveling-wave and resonant metamaterial antennas", *Antenn. Propag. Magazine*, 2008; 50(5): 25–39.

[29] Bonache, J., Gil, I., Garcıa-Garcıa, J., and Martın, F., "New microstrip filters based on complementary split rings resonators", *IEEE Trans. Microw. Theory Tech.*, 2006; 54: 265–271.

[30] Niu, J.-X., "Dual-band dual-mode patch antenna based on resonant-type metamaterial transmission line", *Electron. Lett.*, 2010; 46: 266–268.

[31] Y. Sun, G. Wen, P. Wang, Y. Huang, and Z. Du, "A compact printed end-fire antenna for radio frequency identification (RFID) handheld reader", *Appl. Comp. Electro. Soc. (ACES) J.*, 2013; 28(1): 71–76.

[32] C.-Y. Pan, T.-S. Horng, W.-S. Chen, and C.-H. Huang, "Dual wideband printed monopole antenna for WLAN/WiMAX applications", *IEEE Antenn. Wirel. Propag. Lett.*, 2007; 6: 149–151.